Quantum Structural Studies

Classical Emergence from the Quantum Level

Quantum Structural Studies

Classical Emergence from the Quantum Level

Edited by

Ruth E Kastner
University of Maryland, College Park, USA

Jasmina Jeknić-Dugić
University of Niš, Serbia

George Jaroszkiewicz
The University of Nottingham, UK

World Scientific

NEW JERSEY • LONDON • SINGAPORE • BEIJING • SHANGHAI • HONG KONG • TAIPEI • CHENNAI • TOKYO

Published by

World Scientific Publishing Europe Ltd.

57 Shelton Street, Covent Garden, London WC2H 9HE

Head office: 5 Toh Tuck Link, Singapore 596224

USA office: 27 Warren Street, Suite 401-402, Hackensack, NJ 07601

Library of Congress Cataloging-in-Publication Data

Names: Kastner, Ruth E., 1955– editor. |Ieknić-Dugić, Jasmina, editor. |
Jaroszkiewicz, George, editor.
Title: Quantum structural studies : classical emergence from the quantum level /
edited by, Ruth E. Kastner (University of Maryland, College Park, USA),
Jasmina Jeknić-Dugić (University of Niš, Serbia),
George Jaroszkiewicz (The University of Nottingham, UK).
Description: Hackensack, NJ : World Scientific, [2016] |
Includes bibliographical references and index.
Identifiers: LCCN 2016032484| ISBN 9781786341402 (hc ; alk. paper) |
ISBN 1786341409 (hc ; alk. paper)
Subjects: LCSH: Quantum systems. | Quantum theory. | Degree of freedom. | Atomic structure.
Classification: LCC QC174.13 .Q35 2016 | DDC 530.12--dc23
LC record available at https://lccn.loc.gov/2016032484

British Library Cataloguing-in-Publication Data

A catalogue record for this book is available from the British Library.

Desk Editors: Herbert Moses/Mary Simpson

Typeset by Stallion Press
Email: enquiries@stallionpress.com

Printed in Singapore

About the Editors

Ruth E. Kastner received her BS, *cum laude* with high honors in physics, and after picking up an MS in physics from the University of Maryland, College Park, in 1992, decided to pursue her interest in the foundational aspects being studied at the UMCP Philosophy Department. She received her PhD in Philosophy (History and Philosophy of Science) with Jeffrey Bub as her dissertation advisor in 1999. She has won two National Science Foundation awards for her study of interpretational issues in quantum theory and is Research Associate in the Foundations of Physics Group at UMCP. She wrote the first book on the length treatment of the Transactional Interpretation (*The Transactional Interpretation of Quantum Mechanics: The Reality of Possibility*, Cambridge University Press, 2012) and followed that with a conceptual presentation of TIQM for the general reader (*Understanding Our Unseen Reality: Solving Quantum Riddles*, Imperial College Press, 2015). She has over two dozen peer reviewed publications and regularly attends international conferences on Foundations of Physics, where she is often an invited speaker. She currently resides in upstate New York.

Jasmina Jeknić-Dugić received her BS degree in physics in 2000 when she decided to venture into nuclear sciences and technology. She visited numerous nuclear research centers in Europe, notably IRES in Strasbourg (France) and UCL in Louvain-la-Neuve (Belgium) that resulted in obtaining her MS degree in physics and in starting her academic career as a Teaching Assistant at Department of

Physics, Faculty of Science and Mathematics, University of Niš, Serbia. While working in applied field of nuclear science, numerous open questions and challenges emerged and established her future permanent interest in the foundational topics and issues of quantum theory, quantum decoherence and open quantum systems theory. The research work and scientific publications in these fields earned her position of Associate Professor at Department of Physics, University of Niš, Serbia. With collaborators, she published (2013) the first book on the top-down, minimalist and non-interpretational approach to quantum structuralism with an emphasis on the team's original results, and currently pursues a specific paradigm of local time in the foundations of quantum theory.

George Jaroszkiewicz earned a degree in Mathematical Physics from the University of Edinburgh and then completed his PhD in high energy particle physics with Dr. Peter Landshoff at the Department of Applied Mathematics and Theoretical Physics (DAMTP), University of Cambridge, UK. Awarded a British Council Fellowship, he spent most of 1977 at the Theoretical Physics Department, University of Warsaw, Poland, working with Prof. Krolikowski on the colored quark model. On his return to Britain in 1977, Jaroszkiewicz spent two years as a Post-Doctoral Fellow at the Physics Department, University of Kent at Canterbury, England, working on theoretical end experimental spin-echo nuclear magnetic resonance with John Strange. This was followed by two years as a Post-Doctoral Fellow in the Department of Physical Chemistry, University of Oxford, working on the simulation of dilute aqueous solutions, with Graham Richards. In 1981, Jaroszkiewicz was appointed lecturer in the Department of Mathematics (now School of Mathematical Sciences), University of Nottingham, UK. Over the next three decades, Jaroszkiewicz worked on discrete time mechanics and quantized detector networks (an observer-centered approach to quantum mechanics). A major theme throughout has been his preoccupation with the time concept. He has published two books: *Principles of Discrete Time Mechanics* (CUP, 2014) and *Images of Time* (OUP, 2016). Current projects include a monograph on quantized detector networks, currently scheduled for April 2017 with Oxford University Press.

Contents

Chapter 1

Quantum Structures: An Introduction

R. E. Kastner*, J. Jeknić-Dugić† and G. Jaroszkiewicz‡

**Foundations of Physics Group,*
University of Maryland, College Park, USA
†Department of Physics,
Faculty of Science and Mathematics, University of Niš, Serbia
‡School of Mathematical Sciences,
The University of Nottingham, UK

1. Introduction

Quantum mechanics offers a striking, genuinely novel observation: it is possible to obtain more information about a closed composite system than about the subsystems constituting that system. This is symptomatic of the highly nontrivial concept of a *quantum subsystem*: specifically, there is much more to the "subsystem" concept in quantum mechanics than there is in classical physics.

In classical physics, complete knowledge about the total state of a system is equivalent to a complete knowledge about the state of every constituent subsystem. This accounts for the primary role of subsystems in classical physics, and the fact that the lack of knowledge about subsystems is subjective, i.e. observer–dependent. There are no fundamental physical limits in this regard, since there is no classical information limit at the fundamental physical level. This view of the classical world is typically interpreted as follows: every single classical system exists in space independently of any other physical system at every instant of time. Physical subsystems ("constituting particles") are as physically realistic as the physical objects they build.

Mathematically, this idea is represented by the Cartesian product of a set of ontologically fundamental degrees of freedom (and analogously for continuous fields). In addition, in classical physics, useful artificial degrees of freedom can be defined; for example, the center of mass of an extended composite system. Mathematical manipulations with such constructed degrees of freedom do not directly describe the behavior of a realistic physical object. Thus, a description based on such mathematical degrees of freedom is typically incomplete and approximate, but is assumed ultimately to be reducible to the dynamics of the fundamental degrees of freedom.

This reductionistic attitude is sometimes criticized even in the context of classical physics: there is a conceptual and formal gap between the fundamental and the apparently emergent degrees of freedom [6]. Nevertheless, the primary role of the fundamental degrees of freedom is rarely, if at all, challenged in the physics literature: the existence of fundamental degrees of freedom is assumed to be a *necessary* condition for the emergent behavior that is observed.

However, this view face a serious challenge in the quantum mechanical context for at least the following two reasons. First, knowledge of a composite system's state does not imply, or assume or require knowledge about, the subsystems' states. If the total system is in a pure state $|\Psi\rangle$, the subsystems are statistically described by "reduced density matrices" ("reduced statistical operators") that are sometimes referred to as "improper mixtures". For a bipartite $1+2$ system, a pure state, $|\Psi\rangle$, can be always represented by a Schmidt canonical form, $|\Psi\rangle = \sum_i c_i |i\rangle_1 \otimes |i\rangle_2$, $\sum_i |c_i|^2 = 1$; with the orthonormalized bases, ${}_\alpha\langle i|j\rangle_\alpha = \delta_{ij}$, $\alpha = 1, 2$. Provided $|c_i| < 1, \forall i$, the reduced (subsystems) mixed states $\hat{\rho}_\alpha$ are obtained via the tracing out operation such that $\hat{\rho}_\alpha^2 \neq \hat{\rho}_\alpha$, $\alpha = 1, 2$. The point is that $|\Psi\rangle\langle\Psi| \neq \hat{\rho}_1 \otimes \hat{\rho}_2$, i.e. that the αth subsystem cannot be assumed to be in the $\hat{\rho}_\alpha$ state. Rather, the subsystems' states are not well defined.

Second, there is no unique ensemble description of a mixed quantum state $\hat{\rho}$. The formalism of quantum mechanics allows a non-unique ensemble interpretation of $\hat{\rho}$, which is a positive semidefinite Hermitian operator of trace one. Typically, there is a possibility of choosing different ensemble-representations pertaining to the same mixed state $\hat{\rho}$: $\sum_i p_i \hat{\rho}_i = \hat{\rho} = \sum_m w_m \hat{\sigma}_m$, while $\sum_i p_i = 1 = \sum_m w_m$ and $\mathrm{tr}\hat{\rho}_i = 1 = \mathrm{tr}\hat{\sigma}_m, \forall i, m$. Therefore, the basic interpretational tool of classical statistical physics as embodied in the concept of "Gibbs ensembles" is not necessarily useful in quantum theory.

On the other hand, a closed quantum system [subjected to the Schrödinger unitary dynamics] in a pure state $|\Psi\rangle$ provides all the possible information (predictions) about the system *and* about its subsystems that is allowed by standard quantum theory.[a] Hence, there is a kind of reversal of the above-described reductionist classical thinking that, in the view of the editors, merits further critical attention. This motivates the following questions, to which contributors to this volume have applied themselves:

1. Is there a preferred structure (i.e. decomposition into subsystems) of a composite quantum system? If so, how is it defined and what are the conditions that give rise to it?
2. How successful are extant accounts of quantum/classical correspondence, and/or emergence of classically recognizable structures from a quantum level of description?
3. Can classically familiar degrees of freedom (such as the 'center-of-mass') be considered quantum mechanically realistic, and in what sense — ontologically, epistemologically, operationally?
4. What kinds of alternative structures might be relevant for the transition from a quantum description to a classical one?
5. How, and to what extent, might the answers to the above describe phenomenology, including proposals for new experiments?

This introduction aims to set the stage for the answers offered by the contributors, who present diverse positions. The goal of the Editors of this volume is two-fold: (i) to collect the current perspectives of authors who have contributed to the topic in the

[a]The possible existence of "hidden variables" does not change these conclusions, except possibly on the interpretational level.

past; (ii) to place side-by-side diverse views and approaches to these questions concerning the nature of fundamental quantum structures. Thus, our idea is *not* to ask or look for consensus, but rather to emphasize the wealth of ideas and approaches on the topic, as well as to initiate a discussion that could provide a fruitful foundation for progress in theory and experiment devoted to this topic. The contributors, in their independent research, have come to the point at which they realized that the question "How do [quantum] components relate to a composite?" cannot be ignored or treated as a secondary issue.

A significant aspect of this book's mission is two-fold: (1) to present (as comprehensively as possible given space limitations) and extend the historical perspective and (2) to highlight the ubiquity of our topic of quantum structures in modern quantum physics research. Our presentation is general; further details can be found as indicated in the references list. Our choice of the relevant results is by no means exhaustive and is certainly not the only one possible. In addition, not all those invited have been willing or able (due to time or other constraints) to contribute. Nevertheless, we offer a beginning that, hopefully, will be extended and enriched in the near future.

2. Quantum Structures: A Brief Review

Quantum structural studies (QSS) cannot be given a definite chronological origin, nor can breakthrough results be non-arguably distinguished. Rather, a re-discovery of certain basic insights and their diverse (often implicit) understandings and interpretations are presented in this collection. The resurgence of interest in the issue of fundamental structure is not surprising, bearing in mind that the concept of "structure", i.e. of "decomposition into parts", is omnipresent in physical theory, in its mathematical formulation, and in its applications. To this end, arguably the most fundamental endeavor in science is investigation of the finite decomposability of physical matter, i.e. the existence and nature of elementary particles. Hence, one may argue that QSS are as old as physics (and its mathematical foundations) itself.

Subsystems of a *closed* composite system are typically regarded as ontologically realistic in an approach that views quantum theory as incomplete [11, 37], in the unitary-only Everettian interpretation of quantum theory [97, 107], in certain category structural studies [68]; in the context of non-unitary collapse theories [7, 46] and the 'Schrödinger–Newton' theory [27, 28], Transactional Interpretation [22, 62]; in the so-called "primitive ontology" (PO) approach [1, 2]; and in a non-interpretational context likewise [52, 53].

In contrast to this, a denial of the ontologically realistic status of quantum subsystems ('particles') can be found in a quantum-decoherence/open systems discourse assuming a unitary-only, complete quantum theory [57, 106], in the extreme operational context [19] and Modal interpretation [15], and also in the context of category structural studies [8] and quantum information [96].

In a unitary-only context, the absence of a preferred structure (decomposition into "parts") for a closed system without external interactions has been emphasized [15–17, 32, 33, 51, 104]. As a corollary, a "subsystem" can only be a relational concept — i.e. it can be defined only in relation to the rest of the closed system that is often termed "environment". In this context, preferred structure of a closed system can only be introduced by an additional assumption, e.g. to be postulated [107] or conjectured to exist due to additional symmetry-based requirements [50, 104, 105].

In the same context, a preferred structure of an (relationally defined) open system in interaction with its environment is often conjectured [47] and sometimes supported [4, 84, 90] based on the environmental influence. As a corollary, quantum decoherence, as it has been formulated in the literature (see, e.g. [47]), appears also to be relational rather than absolute: the occurrence of decoherence for mutually incompatible global structures of a closed system is implied [34], and thus creates a challenge for both old and modern Everett interpretation [58] as well as for Quantum Darwinism [63, 64] (see also [4, 5, 28, 47, 87, 90, 97]).

Certain collapse models must assume a preferred local system structure [7, 46] and the Schrödinger–Newton theory [27–29]. On the other hand, a possible "bottom-up" emergence of structure can be based on identifying bound states as composite systems whose entanglement is fundamental rather than relative (Kastner [108], this volume), and on finding asymmetry between observables at the relativistic level, where the dynamical variables corresponding to conserved quantities are arguably more physically fundamental than the spacetime parameters ([62] and [109], this volume).

Much of the research thus far addressing our topical questions has been conducted under the assumption that all quantum dynamics is unitary (and quantum theory is complete; i.e. no hidden variables). Below, we offer a critical review of the unitary-only program as it has been applied to the issue of classical emergence from the quantum level.

The idea that unitary-only dynamics can lead naturally to preferred observables such that decoherence suffices to explain emergence of classical phenomena seems to be inadequate. (The basic "preferred basis" problem is laid out especially clearly in [13].) Such 'Quantum Darwinism'-type arguments depend on assuming special initial conditions of separable degrees of freedom which amount to 'seeds' of classicality from the outset, in which case the explanation of classical emergence becomes circular. That is, the argument becomes:

1. if the quantum dynamics is unitary-only, and
2. if the universe has initially separable (localizable) degrees of freedom such as distinguishable atoms and photons, and
3. if those degrees of freedom interact by Hamiltonians that do not re-entangle them, then
4. decoherence occurs and classicality emerges.

For decoherence to (even approximately) explain the emergence of classicality under the assumption of unitary-only (U-O) evolution, all four premises must hold. However, classicality is implicitly contained in (2) and (3) through the partitioning of the universal degrees of freedom into separable, localized substructures interacting via Hamiltonians that do not re-entangle them, so (given U-O) one has to put in classicality to get classicality out. Premises (2) and (3) are special initial conditions on the early universe that may not hold — certainly they are not the most general case for a quantum-only initial universe. Yet it seems common for researchers assuming U-O to assert that (2), and (3) also must hold without question; i.e. that they can (or should) be taken as axiomatic. This actually amounts to the fallacy of affirming the consequent, as follows: one observes that we have an apparently classical world (affirm 4), and then one asserts that (2) and (3) must hold.

The insistence on (2) appears, for example, in Wallace's invocation of "additional structure on the Hilbert Space" as ostensibly part of the basic formalism ([97], pp. 14–15). Such additional structure — preferred sets of basis vectors and/or a particular decomposition of the Hilbert space — is imposed when quantum theory is applied to specific situations in the laboratory. However, what we observe in the laboratory is the already-emergent classical world, in which classical physics describes our macroscopic measuring instruments and quantum physics is applied only to prepared quantum systems that are not already entangled with other degrees of freedom.

If the task is to explain how we got to this empirical situation from an initially quantum-only universe, then clearly we cannot assume what amounts to the explanandum; i.e. that the universe began with quasi-localized quantum systems distinguishable from each other and their environment, as it appears to us today. Wallace includes this auxiliary condition imposing structural separability under a section entitled "The Bare Formalism" (by which he means U-O), despite noting that we assign the relevant Hilbert space structures "in practice" to empirical laboratory situations. However, such imposed structures are part of the application of the theory to a particular empirical situation; they constitute contingent information and are therefore not aspects of the "bare formalism," (any more than, for example, field boundary conditions are part of the bare theory of electromagnetism). These separability conditions are auxiliary hypotheses to which we cannot help ourselves, especially since the most general state of an early quantum universe is not one that comes with preferred basis vectors and/or distinguishable degrees of freedom. Thus, the addition of this condition amounts to asserting (2), and becomes part of the circular reasoning used to support the claim that quasi-classical world 'branches' naturally appear in an Everettian (U-O) picture.

Now, of course U-O theorists assume that (1) is not subject to question and is just a background fact. If one insists on (1) in this way, then (2) and (3) are *required* in order to arrive via a decoherence-type argument at our current apparently classical world. Thus, they assume that the logical structure of the argument is: (2) and (3) iff (4). So, rather than reject the argument based on its circularity, it is commonly assumed under U-O that the consequent is evidence for the truth of premises (2) and (3). The possibility that the dynamics may not be wholly unitary — the falsity of the unitary-only assumption — is not considered. However, it may be false.

3. Applications: Theoretical and Experimental

Entanglement relativity (ER) has been repeatedly discovered under the unitary-only assumption [24, 32, 33, 38, 50, 92, 94, 99]. More general quantum correlations (such as "quantum discord") are subject to the same kind of relativity as a corollary of ER, and are referred to as "quantum correlations relativity" (QCR) *rule* [35]. QCR implies the existence of virtually any kind of quantum correlations for an arbitrary composite system's state; the correlations are linked with *structure.*

The so-called "entanglement swapping" procedure involves direct application of ER, i.e. the fact that entanglement of two or more degrees of freedom is generally definable only with respect to a particular observable. An important paradigm demonstrating this is quantum teleportation [10, 12]: a set of three qubits denoted $1, 2$ and 3 is initially prepared in a tensor-product pure state for the bipartition $1+(2+3)$, where the pair $2+3$ is in a maximally entangled state. Regarding the $(1+2)+3$ bipartition, this state [in the same instant of time] is entangled: the entangled pair $1 + 2$ is, as a whole, entangled with the qubit 3 — although not yet in Schmidt canonical form. Then, collective operations performed on the whole $1+2$ with the aid of a classical channel allow for quantum teleportation — the "transmission" of unknown information about the initial state of the 1 qubit to the qubit 3. The extensive use of ER via entanglement swapping in a four-qubits system has been discussed [72]. In quantum key distribution, entanglement swapping has been used in [44] as well as a basis for the method (protocol) of quantum information hiding in [102].

"Entanglement monogamy" [67] is in intimate relation with ER and imposes limitations on the operational use of entanglement swapping. It concerns, for example, optimal cloning [65], frustration in a quantum Ising spin system [83], quantum steering [75] and particle-by-particle measurements on a macroscopic quantum state [9].

"Entanglement monogamy" proves essential for non-signaling theories [95] as well as for the device-independent test of gravitational decoherence [79]. Conditions for quantum correlations stability are presented in [41, 42]. ER restricts the use of master equations for the "alternative" degrees of freedom of a closed composite system [5]. Methods and models have been developed for: avoiding decoherence [56, 82] and dissipation [74], isolation of systems [61, 103], and decoupling [40].

There is current interest in the operational importance of "collective" degrees of freedom e.g. in gravity-induced decoherence [28, 79], the collective effect of "superradiance", e.g. [49, 59], and "subradiance" [89], as well as of the local environmental structure [34, 54, 70]. A kind of divergence between the local and collective degrees of freedom is reported in [73].

On the other hand, models of quantum phase transitions can arguably not even be formulated without suitable variable transformations [69, 86, 99]. In addition, so-called entanglement renormalization [40] has been proposed to deal with many-body situations assuming a unitary-only dynamics.[b] There is no doubt that QSS will be an important aspect of the emerging field of quantum thermodynamics [25, 48, 49, 100].

The prospects of operational manipulation with constructed degrees of freedom open nontrivial technological and engineering possibilities and challenges. Such "artificial" degrees of freedom are

---

[b]It is possible however that in a suitable collapse model, disentanglement arises without the need for renormalization techniques.

experimentally observed and/or manipulated in [14, 30, 43, 60, 77, 93, 101], as well as theoretically investigated in [3, 23, 45, 66, 85, 88].

4. Overview of the Contributions

Peter Holland surveys the so-called trajectory conception of quantum dynamics with the aim to exhibit a close relation between the wave-mechanical and trajectory pictures of quantum theory that emerges from a canonical transformation performed on a suitable phase space coordinates.

Nathan Harshman considers the Hamiltonian-symmetry induced tensor-product structures for a model of three one-dimensional particles. The related [but non-equivalent] notions of integrability, separability, and solvability of the equations of motion typically distinguish certain collective observables that provide a "natural" description of a composite system. This provides observation-accessible degrees of freedom of the system and a specific top-down approach to structuring quantum systems.

Eliahu Cohen analyzes quantum measurement from the so-called Two-State Vector Formalism (the "weak measurement" theory) perspective with an emphasis on "macroscopic robustness under time-reversal". It is shown that imposing a boundary condition for the postselected state, it is possible to have [from our local perspective] non-unitary evolution of the measured microstate and a deterministic dynamics of the related measurement reading.

Sebastian Fortin and Olimpia Lombardi propose a specific *top-down* approach to quantum decoherence and to the hence-derived classical limit of quantum mechanics. The reduced state of a closed system is introduced via projection ("coarse-grained"), which discards certain information about the system and hence gives rise to non-unitary dynamics of the closed system.

Christian Baumgarten provides an alternative approach to foundational questions of quantum and relativistic theory from a specific mathematical perspective. The concept of physical time is defined as a measure of change of physical systems on the most fundamental level; thus time supervenes on physical systems and is an emergent quantity. Along with time, standard Euclidean space also proves to be emergent. Thus, neither space nor time is viewed as ontologically fundamental, and therefore the spacetime manifold is not subject to quantization.

Momir Arsenijveić, Jasmina Jeknić-Dugić and Miroljub Dugić employ Entanglement Relativity (Section 3) in an analysis of the Stern–Gerlach (SG) experiment in the context of the standard and on Bohm's version of quantum theory. After a short description of a minimalist *top-down* approach to the standard quantum theory, the role of ER is distinguished in both contexts. It is concluded that, according to the standard model of the SG effect, the magnetic field does not see either the atomic proton or electron but couples locally to the atomic center of mass system.

Tongdong Bai and John Stachel, offer a critique of some of the literature asserting that Bohr's philosophy was positivist, based on Bohr's reply to the 1935 Einstein–Podolsky–Rosen (EPR) paper. In addition, Stachel critically examines some discussions of the Copenhagen Interpretation, arguing that there are important distinctions between Bohr's Complementarity and Heisenberg's philosophical approach concerning the role of measurement in interpreting quantum theory.

Ruth Kastner provides two contributions. She critiques Bohr's "Complementarity" as being inadequate to the task of accounting for the emergence of classicality from the quantum level of description. She proposes that bound states present intrinsically preferred composite degrees of freedom due to their status as emergent structures. Thus, in the case of bound states, the entanglement of component degrees of freedom becomes absolute rather than relative.

Christian de Ronde argues against trying to interpret quantum theory in terms of classical concepts, a central assumption of Bohr's approach. He provides a criterion for a representational realist account of quantum theory which notes that what is considered as "observed" is always a theory-laden and metaphysically laden construct. He argues that the task in interpreting quantum theory is to develop a new set of concepts that provide physical referents for the mathematical elements of the theory; and that accounting for the quantum-to-classical limit is inherently an ontological problem that cannot be skirted by instrumentalist or epistemic approaches.

George Jaroszkiewicz points out that classical physics is based on a reductionist attitude that is revealed as in need of critical reconsideration, based on the failure of reductionist classical physics in the microscopic domain. His response to this failure of the classical description is to make the observer a primary, primitive concept and to relate physical theory to that primary observer. His proposed resolution to the apparent conflict between classical and quantum theory is that the two theories have separate, complementary roles in physics.

Valia Allori discusses the PO approach to interpreting quantum theory. PO distinguishes different roles for theoretical structures: some structures refer ontologically, where others refer nomologically. She identifies the quantum state and its evolution as playing a nomological, rather than ontological, role. A defining assumption of PO is that the fundamental ontological referents must be spacetime objects.

Leonardo Chiatti and Ignazio Licata propose an alternative view of the standard Bohmian conception of a semi-classical structure for the world (i.e. localized particles on spacetime trajectories, guided by a non-material "guiding wave." Instead of the concept of a semi-classical particle as a persistent object, they regard the "particle" as an aspect of the dialectic between the quantum background and determinate events, and investigate whether this picture can help to reconcile non-locality with relativity. Under this scheme, the determinate events occurring in a quantum whole gain the primary conceptual role and appear to form the substance of the physical world.

Needless to say, this is an area of active interest and multiple competing approaches, as exemplified not only by the diversity of views expressed by the contributors but by the diverse views among the editors themselves. We invite the reader to make his or her own assessment as to what approach(es) may be most fruitful.

References

[1] Allori, V. (2007). Fundamental Physical Theories: Mathematical Structures Grounded on a Primitive Ontology. Ph.D. Thesis, Department of Philosophy, Rutgers University.

[2] Allori, V., Goldstein, S., Tumulka, R. and Zanghi, N. (2008). On the common structure of bohmian mechanics and the Ghirardi–Rimini–Weber theory. *The British Journal for the Philosophy of Science*, **59**: 353.

[3] Ardenghi, J. S. (2015). Entanglement entropy between real and virtual particles in φ^4 quantum field theory. *Physics Review D*, **91**: 085006.

[4] Arsenijević, M., Jeknić-Dugić, J. and Dugić, M. (2013). Asymptotic dynamics of the alternate degrees of freedom for a two-mode system: an analytically solvable model. *Chinese Physics B*, **22**: 020302.

[5] Arsenijević, M., Jeknić-Dugić, J., Todorović, D. and Dugić, M. (2015). Entanglement relativity in the foundations of the open quantum systems theory, in *New Research on Quantum Entanglement*, Lori Watson (ed.), Nova Science Publishers: New York, pp. 99–116.

[6] Auynag, S. (1998). *Foundations of Complex-system Theories.* Cambridge University Press: Cambridge.

[7] Bahrami, M., Paternostro, M., Bassi, A. and Ulbricht, H. (2014). Proposal for a noninterferometric test of collapse models in optomechanical systems. *Physics Review Letters*, **112**: 210404.

[8] Bain, J. (2013). Category-theoretic structure and radical ontic structural realism. *Synthese*, **190**: 1621.

[9] Beduini, F. A., Zielinska, J. A., Lucivero, V. G., de Icaza Astiz, Y. A. and Mitchell, M. W. (2015). A macroscopic quantum state analysed particle-by-particle, *Physics Review Letters*, **114**: 120402.

[10] Bennett, C. H., Brassard, G., Crépeau, C., Jozsa, R., Peres, A. and Wootters, W. K. (1993). Teleporting an unknown quantum state via dual classical and Einstein–Podolsky–Rosen channels. *Physics Review Letters*, **70**: 1895.

[11] Bohm, D. (1951). *Quantum Theory.* New York: Prentice Hall.

[12] Bouwmeester, D., Pan, J.-W., Mattle, K., Eibl, M., Weinfurter, H. and Zeilinger, A. (1997). Experimental quantum teleportation. *Nature*, **390**: 575.

[13] Bub, J., Clifton, R. and Monton, B. (1998). The Bare theory has no clothes, in *Quantum Measurement: Beyond Paradox.* Healey, R. A. and Hellman, G. (eds.), Minnesota studies in the philosophy of science, Vol. 17, University of Minnesota Press: Minneapolis, pp. 32–51.

[14] Casabone, B., Friebe, K., Brandstltter, B., Schüppert, K., Blatt, R. and Northup T. E. (2015). Enhanced quantum interface with collective Ion-Cavity coupling. *Physics Review Letters*, **114**: 023602.
[15] Castagnino, M. and Lombardi, O. (2003). The self-induced approach to decoherence in cosmology. *International Journal of Theoretical Physics*, **42**: 1281.
[16] Castagnino, M. and Ordonez, A. (2004). Algebraic formulation of quantum decoherence. *International Journal of Theoretical Physics*, **43**: 695.
[17] Castagnino, M. and Lombardi, O. (2005). Decoherence time in self-induced decoherence. *Physical Review A*, **72**: 012102.
[18] Castagnino, M., Fortin, S. and Lombardi, O. (2009). Decoherence as a relative phenomenon: a generalization of the spin-bath model, arXiv:0907.1933 [quant-ph].
[19] Caves, C. M., Fuchs, C. A. and Schack, R. (2002). Quantum probabilities as Bayesian probabilities. *Physics Review A*, **65**: 022305.
[20] Ciancio, E., Giorda, P. and Zanardi, P. (2006). Mode transformations and entanglement relativity in bipartite Gaussian states. *Physics Letters A*, **354**: 274.
[21] Cooney, T., Junge, M., Navascués, M., Perez-Garcia, D. and Villanueva, I. (2013). Joint system quantum descriptions arising from local quantumness. *Communication in Mathematical Physics*, **322**: 501.
[22] Cramer, J. G. (1986). The transactional interpretation of quantum mechanics, *Reviews of Modern Physics*, **58**: 647–688.
[23] Dalton, B. J., Barnett, S. M. and Garraway, B. M. (2001). Theory of pseudomodes in quantum optical processes. *Physical Review A*, **64**: 053813.
[24] De la Torre, A. C., Goyeneche, D. and Leitao, L. (2010). Entanglement for all quantum states. *European Journal of Physics*, **31**: 325.
[25] Del Rio, L., Hutter, A., Renner, R. and Wehner, S. (2014). Relative thermalization, arXiv:1401.7997 [quant-ph].
[26] Dieks D. (1998). Preferred factorizations and consistent property attribution, *Minnesota Stud. Phil. Sci.*, **17**: 144.
[27] Diósi, L. (1984). Gravitation and quantum mechanical localization of macroobjects. *Physics Letters A*, **105**: 199.
[28] Diósi, L. (2013). Gravity-related wave function collapse: mass density resolution. *Journal of Physics: Conference Series*, **442**: 012001.
[29] Diósi, L. (2014). Gravity-related spontaneous collapse in bulk matter. *New Journal of Physics*, **16**: 105006.
[30] Dubessy, R., De Rossi, C., Badr, T., Longchambon, L. and Perrin, H. (2014). Imaging the collective excitations of an

ultracold gas using statistical correlations. *New Journal of Physics*, **16**: 122001.

[31] Dugić, M. (1999). What is "system": The arguments from the decoherence theory, arXiv:quant-ph/9903037.

[32] Dugić, M. and Jeknić, J. (2006). What is "System": Some decoherence-theory arguments, *International Journal of Theoretical Physics*, **45**: 2215.

[33] Dugić, M. and Jeknić-Dugić, J. (2008). What is "System": The information-theoretic arguments, *International Journal of Theoretical Physics*, **47**: 805.

[34] Dugić, M. and Jeknić-Dugić, J. (2012). Parallel decoherence in composite quantum systems. *Pramana*, **79**: 199.

[35] Dugić, M., Arsenijević, M. and Jeknić-Dugić, J. (2013). Quantum correlations relativity. *Science China Physics, Mechanics Astronomics*, **56**: 732.

[36] Dür, W., Vidal, G. and Cirac, J. I. (2000). Three qubits can be entangled in two inequivalent ways. *Physical Review A*, **62**: 062314.

[37] Dürr, D., Goldstein, S. and Zangh, N. (2013). *Quantum Physics Without Quantum Philosophy*. Springer: Berlin.

[38] Eakins, J. and Jaroszkiewicz, G. (2003). Factorization and entanglement in quantum systems. *Journal of Physics A: Mathematical and General*, **36**: 517.

[39] Eilenberg, S. and Mac Lane, S. (1945). General theory of natural equivalences. *Transactions of American Mathematical Society*, **58**: 231.

[40] Evenbly, G. and Vidal, G. (2014). Real-space decoupling transformation for quantum many-body systems. *Physics Review Letters*, **112**: 220502.

[41] Fel'dman, E. B. and Zenchuk, A. I. (2014a). Systems with stationary distribution of quantum correlations: open spin-1/2 chains with XY interaction. *Quantum Information Processing*, **13**: 201.

[42] Fel'dman, E. B. and Zenchuk, A. I. (2014b). Robust stationary distributed discord in Jordan–Wigner fermion system under perturbations of initial state. *Journal of Experimental and Theoretical Physics*, **119**: 412.

[43] Feng, M., Zhong, Y. P., Liu, T., Yan, L. L., Yang, W. L., Twamley, J. and Wang, H. (2015). Exploring the quantum critical behaviour in a driven Tavis–Cummings circuit. *Nature Communications*, **6**: 7111.

[44] Gao, F., Guo, F.-Z., Wen, Q.-Y. and Zhu, F.-C. (2010). Quantum key distribution by constructing nonorthogonal states with Bell states. *International Journal of Modern Physics B*, **24**: 4611.

[45] Gavrilik, A. M. and Mishchenko, Yu. A. (2015). Entanglement entropy of composite fermions realized by (deformed) fermions vs. that of composite bosons, arXiv:1504.00529 [quant-ph].
[46] Ghirardi, G. C., Rimini, A. and Weber, T. (1986). Unified dynamics for microscopic and macroscopic systems. *Physics Review D*, **34**: 470.
[47] Giulini, D., Joos, E., Kiefer, C., Kupsch, J., Stamatescu I.-O. and Zeh, H. D. (1996). *Decoherence and the Appearance of a Classical World in Quantum Theory.* Spinger: Berlin.
[48] Goold, J., Huber, M., Riera, A., del Rio, L. and Skrzypzyk, P. (2015). The role of quantum information in thermodynamics, arXiv:1505.07835 [quant-ph].
[49] Hardal, A. U. C. and Mustecaplioglu, U. E. (2015). Superradiant quantum heat engine. *Scientific Reports*, **5**: 12953.
[50] Harshman, N. L. and Ranade, K. S. (2011). Observables can be tailored to change the entanglement of any pure state. *Physical Review A*, **84**: 012303.
[51] Hartle, J. (2010). Quasiclassical realms, in *Many-Worlds? Everett, Quantum Theory, and Reality*, S. Saunders, J. Barrett, A. Kent, D. Wallace (eds.), Oxford University Press: Oxford, pp. 73–98.
[52] Holland, P. (2005). Computing the wave function from trajectories: particle and wave pictures in quantum mechanics and their relation. *Annals of Physics (NY)*, **315**: 505.
[53] Holland, P. (2005). Hydrodynamic construction of the electromagnetic field. *Proceedings of the Royal Society of London A*, **461**: 3659.
[54] Huh, J., Mostame, S., Fujita, T., Yung, M.-H. and Aspuru-Guzik, A. (2014). Linear-algebraic bath transformation for simulating complex open quantum systems. *New Journal of Physics*, **16**: 123008.
[55] Janotta, P. and Hinrichsen, H. (2014). Generalized probability theories: what determines the structure of quantum theory? *Journal of Physics A: Mathematical and Theoretical*, **47**: 323001.
[56] Jeknić-Dugić, J. and Dugić, M. (2008). Multiple system decomposition method for avoiding quantum decoherence. *Chinese Physics Letters*, **25**: 371.
[57] Jeknić-Dugić, J., Arsenijević, M. and Dugić, M. (2013). *Quantum Structures: A View of the Quantum World.* LAP Lambert Scientific Publishing: Saarbrücken.
[58] Jeknić-Dugić, J., Dugić, M. and Francom, A. (2014). Quantum structures of a model-universe: questioning the Everett interpretation of quantum mechanics. *International Journal of Theoretical Physics*, **53**: 169.

[59] Jen, H. H. (2015). Collective Lamb shift of superradiant cascade emissions in an atomic ensemble, arXiv:1501.05099 [quant-ph].
[60] Jurcevic, P., Lanyon, B. P., Hauke, P., Hempel, C., Zoller, P., Blatt, R. and Roos, C. F. (2014). Quasiparticle engineering and entanglement propagation in a quantum many-body system. *Nature*, **511**: 202.
[61] Kanada-En'yo, Y. (2015). Analysis of delocalization of clusters in linear-chain a-cluster states with entanglement entropy. *Physics Review C*, **91**: 034303.
[62] Kastner, R. E. (2012). *The Transactional Interpretation of Quantum Mechanics: The Reality of Possibility*. Cambridge University Press.
[63] Kastner, R. E. (2014). Einselection of pointer observables: The new H-theorem? *Studies in History and Philosophy of Science Part B: Studies in History and Philosophy of Modern Physics*, **48**: 56.
[64] Kastner, R. E. (2015). Classical selection and quantum Darwinism. *Physics Today*, **68**: 8.
[65] Kay, A., Kaszlikowski, D. and Ramanathan, R. (2009). Optimal cloning and singlet monogamy. *Physics Review Letters*, **103**: 050501.
[66] Knott, P. A., Sindt, J. and Dunningham J. A. (2013). Detecting measurement-induced relative-position localization. *Journal of Physics B: Atomic, Molecular and Optical Physics*, **46**: 095501.
[67] Koashi, M. and Winter, A. (2004). Monogamy of entanglement and other correlations. *Physics Review A*, **69**: 022309.
[68] Lal, R. and Teh, N. J. (2014). Categorical generalization and physical structuralism, arXiv:1404.3049 [physics.hist-ph].
[69] Lee, T. E., Chan, C.-K. and Yelin, S. F. (2014). Dissipative phase transitions: independent vs. collective decay and spin squeezing. *Physics Review A*, **90**: 052109.
[70] Lim, J., Tame, M., Yee, K. H., Lee, J.-S. and Lee, J. (2014). Comment on 'Energy transfer, entanglement and decoherence in a molecular dimer interacting with a phonon bath? *New Journal of Physics*, **16**: 018001.
[71] Liu, X. F. and Sun, C. P. (2004). On the relative quantum entanglement with respect to tensor product structure, arXiv:quant-ph/0410245.
[72] Ma, X.-S., Zotter, S., Kofler, J., Ursin, R., Jennewein, T., Brukner, C. and Zeilinger, A. (2012). Experimental delayed-choice entanglement swapping. *Nature Physics*, **8**: 480.
[73] Manrique, P. D., Rodriguez, F., Quiroga, L. and Johnson, N. F. (2015). Non-equilibrium quantum systems: Divergence between

global and local descriptions. *Advances in Condensed Matter Physics*, **2015**: 615727.

[74] Manzano, G., Galve, F. and Zambrini, R. (2013). Avoiding dissipation in a system of three quantum harmonic oscillators. *Physics Review A*, **87**: 032114.

[75] Milne, A., Jevtic, S., Jennings, D., Wiseman, H. and Rudolph, T. (2014). Quantum steering ellipsoids, extremal physical states and monogamy. *New Journal of Physics*, **16**: 083017.

[76] Mitsch, R., Sayrin, C., Albrecht, B., Schneeweiss, P. and Rauschenbeutel, A. (2014). Exploiting the local polarization of strongly confined light for sub-micrometer-resolution internal state preparation and manipulation of cold atoms. *Physics Review A*, **89**: 063829.

[77] Nadj-Perge, S., Drozdov, I. K., Li, J., Chen, H., Jeon, S., Seo, J., MacDonald, A. H., Bernevig, B. A. and Yazdani, A. (2014). Observation of Majorana fermions in ferromagnetic atomic chains on a superconductor. *Science*, **346**: 602.

[78] Penrose, R. (1996). On gravity's role in quantum state reduction. *General Relativity and Gravitation*, **28**: 581.

[79] Pfister, C., Kaniewski, J., Tomamichel, M., Mantri, A., Schmucker, R., McMahon, N., Milburn, G. and Wehner, S. (2015). Understanding nature from experimental observations: A theory independent test for gravitational decoherence, arXiv:1503.00577 [quant-ph].

[80] Piazza, F. and Costa, F. (2007). Volumes of space as subsystems, arXiv:0711.3048 [gr-qc].

[81] Pigeon, S., Xuereb, A., Lesanovsky, I., Garrahan J. P., De Chiara, G. and Paternostro, M. (2015). Dynamical symmetries and crossovers in a three-spin system with collective dissipation. *New Journal of Physics*, **17**: 015010.

[82] Putz, S., Krimer, D. O., Amsüss, R., Valookaran, A., Nöbauer, T., Schmiedmayer, J., Rotter, S. and Majer, J. (2014). Protecting a spin ensemble against decoherence in the strong-coupling regime of cavity QED. *Nature Physics*, **10**: 720.

[83] Rao, K. R. K., Katiyar, H., Mahesh T. S., De, A. S., Sen, U. and Kumar A. (2013). Monogamy of quantum correlations reveals frustration in a quantum Ising spin system: Experimental demonstration. *Physics Review A*, **88**: 022312.

[84] Rau, A. V., Dunningham, J. A. and Burnett, K. (2003). Measurement-induced relative-position localization through entanglement. *Science*, **301**: 1081.

[85] Röntynen, J. and Ojanen, T. (2015). Topological superconductivity and high Chern numbers in 2D ferromagnetic Shiba lattices. *Physics Review Letters*, **114**: 236803.

[86] Sachdev, S. (2006). Quantum phase transitions, in *The New physics for the Twenty-First Century*, G. Fraser (ed.), Cambridge University Press: Cambridge.

[87] Schlosshauer, M. (2004). Decoherence, the measurement problem, and interpretations of quantum mechanics. *Review of Modular Physics*, **76**: 1267.

[88] Schönleber, D. W., Croy, A. and Eisfeld, A. (2015). Pseudomodes and the corresponding transformation of the temperature-dependent bath correlation function. *Physics Review A*, **91**: 052108.

[89] Scully, M. (2015). Single photon subradiance: Quantum control of spontaneous emission and ultrafast readout, arXiv:1505.03073 [quant-ph].

[90] Stokes, A., Kurcz, A., Spiller T. P. and Beige, A. (2012). Extending the validity range of quantum optical master equations. *Physics Review A*, **85**: 053805.

[91] Tay, B. A. (2014). Attenuation of excitation decay rate due to collective effect. *Physics Review E*, **90**: 022142.

[92] Thirring, W., Bertlmanna, R. A., Kohler, P. and Narnhofer, H. (2011). Entanglement or separability: The choice of how to factorize the algebra of a density matrix, *European Physical Journal D*, **64**: 181.

[93] Tokunaga, Y., Yu, X. Z., J. White, S., Rønnow, H. M., Morikawa, D., Taguchi, Y. and Tokura, Y. (2015). A new class of chiral materials hosting magnetic skyrmions beyond room temperature. *Nature Communications*, **6**: 7638.

[94] Tommasini, P., Timmermans, E. and de Toledo Piza, A. F. R. (1998). The hydrogen atom as an entangled electron-proton system. *American Journal of Physics*, **66**: 881.

[95] Toner, B. (2009). Monogamy of non-local quantum correlations. *Proceedings of the Royal Society of London A*, **465**: 59.

[96] Vedral, V. (2010). *Decoding Reality*. Oxford University Press: Oxford.

[97] Wallace, D. (2012). *The Emergent Multiverse: Quantum Theory According to the Everett Interpretation*. Oxford University Press: Oxford.

[98] Wang, Z. H., Ji, Y. J., Li, Y. and Zhou D. L. (2014). Dissipation and decoherence induced by collective dephasing in coupled-qubit system with a common bath, arXiv:1409.0172 [quant-ph].

[99] Wichterich, H. C. (2011). *Entanglement between Noncomplementary Parts of Many-Body Systems*. Springer Theses, Springer: Berlin.

[100] Woods, M. P., Ng, N. and Wehner, S. (2015). The maximum efficiency of nano heat engines depends on more than temperature, arXiv:1506.02322 [quant-ph].

[101] Xu, S.-J., Chen, X.-B., Wang, L.-H., Niu, X.-X. and Yang, Y.-X. (2015). A novel quantum information hiding protocol based on entanglement swapping of high-level Bell states. *Chinese Physics B*, **24**: 050306.

[102] Xu, S.-Y., Belopolski, B., Alidoust, N. *et al.* (2015). Discovery of a Weyl Fermion semimetal and topological Fermi arcs. *Science*, **349**: 613.

[103] Zagoskin, A. M., Chipouline, A., Il'ichev, E., Johansson, J. R. and Nori F. (2014). Toroidal qubits: naturally-decoupled quiet artificial atoms, arXiv:1406.7678 [quant-ph].

[104] Zanardi, P. (2001). Virtual quantum subsystems. *Physics Review Letters*, **87**: 077901.

[105] Zanardi, P., Lidar D. A. and Lloyd, S. (2004). Quantum tensor product structures are observable induced. *Physics Review Letters*, **92**: 060402.

[106] Zeh, H. D. (1993). There are no quantum jumps, nor are there particles! *Physics Letters A*, **172**: 189.

[107] Zurek, W. H. (2003). Decoherence, einselection, and the quantum origins of the classical. *Review of Modular Physics*, **73**: 715.

[108] Kastner, R. E. (2016a). Bound states as fundamental quantum structures. This volume, p. 427.

[109] Kastner, R. E. (2016b). Beyond Complementarity. This volume, p. 77.

Historical Aspects

Chapter 2

Bohr's Diaphragms

Tongdong Bai*,‡ and John Stachel†

*Department of Philosophy, Xavier University
5 University Drive, Cincinnati, OH 45207

†Center for Einstein Studies, Boston University
‡bai@xavier.edu

In his response to EPR, Bohr introduces several ideal experimental arrangements that often are not understood correctly, and his discussion about them is given a positivist reading. Our analysis demonstrates the difference between such a reading and Bohr's actual position, and also clarifies the meaning of several of Bohr's key physical and philosophical ideas:

The role of the quantum of action in the distinction between classical and quantum systems;
The criterion of measurability for theoretically defined concepts;
The freedom in placement of the "cut" between measuring instrument and measured system;
The non-visualizability of the quantum formalism; and Bohr's concepts of phenomenon and complementarity.

1. Introduction

In EPR [16], Albert Einstein, Boris Podolsky, and Nathan Rosen challenged the completeness of the quantum mechanical description of a physical system. Their argument may be summarized as follows:

Let two systems (hereafter I and II) initially be non-interacting, so that the initial wave function of the total system (hereafter III) is the product of separate wave functions for I and II. According to quantum mechanics (QM), if they subsequently interact during a finite time interval, then they become a single composite system III. Even after the interaction has ceased, I and II no longer have individual wave functions, and III is represented by a single joint wave function.

As an example, EPR consider a composite system III such that, after the interaction, the difference in the positions $x_1 - x_2$ and the sum of the momenta $p_1 + p_2$ of systems I and II are eigenvalues of the joint wave function III.[a]

After the two systems cease to interact, a subsequent sharp measurement of the coordinate x_1 of system I at a certain time[b] thus can be used, together with the information about the joint wave function III, to predict with certainty the coordinate value x_2 of system II at the same time. EPR argue that such a determinate coordinate value x_2 should be taken as an element of reality? belonging to system II immediately before the measurement made on system I:

> [A]t the time of measurement, the two systems no longer interact, [and] no real change can take place in the second system in consequence of anything that may be done to the first system.

If this argument were accepted, it already would show the incompleteness of the QM description: according to QM, x_2 does not have a determinate value just before the sharp measurement of x_1. But EPR did not base their argument for the incompleteness of QM on this observation. They argue that, instead of measuring the position of system I, one could have chosen to measure another physical quantity, represented by an operator that does not commute

[a]This is possible because the corresponding QM operators are easily seen to commute. For simplicity, here we consider only one spatial dimension of the system. In the next section, the other spatial dimensions are included in the discussion of diaphragms.

[b]Later, we shall return in some detail to the question of why a certain moment of time must be chosen.

with its coordinate operator,[c] in particular, its momentum p_1. Measurement of p_1, together with the prior information about the joint wave function, could have been used to predict with certainty a determinate value p_2 of system II's momentum at the same time as that, at which the position measurement of system I was actually made.

From this observation, EPR conclude that system II must have had *both* a determinate position x_2 and momentum p_2 just before any measurement is made on system I (see the quotation given above). But QM precludes the possibility of a description ascribing to a system at a certain time determinate values of a pair of non-commuting variables; in particular, the components of position and momentum in some given direction. Thus, the QM description of a system cannot be complete.

Bohr's response [7] introduces several new ideal experimental arrangements. Although many commentaries on Bohr's views exist, detailed analyses of his use of these diaphragms are rare, and accurate ones even rarer.[d] In particular, the argument based on these ideal experiments has been given a positivist reading, leading to a positivist interpretation of Bohr's reply to the EPR paper, and more generally of Bohr's philosophical viewpoint.[e]

In this chapter, we present a reading of these arrangements that we believe coincides substantially with Bohr's own. In the next section, we shall explain in detail these ideal experimental arrangements based on the use of one or more diaphragms. The third

[c]Sometimes, for brevity, we shall say that two quantities do or do not commute, but this is always what we really mean.

[d]We have chosen as prime examples of inaccuracy, [1, 2] Chapter 7 (our criticisms of the book are confined to this chapter), and Fine [18]. One of us (J.S.) was a close friend of Mara Beller, whose untimely death he will never cease to mourn. But she and he shared a sense of philosophy of science as a search for clarity through dialogue, and in their correspondence he had explained his criticisms to her.

[e]Beller and Fine [2] speak of "the positivist perspective that Bohr eventually adopted" and "his positivist solution to EPR." See for a critique of some aspects of Beller and Fine's account of Bohr's position. Howard [21] asserts: "Bohr was ... in no way a positivist," a position with which we agree.

section explains the role of the quantum of action and of the "cut", both crucial to understanding Bohr's response. In the fourth section, we explain Bohr's concept of phenomenon, which is closely related to the discussion in the third section, and crucial to the difference between Bohr's position and a positivist one. Finally, the last section discusses a few difficulties with Bohr's experimental arrangements. The conclusion summarizes the differences between Bohr's position and a positivist one.

2. Uses of Diaphragm(s) in Different Experimental Arrangements in Bohr's Response

Bohr introduces several new ideal experimental arrangements using diaphragms. Central to his analysis of all such experiments is the presence of a [material] support which defines the space frame of reference, tacitly assumed to be inertial.[f] The coordinates and momenta are defined with respect to this inertial frame, and the instruments for measuring these quantities are located, spatially and temporally, with respect to it. It is also assumed that the material support is so massive that, the momentum exchanged between the particle and the diaphragm will, together with the reaction of the particle on the other bodies [rigidly attached to the support], pass into this common support [7, p. 697] without significant effect on its state of motion.

A diaphragm is a large screen (idealized as two-dimensional) that is opaque to a beam of particles or radiation impinging upon it, except for the part of the beam that passes through one or more slits in it.[g] The width of the slits is assumed large compared to the

[f]Bohr refers to "some ultimate measuring instruments, like the scales and clocks which determine the frame of spacetime coordination — on which, in the last resort, even the definitions of momentum and energy quantities rest..." ([14], p. 104). Since the discussion is confined to non-relativistic QM, we may confine our consideration of time to the Newtonian absolute time.

[g]We speak of a beam, even though the particles or quanta of radiation may be so spread out in time that (on the average) only one at a time passes through the diaphragm. QM can only treat ensembles of particles produced by some (ideal) device, even if the ensembles are temporal rather than a spatial, and a virtual rather than real (see, e.g. [28, 29]).

(average) wavelength of the particles or photons in the incident beam; so that the slit width may be taken as the uncertainty Δq of the position of the particle relative to the diaphragm, in the direction perpendicular to the slit (*ibid.*, p. 697).[h]

One type of diaphragm (D1) has a single slit in it,[i] and may form part of some more or less complicated experimental arrangement? (*ibid.*, p. 697). But let us dwell on D1 for a moment. It can be used for measuring one component of the position (q_y) of a particle passing through the slit *relative to the plane of the diaphragm* and to the direction ($\boldsymbol{y}$) perpendicular to the slit. An indeterminateness[j] or uncertainty in position (Δq_y) is associated with this measurement and, in accord with Heisenberg's general principle $\Delta p \Delta q \sim h$ (*ibid.*, p. 697), an uncertainty of the particle's momentum (Δp_y) is associated with it:

> Obviously, the uncertainty Δp is inseparably connected with the possibility of an exchange of momentum between the particle and the diaphragm; and the question of principle interest for our discussion is now to what extent the momentum thus exchanged can be taken into account in the description of the phenomenon to be studied by the experimental arrangement concerned, of which the passage of the particle through the slit [D1] may be considered as the initial stage (*ibid.*, p. 697).

Bohr considers two such arrangements. In the first (A1), D1 is rigidly fixed to the frame of reference. When a particle passes through such a slit:

> [t]hen the momentum exchanged between the particle and the diaphragm will . . . pass into this common support, and we have thus voluntarily cut ourselves off from any possibility of taking these reactions separately into account in predictions regarding the final outcome of the experiment . . . (*ibid.*, p. 697).

---

[h]We assume that, if the diaphragm has more than one slit, the width of each is small compared to the distance between slits.

[i]The designations of the diaphragms as D1, D2, and of the experimental arrangements as A1, A2, etc., are ours, introduced for clarity in our references to them.

[j]Heisenberg's "Unbestimmtheit" is better translated "indeterminateness," but "uncertainty" seems impossible to replace [19].

But, just because the momentum exchange between D1 and the particle is indeterminate, the position component q_y *relative to the support*[k] can be measured.

In the second arrangement (A2), D1 is not rigidly fixed to the support, but connected to it, by a system of springs or the like, that allows D1 to recoil when a particle passes through it, relative to the support. In arrangement (A2), one still has a free choice *after* the particle passes the slit. One can *either*[l]:

(1) add an arrangement (A2-1) to the apparatus, allowing measurement of the Y-component of the *position* of the slit in D1 relative to the support immediately after the particle passes through it; and hence deduce the y-component of the position of the particle at that time; or
(2) add a different arrangement (A2-2) to the apparatus, allowing measurement of P_y, the Y-component of the *momentum* of D1 relative to the support after the passage of the particle through it; and, using conservation of momentum, deduce p_y, the y-component of the momentum of the particle, as it passed through the slit in D1.[m]

One actually measures the change in the momentum of D1, and hence in that of the particle. Calculating the total momentum of each

[k]From now on, we use "support" and "frame of reference" interchangeably.

[l]In an arrangement suited for measurements of the momentum of the first diaphragm [i.e. A2], it is further clear that even if we have measured this momentum before the passage of the particle through the slit, we are after this passage still left with a *free choice* whether we wish to know the momentum of the particle or its initial position [i.e. at the moment of passing through D1] relative to the rest of the apparatus [7, p. 698].

[m]Note that a measurement of the position of the slit must be made immediately after passage of the particle through it, since the position of the particle does not remain sharp (i.e. position measurements on it made a finite time after passage through the slit would give a spread of results). On the other hand, assuming that particle and diaphragm are not subject to any external forces after passage through the slit, the y-components of their momenta remain constant (i.e. a momentum measurement made a finite time after passage through the slit would give the same sharp result). This point is discussed in greater detail below.

presupposes that the y-components of their initial momenta have determinate values just before the particle passes through the slit. D1 is usually taken to be at rest initially, so that its initial $P_y = 0$. If it is assumed that the beam of particles was prepared in a sharp momentum state, with the total momentum of each particle in it in the (x) direction of propagation of the beam; then the initial $p_y = 0$ and the change in p_y is equal to p_y.

Bohr asserts that in QM sub-arrangements A2-1 and A2-2 are mutually exclusive,[n] and below we shall go into the important reasons for this assertion. But first, a few other points need clarification. First of all, we emphasize that D1 plays a quite *different* role in arrangements A1 and A2. In A1, D1 is rigidly attached to the frame of reference and so forms part of the measuring apparatus. The phenomenon[o] that started with the preparation of the particle in a definite momentum state is complete when the position of the particle at the time of its passage through the slit is registered. A clock, rigidly fixed to the frame of reference, would be needed to actually record this time.[p] In arrangement A1, no further analysis of the behavior of D1 is permitted; or better put: if one were to attempt it, then one would no longer be dealing with the same phenomenon.

Arrangement A2 provides an example of such an alternative attempt. Here, D1 is not part of the measuring apparatus, but is part of another, composite system, consisting of the particle plus D1. Now one can perform a further measurement on D1 — either A2-1

[n] In the first eventuality, we only need to make a second determination of the momentum of the diaphragm [D1], leaving unknown forever its exact position when the particle passed [through it]. In the second eventuality, we need to only determine its position relative to the space frame with the inevitable loss of the knowledge of the momentum exchanged between the diaphragm and the particle [7, p. 698].

[o] For Bohr, a phenomenon is a complete process, consisting of an initial preparation of a system, its subsequent interactions, and a final registration of the result of some measurement on it. Any attempt to physically subdivide a phenomenon (for example, by an intermediate measurement) results in a different phenomenon. See the final section for further discussion of Bohr's concept.

[p] Elsewhere, Bohr made this point explicitly: "If we want to use the idea of spacetime, we must have watches and rods which are outside and independent of the object under observation...".

or A2-2 — and use the result to deduce corresponding information about D1's partner in the new composite system: the particle.[q] Bohr emphasizes that, if A2-2 is used, the possibility of also using the output of D1 to produce interference patterns by subsequently passing this output through another diaphragm with two or more slits in it, no longer exists:

> [E]ven the minimum uncertainty of the position of the first diaphragm [D1 associated with the sharp measurement of P_y] . . . will imply the total wiping out of any interference effect . . . to which the presence of more than one slit in the second diaphragm would give rise in case the positions of all apparatus are fixed relative to each other [as in case A1, discussed above] [7, p. 698].

As far as questions of principle are concerned, arrangement A2 is really quite analogous to the EPR arrangement: two systems, the particle and diaphragm D1, initially non-interacting before the particle's passage through the slit in D1, interact for the limited time interval it takes the particle to pass through the slit; and remain entangled even though the particle and D1 do not interact physically afterwards. In order to extract information about the particle from D1, a further measurement on it is necessary; and here the alternate arrangements A2-1 and A2-2 enter the story. Although Bohr does not stress the point, the analogy with the two particles of EPR is really quite striking.[r]

With detailed discussion of the one-particle case behind him, Bohr is able to make short work of the two-particle EPR case by means of a third experimental arrangement (A3), utilizing a two-slit

[q]The principle difference between the two experimental arrangements . . . is . . . that in the arrangement suited for the control of the momentum of the first diaphragm [i.e. D1 in arrangement A2], this body can no longer be used as a measuring instrument for the same purpose as in the previous case [i.e. D1 in arrangement A1], but must, as regards its position relative to the rest of the apparatus [i.e. the frame of reference], be treated, like the particle traversing the slit, as an object of investigation. . . [7, p. 698].

[r]Of course, if one does not see the force of this analogy, it may appear that Bohr has never really came to grips with the EPR thought-experiment; we discuss this objection below.

diaphragm (D2) that (like the one-slit diaphragm in arrangement A2) is free to move. Bohr describes A3 as:

> a simple experimental arrangement, comprising a rigid diaphragm with two parallel slits [D2], which are very narrow compared to their separation, and through each of which one particle with given initial momentum passes independently of the other. If the momentum of this [mobile] diaphragm is measured accurately before as well as after the passing of the particles, we shall in fact know the sum of the components [of the particle momenta] perpendicular to the slits [7, p. 699].

Some further explication may be helpful. The purpose of A3 is to realize, by means of an ideal experimental arrangement, the wave function initially introduced by EPR. By making measurements of the y-components of the two particles' momenta before passage through D2, and of the y-component of D2's momentum before and after their passage; and then invoking conservation of momentum, we are able to obtain the y-component of the total momentum of the two particles after passage through the slits. Or to put it more accurately, this y-component is QM-determinate if an apparatus is subsequently introduced to complete the phenomenon by its measurement.

Since they passed through the two slits, the difference in their q_y's — but not their individual q_y's, since D2 is mobile — just after passage through the slit is also determinate.[s] But this is just a physical realization of the state described by the EPR wave function, in which both the y-component of their total momentum and the difference between the y-components of their positions are determinate.

A source of possible misunderstanding is to assume that Bohr is treating arrangements A1 and A2 symmetrically; i.e. that, just as A1 can be used only to measure the position q_y of the particle, so A2 can be used only to measure its momentum p_y — in other

[s]The requirement that slit widths be small compared to the distance between them assures that the Δq_y's are small compared to the difference between the two q_y's.

words, to assume that Bohr treats A1 and A2 as mutually exclusive, complementary arrangements.[t]

As we have just seen, this is not the case. Arrangement A1 can be used *only* to measure a position component since D1 is part of the measuring apparatus, and thus one does not have any choice of how to use it after the particle passes the slit. In order to measure a momentum component, one must have recourse to another arrangement. But, instead of a *complementary* arrangement that would enable *only* a measurement of p_y, Bohr introduces the second arrangement A2, in which D1 itself can be used subsequently *either for* a position *or* a momentum measurement. A2-1 and A2-2 are *truly* complementary to each other; but involve the introduction of *additional* arrangements (such as D2 in the EPR case).[u]

Of course, Bohr [7] bears some responsibility for the possibility of this misinterpretation. It might *seem* that he regards the two arrangements A1 and A2 as symmetrical; although, if the text is examined carefully, it is readily seen that they are not — and nowhere does he imply that they are. When he does talk about two arrangements, one of which can be used only to measure position and the other only to measure momentum, rather than A1 and A2,

---

[t]Indeed, Beller captions a reproduction of Bohr's drawings: "The rigidly bolted apparatus (*center*) is for position measurement. The apparatus suspended by weak springs (*top*) is for momentum measurement" [1, p. 148]; and elaborates in the text: "In addition to the first two-slit diaphragm, we employ a second diaphragm suspended by weak springs or rigidly bolted, depending on whether we intend to measure the position or momentum of the first particle.... For the measurement of P_1 we use the movable diaphragm, so we exclude in principle the possibility of measuring Q_1..." [1, p. 147]. Not only Bohr's critics commit this error. In his defense of Bohr against charges of positivism, Hooker [20] writes: "[W]hen the diaphragm is not rigidly attached to the coordinate framework, there is no question of applying the concept of position (in the direction perpendicular to [particle] A's trajectory) to the A component's passage through the slit. Hence, there is no question of applying this concept of position to the situation as a whole at all" (p. 223). The same misunderstanding can also be found in [22, pp. 95 and 196]. In contrast, Murdoch [17, p. 308]; [25, pp. 168–170] states it correctly.

[u]Presumably Bohr wanted to demonstrate that D1 could also be used for a momentum measurement, but could not think of any way of using it *exclusively* for a momentum measurement. At any rate, *we* have not been able to do so.

he is always referring to the two sub-arrangements A2-1 and A2-2 (otherwise, his either... or, would not be exclusive). Such possibly ambiguous expressions aside, Bohr does clearly and explicitly point out that A2 can be used for *both* kinds of measurements, and that one still has freedom of choice *after* passage of the particle through D1.[v]

Bohr returned to the EPR paper in his account of discussions with Einstein [11], and this later version is often much clearer and more explicit. He writes:

> In fact, after a preliminary measurement of the momentum of the diaphragm, we are in principle offered the choice, when an electron or photon has passed through the slit, either to repeat the momentum measurement or to control the position of the diaphragm, and, thus, to make predictions pertaining to alternative subsequent observations. [11, p. 57]

Later, he repeats this point [11, p. 60].[w] It is clear that, for Bohr [7, 11], the movable diaphragm can be used for either a momentum-component or a position-component measurement, and that one has free choice of which to measure *after* passage of the particle through the diaphragm.

This point is important for Bohr's reply to the EPR paper. Freedom of choice *after* passage of the particle is closer to EPR's alternative measurement requirement (i.e. one could decide to make either a position or a momentum measurement) than would be such freedom *before* passage of the particle. More importantly, assuming that A1 and A2 are symmetrical is simply wrong physics. Both in the case of classical mechanics (CM) and QM, in arrangement A1, one can measure only a position component with D1; while in

[v]See the citation in Note 12.

[w]After presenting her misinterpretation of Bohr's argument (see note 20), Beller goes on to cite these lines of Bohr [1, p. 147], suggesting that: "Two different, even incompatible, answers are concurrently present in Bohr's response to EPR" [1, p. 146]. There are indeed "two different, even incompatible answers" in her text, but only one of them is Bohr's. See for further comments on Beller's treatment of Bohr.

both CM and QM, in arrangement A2, one can make a position and a momentum component measurement with D1. In this case, the "only" difference between CM and QM is that, according to the former, we can make *both* measurements using one experimental arrangement; while, according to the latter, we must make a choice between either A2-1 or A2-2, the possibility of making the other being thereby excluded.

3. The Role of Quantum of Action, the Idea of the Cut, and an Implicit Use of Entanglement

Now we come to the crucial question: Why this difference between CM and QM? Is it merely that a measurement of one quantity so disturbs the measured system that it is unusable for a second measurement of a different quantity? Indeed, even in CM, a classical-mechanical disturbance during a measurement is often inevitable, so that subsequent measurement of a second quantity gives a different result than if that quantity had been measured first. But such a classical disturbance can be controlled in the sense that: either the effect of the disturbance can be made negligible by some appropriate modification of the first measurement procedure; or, with the help of physical theories, the quantitative effect of this disturbance can be calculated and subtracted from the second measured result to obtain the answer that would have been obtained if the first measurement had not been made. So, classical, controllable disturbances cannot be the reason for this difference.

According to Bohr, what is unique to the case of QM is that the disturbance is uncontrollable in principle. Indeed, he uses the word "uncontrollable" again and again throughout his writings. According to Bohr, it is this uncontrollable nature of the quantum exchange between the measured system and the measuring instrument that makes them into a whole that cannot be further subdivided.

The reason that the quantum exchange is uncontrollable is not a question of the accuracy of measurements, but lies in the existence of Planck's quantum of action h [13, p. 5]. *Because of the existence of the quantum of action*, the exchange between system and measuring instrument becomes uncontrollable.[x,y] The quantum of action cannot be explained by or derived from a classical or quasi-ordinary disturbance picture. Rather, the existence of h is responsible for, is revealed in, and is presupposed by, the nature and limits of possible measurements in QM. Similarly, the indeterminacy relations and other QM characteristics, such as the necessity of a final renunciation of the classical ideal of causality and a radical revision of our attitude towards the problem of physical reality, also arise from the existence of the quantum of action [7, p. 697]. For Bohr, the existence and efficacy of the quantum of action is the most fundamental feature of QM, which cannot be reduced to any other empirical facts. Bohr states that one must treat the quantum of action as an element evading customary explanation ... similar to the role of the velocity of light in relativity theory as a maximal speed of signals [13, p. 12].

In the case of Bohr's diaphragms, the uncontrollable, finite, interaction is therefore the result of the existence of the quantum of action. Applying this general principle to his experimental arrangements, Bohr argues:

> The principle difference between the two experimental arrangements under consideration is, however, that in the arrangement suited for the control of the momentum of the first diaphragm [i.e. the second arrangement A2], this body can no longer be used as a measuring instrument for the same purpose as in the previous case, but must, as regards its position relative to the rest of the apparatus, be treated, like the particle traversing

[x][T]he *finite interaction between object and measuring agencies* conditioned by the very existence of the quantum of action entails because of the impossibility of controlling the reaction of the object on the measuring instruments if these are to serve their purpose — the necessity of a final renunciation of the classical ideal of causality [7, p. 697].

[y]In a highly problematic presentation of the EPR debate (we only point out a few problems closely related to our topic), Fine [18] keeps criticizing Bohr's use of the language of disturbance. However, he fails to make clear the essential distinction Bohr makes between controllable classical disturbances and uncontrollable quantum disturbances; and hence does not explain what makes quantum disturbances uncontrollable (see Section 4).

> the slit, as a system of [i.e. under] investigation, in the sense that the quantum-mechanical uncertainty relations regarding its position and momentum must be taken explicitly into account. In fact, even if we knew the position of the diaphragm relative to the space frame [of reference] before the first measurement of its momentum, and even though its position after the last measurement can be accurately fixed, we lose, on account of the uncontrollable displacement of the diaphragm during each collision process with the test bodies, the knowledge of its position when the particle passed through the slit. The whole arrangement is therefore obviously unsuited to study the same kind of phenomena as in the previous case. [7, p. 698]

Taken in isolation, the last sentence could be taken to mean that the second arrangement can be used only to study momentum, but not position. However, continuing with Bohr's own words (parts of which were quoted above in the discussion of D2), one sees that the different phenomena Bohr is talking about are the presence of interference effects, or the lack of them, if a diaphragm with several slits (D2) and a photographic plate are placed behind D1. In short, Bohr is arguing here that the first diaphragm D1 in the second arrangement A2 has to be treated as part of the quantum system, and hence the uncertainty relations applied to D1.

His argument involves several interesting points: first, it is wrong to assume that the role of an object as part of the measuring instrument or the measured system is uniquely fixed. In the arrangement A1, diaphragm D1 is part of the measuring instrument; while in A2, the same diaphragm D1 is part of the quantum system and subject to measurement by other instruments in other possible experimental arrangements. He gives a clear physical criterion for when an entity may be treated as a measuring instrument and when it must be considered part of the quantum system: it depends on whether the quantum-mechanical uncertainty relations — or more generally, the effects of the quantum of action — may be overlooked or must be applied to the treatment of the entity in question.

> In fact, it is an obvious consequence of the above argumentation that in each experimental arrangement and measuring procedure we have only a free choice of this place [where the discrimination

is made between object and measuring apparatus] within a region where the quantum-mechanical description of the process concerned is effectively equivalent with the classical description [7, p. 702].

The measuring apparatus as a whole must be sufficiently complex to record an irreversible change as a result of its interaction with the quantum system;[z] but this does not imply anything about the size of an entity that forms a part of the total measuring apparatus: even a small micro-system can be included as a measuring instrument if, in the phenomenon under investigation, one does not have to take quantum effects into account when dealing with this micro-system. Conversely, if such quantum effects have to be taken into account, then an object that is usually taken as a measuring instrument — even one that is macroscopic in size and mass, as is the case with D1[aa] — has to be considered part of the quantum system.[bb]

Second, Bohr argues that, after passage of the particle, one still has freedom of choice to determine either a position or a momentum component of the particle by measuring either the position or momentum component of the diaphragm. In "state" language, this implies that a measurement on the diaphragm immediately determines the state of the particle, which may have separated completely from the diaphragm. In more contemporary terminology, after its passage through the diaphragm, particle and diaphragm are *entangled*.[cc] As we have seen, the concept of quantum entanglement is exactly what is at issue for EPR. Although Bohr does not use the word entanglement, he was nevertheless well aware of the concept, which is implicit in his definition of a phenomenon (see the next

[z] "Every atomic phenomenon is closed in the sense that its observation is based on registrations obtained by means of suitable amplification devices with irreversible functioning such as, for example, permanent marks on a photographic plate, caused by the penetration of electrons into the emulsion" [11, p. 73].
[aa] See Bohr's drawing of it.
[bb] Hence, Beller and Fine are not quite accurate, when they state: for him [Bohr] the measuring device must — in principle — be heavy and classical [2, p. 27]. Fine [18] makes a similar error, claiming that the picture of Bohr's diaphragms is a tiny object banging into a big apparatus.
[cc] The term entanglement was introduced in [27].

section).[dd] Indeed, as has been seen, he takes this concept for granted and utilizes it in his response. One may criticize Bohr for not expressing this awareness more explicitly and answering the EPR challenge more straightforwardly and in more detail; but one cannot justly claim that he was unaware of it.

Beller and Fine [2] challenge the analogy between the Bohr's example of particle–diaphragm interaction and EPR's example of particle–particle interaction, claiming that:

> Bohr's "wholeness" cannot extend to composite systems consisting of micro-systems, who [sic] might be said to lose their individuality when their state functions become entangled [2, p. 27].

Their argument seems to be: in the former case, one of the entangled systems — the diaphragm — is a macro-system, and Bohr's wholeness criterion is applied to the relation between macro-system and micro-system; while the EPR paper involves a micro-subsystem–micro-subsystem entanglement within a composite micro-system. They conclude that Bohr did not have the conceptual tools to handle the EPR problem.

But, as discussed above, the macro- and micro-distinction is not the same as Bohr's "measuring instrument and system" distinction. The mobile macro-diaphragm is part of a composite quantum system. And the key quantum phenomenon that EPR challenge is entanglement: it does not really matter whether it is entanglement between two micro-systems or between a micro-system and a macro-system.

4. Bohr's Concepts of Phenomenon and Complementarity

However, a question still remains. Did Bohr adopt a quantum disturbance view after all? i.e. did he believe that, although — in contrast to the classical disturbance view — the source of uncontrollable disturbances is quantum-mechanical, a system may

[dd] See also Bohr [7–9, 11].

still have a determinate position *before* a sharp position measurement has been arranged even though QM does not attribute one to it?[ee] Did he maintain that, in spite of this, QM is not incomplete because the existence of the quantum of action makes knowledge of this position inaccessible to human beings, and what counts for physics is only what is measurable by man? If Bohr did espouse this viewpoint, then his answer to EPR would indeed be positivistic.

But Bohr's view is quite different. For him, a quantum system does not have a position unless it is part of a process defining one. Position is a (system-instrument) holistic concept. Separated from such a process, talk about the position of a quantum (macro- or micro-) system makes no sense. Absolutely crucial to an understanding of Bohr's approach is a clear understanding of his concept of a quantum phenomenon; from it follows the role of complementarity in the "proper study of quantum phenomena." Bohr states:

> I advocated the application of the word phenomenon exclusively to refer to the observations obtained under specified circumstances, including an account of the whole experimental arrangement [11, p. 64].

To expatiate on Bohr, a phenomenon consists of a total process, which includes preparation of a system, subsequent interactions, and final registration of some quantity associated with the system. The word "measurement" is often used in connection with the acts of preparation and registration.

> The essential lesson of the analysis of measurements in quantum theory is thus the emphasis on the necessity, in the account of the phenomena, of taking the whole experimental arrangement into consideration, in complete conformity with the fact that all unambiguous interpretation of the quantum mechanical formalism involves the fixation of the external conditions, defining the initial state of the atomic system concerned and the character of the possible predictions as regards subsequent observable properties

[ee]Rather than Bohr's view, this is Heisenberg's (see the last section). Talk of a single "Copenhagen interpretation" has led to frequent conflation of their two differing viewpoints.

> of that system. Any measurement in quantum theory can in fact only refer either to a fixation of the initial state or to the test of such predictions, and it is first the combination of measurements of both kinds which constitutes a well-defined phenomenon [14, p. 101].[ff]

These conditions, including an account of the properties and manipulation of ideal measuring instruments involved in preparation and registration, constitute in fact the only basis for the definition of the concepts used to describe the phenomenon.

Bohr's understanding of phenomena, together with his understanding of the nature of quantum disturbance, his idea of (the limited arbitrariness of) the cut, and his implicit reference to what is now called entanglement, are crucial to an appreciation of Bohr's central argument in response to EPR. According to Bohr, a sharp measurement of system I exerts — *an influence on the very conditions which*... constitute an inherent element of the description of any phenomenon involving system II [7, p. 700].

Fine [18] completely misreads Bohr's crucial argument (also see [2]): First, this conditional disturbance becomes an "informational disturbance", and the notion of phenomenon (he uses the term "relational property") becomes merely a linguistic and semantic one. Second, as seen earlier in this section and the last, the reason why a measurement on system I is a condition of the phenomenon involving system II, is what is now known as *entanglement* or *non-separability* of the two systems, to which Bohr implicitly refers; but, according to Fine, Bohr has to appeal to *non-locality.* In addition to this conflation, Fine also conflates quantum non-separability or entanglement with the kind of harmless classical non-locality shown in the example he offers: becoming the best' when your only competitor — who might be miles away — fails.

[ff]Bohr' concept of "phenomenon" is quite similar to Feynman's concept of "process." Bohr's attitude to the usual state-vector formalism is also close to Feynman's, and Feynman's emphasis on calculation of the probability amplitude for a process as the fundamental task of QM fits well with Bohr's approach.

With the concept of phenomenon so understood, it follows then that any attempt to subdivide a phenomenon physically (for example, by an intermediate measurement) results in a different phenomenon:

> [I]n the phenomena concerned, we are not dealing with an incomplete description characterized by the arbitrary picking out of different elements of physical reality at the cost of sacrificing other such elements, but with a rational discrimination between essentially different experimental arrangements and procedures which are suited either for an unambiguous use of the idea of space location, or for a legitimate application of the conservation theorem of momentum... Indeed we have in each experimental arrangement suited for the study of proper quantum phenomena not merely to do with an ignorance of the value of certain physical quantities, but with the impossibility of defining these quantities in an unambiguous way [7, p. 699].

Fine [18] quotes a related passage by Bohr. Commenting on the double slit experiment, he says:

> If we only imagine the possibility that without disturbing the phenomena we determine through which hole the electron passes, we would truly find ourselves in irrational territory....

What Bohr is arguing against is the idea that we can physically subdivide a phenomenon without altering it (i.e. without replacing the original phenomenon with new ones), asserting that this would indeed land us in irrational territory. But Fine calls this "the language of disturbance" and says "Bohr defends locality and regards the very contemplation of non-locality as irrational and completely incomprehensible.... Fine's interpretation of the whole passage again conflates a form of non-separability (with two slits open, the electron cannot be said to go through either) with non-locality (if the electron *did* go through one of the two slits, it *would* be influenced by whether the other slit is open or not). As mentioned above, he then uses Bohr's apparent denial of non-locality to argue that Bohr cannot endorse a relational view, which is based upon non-locality. But he is wrong, for the relational view is actually based upon non-separability.

To be clear, Bohr did not argue that *all* physical properties have to be defined by way of phenomena. For example, he never challenged the idea that certain physical properties of a quantum system, such as the mass, charge, and spin of the elementary particles, were intrinsic and independent of any phenomenon (or process). This fact poses yet another challenge to Fine's metaphysical reading of Bohr that attempts to universalize (Fine's misunderstanding of) his concept of phenomenon and turns it into a universal positivist ontology that covers all physical properties.

It is sometimes claimed that Bohr's views on these matters only developed after, and in response to, EPR. Of course, some evolution of his views is only natural. Yet, Bohr [11], contains very similar ideas:[gg]

> [T]he unambiguous application of such fundamental concepts as space and time is essentially limited on account of the finite interaction between object and the measuring tools, which, as a consequence of the existence of the elementary quantum [of action], is involved in any measurement.... [W]e must remember that this interaction cannot be taken fully into account in the description of the phenomena, since the very definition of the spacetime-frame implies the negligence of the reaction of the objects on the measuring instruments.... Inversely, every application of conservation theorems...involves an essential renunciation as regards the pursuance in space and time of the individual atomic particles....
>
> [T]he viewpoint of "complementarity" in the description of atomic phenomena is forced upon us by the existence of the quantum of action...[S]pacetime co-ordination and dynamical conservation laws may be considered *as two complementary aspects of ordinary causality* which in this field exclude one another to a certain extent, although neither of them has lost its intrinsic validity [14, pp. 53–55].

[gg]Hooker [20] observes: "[T]here is no suggestion, that I can detect, that EPR did alter Bohr's conception of quantum theory.... It may...be true that, as Bohr himself seems to allow, Einstein's penetrating criticisms of quantum theory served to crystallize the elements of the doctrine of complementarity, giving impetus to a more precise development and to the broadening of their scope" (p. 149).

As for "visualizability," if we understand it, as did Bohr, as the kind of picture that CM offers — a causal description of a process in terms of momentum and energy that also provides information about a system's position in space and time, all of which are determinate — then, we have to say that no classical visualization or picturing of quantum phenomena is possible. QM can only offer such pictures in a complementary fashion. That is, we can either have position as a function of time, i.e. a well-defined spatiotemporal description, defined by the specification of a corresponding experimental arrangement allowing a continuous monitoring of position, at the sacrifice of any application of classical concepts of causality; or we can have well-defined momentum and energy, with their conservation laws obeyed as demanded by classical causality, but only by the complete sacrifice of a spatiotemporal description of this process. No single QM picture can unite both descriptions in a single process, and in this sense the quantum world cannot be visualized. This also explains why Bohr avoided Heisenberg's early disturbance language, especially in Bohr's later writings. He argues that it is misleading to formulate Heisenberg's indeterminacy relations by such a statement as: the position and momentum of a particle cannot simultaneously be measured with arbitrary accuracy. According to such a formulation, it would appear as though we had to do with some arbitrary renunciation of the measurement of either the one or the other of the two well-defined attributes of the object, which would not preclude the possibility of a future theory taking both attributes into account on the lines of the classical physics. From the above considerations, it should be clear that the whole situation in atomic physics deprives of all meaning such inherent attributes as the idealization of classical physics would ascribe to the object. [8, pp. 292–293]

To restate what was said above, concepts such as position are not applicable to an "isolated" quantum system, i.e. to one treated in isolation from an appropriate process, including some preparation. This position might seem closer to the one held by Pascual Jordan [23], who claimed that a position measurement creates a position.

But Bohr shies away from this kind of language, in fact criticizing this creation view:

> The unaccustomed features of the situation with which we are confronted in quantum theory necessitate the greatest caution as regards all questions of terminology. Speaking, as is often done, of disturbing a phenomenon by observation, or even of creating physical attributes to objects by measuring processes, is, in fact, liable to be confusing, since all such sentences imply a departure from basic conventions of language which, even though it sometimes may be practical for the sake of brevity, can never be unambiguous.[hh] It is certainly far more in accord with the structure and interpretation of the quantum mechanical symbolism, as well as with elementary epistemological principles, to reserve the word "phenomenon" for the comprehension of the effects observed under the given experimental conditions.
>
> These conditions, which include an account of the properties and manipulation of all measuring instruments concerned, constitute in fact the only basis for the definition of the concepts by which the phenomenon is described [14, p. 104].

Fifteen years later, he repeated and amplified the same thought:

> In this context, one sometimes speaks of disturbance of phenomena by observation or creation of physical attributes to atomic object by measurements. Such phrases, however, are apt to cause confusion, since words like phenomena and observation, just as attributes and measurements, are here used in a way incompatible with common language and practical definition. On the lines of objective description, it is indeed more appropriate to use the word phenomenon to refer only to observations obtained under circumstances whose description includes an account of the whole experimental arrangement. In such terminology, the observational problem in quantum physics is deprived of any special intricacy and we are, moreover, directly reminded that every atomic phenomenon is closed in the sense that its observation is based on registrations obtained by means of suitable amplification devices with irreversible functioning

[hh]Fine [18] quotes the passage up to here, but claims that Bohr's (justified) refusal to use either "disturbing" or "creating" shows the latter's difficulty in framing concepts that can be used to derail the EPR challenge.

such as, for example, permanent marks on a photographic plate, caused by the penetration of electrons into the emulsion [11, p. 73].

In summary, we can now see that Bohr's use of diaphragms is not the result of a positivist tendency, as has sometimes been claimed. He does not believe that theoretical concepts like position and momentum are too abstract, and that such theoretical talk should be eliminated by reducing all such concepts to sensations or empirical data, or by defining them in terms of an ordinary experimental set-up. Rather, for Bohr, there is no problem in using such theoretical concepts in either CM or QM. But in the latter case, we must not use them without specifying a phenomenon (in Bohr's sense), because attributes such as position and momentum inherent in classical systems, which may be treated as isolated, are not inherent in quantum systems, which cannot be treated in isolation. In short, the reason Bohr appeals to experimental arrangements, such as diaphragms, is not metaphysical or semantic but *physical*; the challenge to physics posed by the existence of the quantum of action.

5. A Few Final Issues

Having explained and defended Bohr's experimental arrangements, we shall conclude by discussing three further issues raised by his diaphragms and by clarifying his general attitude to the role of experience. First, one might ask: How it is possible that the position of a macro-system such as a diaphragm is indeterminate? At first sight, this might seem counter-intuitive;[ii] but the point is that "position" here means "position with respect to a fixed inertial frame of reference,"[jj] which is associated physically with a

[ii]When one recollects that the closest classical analogue to a quantum system is a canonical ensemble (see, e.g. [28]), then it seems less counterintuitive, since even for a rigid macroscopic body there will be a distribution of positions associated with an ensemble having a sharp momentum, and vice versa.

[jj]See for an extensive discussion of "the role that reference frames play in the definition of physical concepts in quantum theory."

massive system[kk] that is quite independent of the diaphragm. If the diaphragm is rigidly attached to the frame, then its position is indeed fixed in the above sense. But when it is capable of moving relative to the inertial frame, if it is treated as a quantum system its position relative to the frame is subject to effects of the uncertainty relations. Even if its mass is macroscopic, if it is small compared to the mass(es) of the object(s) defining the frame of reference, these effects need not be negligible.

Moreover, it should be remembered that Bohr's experimental arrangements are part of thought experiments, involving idealized apparatus. As he himself explicitly points out, the obvious impossibility of actually carrying out these thought experiments does not clearly affect the theoretical argument, since the procedures in question are essentially equivalent with atomic processes, like the Compton effect, where a corresponding application of the conservation theorem of momentum is well established [6, p. 698].

Bohr and Rosenfeld [15] make a related point about any theory: the need to assure a complete accord between quantities *defined* as observable in a theory and what is measurable *in principle*. This is important, not because the definition of a quantity is based on its measurability (operationalism); but to assure that theory and (ideal) measurement are consistent. Bergmann and Smith 1982 call this criterion "measurability analysis."[ll] "Measurability in principle" implies that the measuring devices introduced need not be realistic, but their construction must be consistent with the theory to which they are being applied. In other words, far from implying

[kk]It must be massive for two reasons: it must be so massive that absorbing uncontrollable recoil energy and momentum do not appreciably effect its own energy and momentum; and so massive that, when comparing it to other inertial frames (in such comparisons it is treated as a system subject to the uncertainty relations), attributing a position and velocity to it does not fall foul of the uncertainty relations: $\Delta V = (\Delta P)/M$, so for sufficiently large M, even a considerable ΔP causes no appreciable change in its velocity.

[ll]They emphasize this criterion is quite distinct from the so-called "measurement problem" in QM, which arises when the state function is interpreted ontologically.

a positivist or operationalist approach, ideal measuring instruments are themselves theoretical tools. For example, in order to show the measurability in principle of the vacuum electric and magnetic field components, Bohr and Rosenfeld [15] invoke massive charged test bodies while totally neglecting their atomic constitution and the quantization of their charge, insisting that:

> In our discussion of the limitations of the measurability of field quantities these difficulties play no role, since for this purpose the atomistic structure of matter is not an essential issue. It is true that the measurement of fields requires the use of material charged test bodies, but their unambiguous application as measuring instruments depends exactly on the extent to which we can treat their response to the fields as well as their influence as field sources on the basis of classical electrodynamics [15, p. 360].

This discussion concerns special-relativistic quantum fields but, suitably modified, their comments apply to non-relativistic QM. We might mention for example, that Bohr [7] assumes the possibility of completely rigid diaphragms and a rigid reference support, and of a completely rigid attachment of the one to the other — all impossible to realize in practice or special relativistic theory; but quite consistent theoretical constructs within Newtonian CM.

Second, as pointed out several times, Bohr's arrangements require that, if the position measurement on one particle is to be used to define the corresponding position coordinate of the other particle, the position measurement be made immediately after the particle's passage through the two-slit diaphragm, which defines the difference in the x-coordinates of the two particles, because this difference is not a conserved quantity. This is indeed a difficulty, but it is not fair to hold Bohr responsible.[mm] It was EPR who introduced the wave function depending on the sum of momenta and difference of positions; and Bohr was just showing the consequences of preparing a system with this wave function. An experimentally feasible arrangement of

[mm]See [2], pp. 14–15.

the EPR type was first proposed by Bohm [4], who replaced these two quantities with two conserved spins.[nn]

Third, there is a real complication, not stressed by Bohr, in applying Bohr's diaphragms or some arrangement of this sort to the EPR case. He states:

> If the momentum of this [double slit] diaphragm is measured accurately before as well as after the passing of the particles, we shall in fact know the sum of the components perpendicular to the slit of the momenta of the two escaping particles, as well as the difference of their initial positional coordinates in the same direction.... In this arrangement, it is therefore clear that a subsequent single measurement either of the position or of the momentum of one of the particles will automatically determine the position or momentum, respectively, of the other particle with any desired accuracy... [7, p. 700].

Bohr does not work out the relation between the determinate quantities. If one does so, an interesting point arises. To illustrate it, we present the details for the momentum case. After the two particles pass through the mobile double slit diaphragm and its recoil momentum has been determined, the two particles are entangled: their total momentum is determinate, as is the distance between them. Suppose particle 1 is subsequently passed through a mobile single slit diaphragm initially at rest, then the momenta of both the diaphragm and of particle 1 after its passage must be measured. The interesting point, which is not mentioned by Bohr, is the need to measure the latter quantity. From this data and conservation of momentum, one can now determine the momenta of both particles *after* they passed through the double slit diaphragm and *before* one of them entered the single slit diaphragm.

[nn]One of us (J.S.) is grateful to Dr. Tilman Sauer of the Einstein Papers for informing him that "I found a version of the EPR incompleteness argument, apparently very late: late 1954 perhaps, on the bottom half of one of his U[nified] F[ield] T[heory] calculations. In it he discusses the argument in terms of spin observables" (private communication).

The calculation goes as follows:

Single slit diaphragm: momentum before passage $= 0$, after passage $= P_D$, both known

Particle 1: momentum before passage $= p_1$, unknown; after passage $= p_{1'}$; this is the quantity needed, but not mentioned in [7].

Particle 2: momentum before $= p2$, unknown

Conservation of momentum after passage of Particle 1 through the single slit diaphragm:

$$p_1 = P_D + p_{1'}.$$

Sum of momentum of two particles, known from measurement of momentum of double slit diaphragm:

$$P = p_1 + p_2,$$

which is conserved and conserved until Particle 1 passes through the single slit diaphragm, which decouples (disentangles) the two particles.

It follows that:

$$p_2 = P - p_1 = P - (P_D + p_1).$$

To recapitulate, *two* measurements *after* passage of Particle 1 through the single slit diaphragm are needed to determine the values of p_1 and p_2 *before* passage through the diaphragm.

Finally, a general comment is in order here. We have already defended Bohr's analysis of the EPR experiment from its characterization as positivist. But some of his more general comments on the role of measurement and experiment might lead people to read him in a positivist way:

> The extent to which an unambiguous meaning can be attributed to such an expression as "physical reality" cannot of course be deduced from *a priori* philosophical conceptions, but — as the authors of the article cited themselves emphasize — must be founded on a direct appeal to experiments and measurements [7, p. 696].

But, as Bohr indicates, in fact it was EPR, the "realists," who introduced this point. EPR claim:

> The elements of the physical reality cannot be determined by *a priori* philosophical considerations, but must be found by an appeal to results of experiments and measurements [16, p. 777].

Moreover, the view Bohr and Einstein share here is *not* that *only* what is measurable exists, the idea that — a physical property of a given system has reality only when it is actually measured — the position Beller and Fine accuse Bohr of holding [2, pp. 9–10, 27 and 29]. Rather, his (and Einstein's) view is that what exists must be measurable, or more accurately, must have measurable consequences. The key point here is that talk of experiments, measurements, instruments, operations, observations, or experiences in general does *not necessarily* imply a positivist position. The issue, rather, is whether experiments and the like are assigned an absolutely privileged position; whether everything in a theory has to be reducible directly to experiences in order to be accepted as real. For Bohr, as well as for EPR, being measurable or experimentable is a sufficient but not a necessary condition for the existence of something or the meaningfulness of the corresponding concept. It may be capable of direct exhibition through experiments, or more generally in experience; but it may only be capable of indirect exhibition. For example, a theoretical posit may be related to other elements of a theory in such a way that it is only the organic whole that is directly connected with observation and experiment, priority being given to the existence of the theoretical entity (or entities) behind the measurement, and not necessarily to the measurement. The theoretical entity is *manifested* in the experiment, but is *not defined or determined* by it. Such a "tempered" emphasis on experience does not, therefore, imply positivism. One can say that Bohr and Einstein are both empirical in their approach to physical reality, but they are not empiricists.

References

[1] Beller, M. (1999). *Quantum Dialogue*. The University of Chicago Press: Chicago.

[2] Beller, M. and Fine, A. (1994). Bohr's response to EPR, in *Niels Bohr and Contemporary Philosophy*, Faye, J. and Folse H. j. (eds.), pp. 1–31.

[3] Bergmann, P. and Smith, G. (1982). Measurability analysis of the linearized gravitational field. *General Relativity and Gravitation*, **14**: 1131–1166.

[4] Bohm, D. (1951). *Quantum Theory*. Prentice-Hall: New York.

[5] Bohr, N. (1932). Chemistry and the quantum theory of atomic constitution. *Journal of the Chemistry Society*, **26**: 349–384.

[6] Bohr, N. (1934). *Atomic Theory and the Description of Nature*. Cambridge University Press: Cambridge.

[7] Bohr, N. (1935). Can quantum-mechanical description of physical reality be considered complete? *Physical Review*, **48**: 669–702.

[8] Bohr, N. (1937). Causality and complementarity. *Philosophy of Science*, **4**(3): 289–298.

[9] Bohr, N. (1939). The Causality Problem in Atomic Physics, in *New Theories in Physics*. Paris: International Institute of Intellectual Co-operation, 1939. Cited from Bohr (1998), pp. 94–121.

[10] Bohr, N. (1948). On the notions of causality and complementarity. *Dialectica*, **2**: 312–319.

[11] Bohr, N. (1949). Discussion with Einstein on epistemological problems in atomic physics, in *Albert Einstein: Philosopher-Scientist*, Paul Arthur Schilpp (ed.), Library of Living Philosophers: Evanston, IL. Cited from Bohr (1958), pp. 32–66.

[12] Bohr, N. (1958). *Atomic Physics and Human Knowledge*. John Wiley & Sons: New York.

[13] Bohr, N. (1963). *Essays 1958–1962 on Atomic Physics and Human Knowledge*. John Wiley & Sons: New York.

[14] Bohr, N. (1998). The Philosophical Writings of Niels Bohr. *Causality and Complementarity: Supplementary Papers*, Vol. 4, Faye, J. and Folse, H. J. (eds.), OxBow Press: Woodbridge, CT.

[15] Bohr, N. and Rosenfeld, L. (1933). Zur Frage der Messbarkeit der elektromagnetischen Feldgrössen, *Kgl. Danske Videnskab. Selskab, Mat.-fys. Med.* 12, no. 8. English translation in *Selected Papers of Leon Rosenfeld*, Cohen, R. S. and Stachel, J. (eds.), D. Reidel Publ. Co.: Dordrecht/Boston/London, pp. 357–400.

[16] Einstein, A., Podolsky, B. and Rosen, N. (EPR) (1935). Can quantum-mechanical description of physical reality to be considered complete? *Physical Review*, **47**: 777–780.

[17] Faye, J. and Folse, H. (1994). *Niels Bohr and Contemporary Philosophy*. Kluwer Academic Publishers: Dordrecht.

[18] Fine, A. (2004). The Einstein–Podolsky–Rosen argument in quantum theory, in *The Stanford Encyclopedia of Philosophy*. Current edition, Edward N. Zalta (ed.), (http://plato.stanford.edu/archives/win2014/entries/qt-epr/).

[19] Heisenberg, W. (1949). *The Physical Principles of the Quantum Theory*. Dover Publications Inc.: New York.

[20] Hooker, C. A. (1972). The nature of quantum mechanical reality: Einstein vs Bohr, in *Paradigms & Paradoxes: The Philosophical Challenge of the Quantum Domain*, Robert Garland Colodny (ed.), University of Pittsburgh Press: Pittsburgh, pp. 167–302.

[21] Howard, D. (2004). Who invented the Copenhagen interpretation? A Study in Mythology. *Philosophy of Science*, **71**: 669–682.

[22] Jammer, M. (1974). *The Philosophy of Quantum Mechanics*. John Wiley & Sons, Inc.: New York.

[23] Jordan, P. (1934). Quantenphysikalische bemerkungen zur biologie und psychologie. *Eikenntnis*, **4**: 215–252.

[24] Meier, C. A. (eds.) (2001). *Atom and Archetype: The Pauli/Jung Letters, 1932–1958*. Princeton University Press: Princeton.

[25] Murdoch, D. (1987). *Niels Bohr's Philosophy of Physics*. Cambridge University Press: Cambridge.

[26] Ruark, A. E. (1935). Is the quantum-mechanical description of physical reality complete? *Physical Review*, **48**: 448–467.

[27] Schrödinger, E. (1935). Discussion of probability relations between separated systems. *Proceedings of the Cambridge Philosophical Society*, **31**: 555–563; **32**: 446–451.

[28] Stachel, J. (1997). Feynman Paths and Entanglement: Is There Any More to the Mystery, in *Potentiality, Entanglement and Passion-at-a-Distance, Quantum Mechanical Studies for Abner Shimony*, Cohen, R. S., Horne, M. and Stachel, J. (eds.), Vol. 2. Kluwer Academic Publishers: Dordrecht/Boston/London, pp. 245–256.

[29] Stachel, J. (2006). Quantum logic, in *The Philosophy of Science: An Encyclopedia*, Sarkar, S. and Pfeifer, J. (eds.), Vol. 2, Routledge: New York and London, pp. 633–649.

Chapter 3

"It Ain't Necessarily So": Interpretations and Misinterpretations of Quantum Theory

John Stachel

Center for Einstein Studies, Boston University

After describing some recent misinterpretations of Bohr's views on quantum theory, largely based on their conflation with those of Heisenberg, a correct account of Bohr's approach is given in his own words. Then some guidelines toward a valid interpretation of quantization are discussed, including: the role of the quantum of action, the primacy of processes over states, the difference between classical and quantum ensembles, and between non-relativistic quantum mechanics and relativistic quantum field theory.

1. What is Bohr's "Copenhagen Interpretation"?

In a recent article, Żurek [39] writes of "quantum weirdness," which he attempts to explain. Basic to his entire approach is the concept of a "quantum state" that "will, when measured, collapse to an eigenstate of the measured observable" [39], p. 45. He sets up a system of five "quantum postulates," but finds a "conflict" between "the randomness inherent in postulates 4 and 5" and "the unitarity of postulate 5." He then asserts that "Niels Bohr insist[ed] that a part of the universe — including measuring devices and observers — must be classical" [39], pp. 44–45. His aim is to avoid this conclusion.

The paper that he cites in support of this conclusion, Bohr [8], was published in 1928; but Bohr's views on quantum theory continued to develop and change in significant ways at least until the end of the 1930s (see, e.g. Camillieri [14] and Bai and Stachel [4]).

In particular, his views about the "quantum state" changed considerably. Bohr's original approach to complementarity placed primary emphasis on the instantaneous, three-dimensional spatial *state* of the system; from this point of view, a four-dimensional spatiotemporal process is just a temporal sequence of changing states of the system. And it is this 1928 approach that Żurek has adopted.

But Bohr's later approach places primary emphasis on four-dimensional *processes* or "*phenomena*" as he calls them. From this point of view, a "state" is just a particular spatial cross-section of a process, and is of quite secondary importance: All such cross-sections are equally valid, and any other sequence of states merely represents a different "perspective" on the same process.

> On the lines of objective description, it is indeed more appropriate to use the word phenomenon to refer only to observations obtained under circumstances whose description includes an account of the whole experimental arrangement. In such terminology, the observational problem in quantum physics is deprived of any special intricacy ... (Bohr [7]).

And the main characteristic of the "measuring devices" is not that they be classical, but that they be sufficiently complex to be capable of irreversible processes. The quotation just broken off continues immediately:

> and we are, moreover, directly reminded that every atomic phenomenon is closed in the sense that its observation is based on registrations obtained by means of suitable amplification devices with irreversible functioning such as, for example, permanent marks on a photographic plate, caused by the penetration of electrons into the emulsion (Bohr [7]).

Bai and Stachel [4] offer a more detailed explanation of Bohr's position:

> First, it is wrong to assume that the role of an object as part of the measuring instrument or the measured system is uniquely fixed. In the arrangement A1, diaphragm D1 is part of the

measuring instrument; while in A2, the same diaphragm D1 is part of the quantum system and subject to measurement by other instruments in other possible experimental arrangements. He gives a clear physical criterion for when an entity may be treated as a measuring instrument and when it must be considered part of the quantum system: It depends on whether the quantum-mechanical uncertainty relations — or more generally, the effects of the quantum of action — may be overlooked or must be applied to the treatment of the entity in question.

> In fact, it is an obvious consequence of the above argumentation that in each experimental arrangement and measuring procedure we have only a free choice of this place [where the discrimination is made between object and measuring apparatus] within a region where the quantum-mechanical description of the process concerned is effectively equivalent with the classical description (Bohr [9]).

The measuring apparatus as a whole must be sufficiently complex to record an irreversible change as a result of its interaction with the quantum system; but this does not imply anything about the size of an entity that forms a part of the total measuring apparatus: even a small micro-system can be included as a measuring instrument if, in the phenomenon under investigation, one does not have to take quantum effects into account when dealing with this micro-system. Conversely, if such quantum effects have to be taken into account, then an object that is usually taken as a measuring instrument — even one that is macroscopic in size and mass, as is the case with D1 — has to be considered part of the quantum system[a] (Bai and Stachel [4]).

2. Other Recent Misinterpretations of Bohr

I am far from suggesting that Żurek is alone in misinterpreting Bohr's views. Unfortunately, he is in good company. Here are some citations from two recent, semi-popular books that discuss quantum mechanics. The first, Stone [37], won the 2014 Phi Beta Kappa Award

[a]Hence, Beller and Fine are not quite accurate, when they state: "for him [Bohr] the measuring device must — in principle — be heavy and classical" (Beller and Fine [5], p. 27). Fine [21] makes a similar error, claiming that the picture of Bohr's diaphragms is "a tiny object banging into a big apparatus".

in Science and was selected as a top science book of 2013 by National Public Radio. Stone's attack on"the Copenhagen interpretation," focuses on:

> Born's probabilistic interpretation of the wave function, Heisenberg's uncertainty principle and Bohr's mysterious complementarity principle [37, p. 281].
>
> Einstein's later critiques of quantum theory focused less on its indeterminacy and more on its strange epistemological status. In quantum mechanics, the actual act of measurement is part of the theory; these magic coins just mentioned exist in a state of (heads, tails)–(tails, heads) uncertainty until they are measured, and then they are forced to "decide" which state they are in [37, p. 285].

Baggott [3] makes a similar assertion:

> Danish physicist Niels Bohr and German Werner Heisenberg argued that particles and waves are merely the shadowy projections of an unfathomable reality into our empirical world of measurement and perception..... This approach to quantum theory became known as the *Copenhagen interpretation*.... At the heart of this interpretation lies Bohr's notion of complementarity, a fundamental duality of wave and particle behavior.

This misconception is so common that Tamsin Shaw [26] repeats it without comment in *The New York Review of Books*:

> Niels Bohr and his colleagues demonstrated that in a quantum system the location of a particle prior to observation could only be described probabilistically, as a wave function .... The Copenhagen interpretation of the quantum phenomenon by Niels Bohr and his associates asserted that the act of observation itself caused the wave function to collapse.

3. Bohr (or) Heisenberg?

This misconception of Bohr's views is so common that it seems worthwhile to discuss it at some length. The authors cited do not seem to be aware that they are giving Heisenberg's interpretation of quantum mechanics, which is quite different from Bohr's interpretation.

You do not have to take my word for this. The difference is well known to those who know it well:

> [O]f the two primary architects of the Copenhagen interpretation, Werner Heisenberg maintained that 'observation of the position will alter the momentum by an unknown and undeterminable amount,' whereas Niels Bohr 'warned specifically against phrases, often found in the physical literature, such as "disturbing of phenomena by observation."' ...The wave function should be regarded as a mathematical tool for calculating the outcomes of observations, not as a physically present entity existing in space such a football, or a nitrogen molecule, or even an electric field (Styer *et al.* [38]).

While I agree completely with this assertion, I should make it clear that, although I have done so elsewhere (see Stachel [34] and [33]), in this chapter I am not primarily arguing *in favor of* Bohr's viewpoint, which indeed evolved over time; but *against its misrepresentation* and for its characterization as a position quite distinct from Heisenberg's, described by Bohr himself as quite non-mystical.

4. Let Bohr Speak for Himself

> The essential lesson of the analysis of measurements in quantum theory is thus the emphasis on the necessity, in the account of the phenomena, of taking the whole experimental arrangement into consideration, in complete conformity with the fact that all unambiguous interpretation of the quantum mechanical formalism involves the fixation of the external conditions, defining the initial state of the atomic system concerned and the character of the possible predictions as regards subsequent observable properties of that system (Bohr [11]).

Bohr is here stating that quantum theory can only deal with open systems, to use a term I have discussed elsewhere (Stachel [28]).

> Any measurement in quantum theory can in fact only refer either to a fixation of the initial state or to the test of such predictions, and it is first the combination of measurements of both kinds which constitutes a well-defined phenomenon.

And here Bohr is asserting that it can only treat a complete process, as Feynman terms it, or phenomenon, the term Bohr employs.

> In this context, one sometimes speaks of "disturbance of phenomena by observation" or ''creation of physical attributes to atomic objects by measurements." Such phrases, however, are apt to cause confusion, since words like phenomena and observation, just as attributes and measurements, are here used in a way incompatible with common language and practical definition. On the lines of objective description, it is indeed more appropriate to use the word phenomenon to refer only to observations obtained under circumstances whose description includes an account of the whole experimental arrangement. In such terminology, the observational problem in quantum physics is deprived of any special intricacy and we are, moreover, directly reminded that every atomic phenomenon is closed in the sense that its observation is based on registrations obtained by means of suitable amplification devices with irreversible functioning such as, for example, permanent marks on a photographic plate, caused by the penetration of electrons into the emulsion (Bohr [13], p. 73).
>
> [I]n order to avoid any misunderstanding concerning the significance of the word "indeterminism",... recall that in quantum effects we were not dealing with behavior independent of the objects, but that the observable phenomena essentially depend upon the interaction of these objects with the measuring instruments which fix the conditions for the experiment. That is the reason why we find ourselves faced by quite a new situation in physics in which the traditional conceptions of determinism or indeterminism are not univocally applicable. It is really wonderful that in spite of this we can, with the help of mathematical abstractions, put so much order into a domain so vast and so rich with experience, in a way that is entirely rational and excludes all mysticism (Bohr [10]).

In 1949, he summed up his position in Bohr [12]:

This point is of great logical consequence, since it is the only circumstance that we are presented with a choice of *either* tracing the path of a particle or observing interference effects, which allows us to escape from the paradoxical necessity of concluding that the behavior of an electron or a photon should depend on the presence of a slit in the diaphragm through which it could be proved not to pass. We have here to do with a typical example of how the complementary phenomena appear under mutually exclusive experimental arrangements and are just faced with the impossibility, in the analysis of quantum effects, of drawing any sharp separation between an independent behavior of atomic objects and their interaction with the measuring instruments which serve to define the conditions under which the phenomena occur.

. . . .

The extent to which renunciation of the visualization of atomic phenomena is imposed upon us by the impossibility of their subdivision is strikingly illustrated by the following example to which Einstein very early called attention and often has reverted. If a semi-reflecting mirror is placed in the way of a photon, leaving two possibilities for its direction of propagation, the photon may either be recorded on one, and only one, of two photographic plates situated at great distances in the two directions in question, or else we may, by replacing the plates by mirrors, observe effects exhibiting an interference between the two reflected wave-trains. In any attempt of a pictorial representation of the behavior of the photon we would, thus, meet with the difficulty: to be obliged to say, on the one hand, that the photon always chooses *one* of the two ways and, on the other hand, that it behaves as if it had passed *both* ways.

It is just arguments of this kind which recall the impossibility of subdividing quantum phenomena and reveal the ambiguity in ascribing customary physical attributes to atomic objects. In particular, it must be realized that besides in the account of the placing and timing of the instruments forming the experimental arrangement all unambiguous use of spacetime concepts in the description of atomic phenomena is confined to the recording of observations which refer to marks on a photographic plate or to similar practically irreversible amplification effects like the building of a water drop around an ion in a cloud-chamber. Although, of course, the existence of the quantum of action is ultimately responsible for the properties of the materials of which the measuring instruments are built and on which the functioning of the recording devices depends, this circumstance is not relevant for the problems of the adequacy and completeness of the quantum-mechanical description in its aspects discussed here.

These problems were instructively commented upon from different sides at the Solvay meeting, in the same session where Einstein raised his general objections. On that occasion, an interesting discussion arose also about how to speak of the appearance of phenomena for which only predictions of statistical character can be made. The question was whether, as to the occurrence of individual effects, we should adopt a terminology proposed by Dirac, that we were concerned with a choice on the part of "nature" or, as suggested by Heisenberg, we should say that we have to do with a choice on the part of the "observer" constructing the measuring instruments and reading their recording. Any such terminology would, however, appear dubious since, on the one hand, it is hardly reasonable to endow nature with volition in the ordinary sense, while on the other hand, it is certainly not possible for the observer to influence the events which may appear under the conditions he has arranged. To my mind, there is no other alternative than to admit that, in this field of experience, we are dealing with individual phenomena and that our possibilities of handling the measuring instruments allow us only to make a choice between the different complementary types of phenomena we want to study.

The epistemological problems touched upon here were more explicitly dealt with in my contribution to the issue of *Naturunssenschaften* in celebration of Planck's 70th birthday in 1929. In this chapter, a comparison was also made between the lesson derived from the discovery of the universal quantum of action and the development which has followed the discovery of the finite velocity of light and which, through Einstein's pioneer work, has so greatly clarified basic principles of natural philosophy. In relativity theory, the emphasis on the dependence of all phenomena on the reference frame opened quite new ways of tracing general physical laws of unparalleled scope. In quantum theory, it was argued, the logical comprehension of hitherto unsuspected fundamental regularities governing atomic phenomena has demanded the recognition that no sharp separation can be made between an independent behavior of the objects and their interaction with the measuring instruments which define the reference frame.

5. What Quantization is — and What it is Not?

First of all, what is *not* being claimed here is that quantization only makes sense when applied to "fundamental" structures or entities. As I noted 30 years, a mystique has been created around quantum mechanics:

> Anything touched by this formalism thereby seems to be elevated — or should it be lowered? — to a fundamental ontological status. The very words "quantum mechanics" conjure up visions of electrons, photons, baryons, mesons, neutrinos, quarks, and other exotic building blocks of the universe. But the scope of the quantum mechanical formalism is by no means limited to such (presumed) fundamental particles. There is no restriction of principle on its application to any physical system. One could apply the formalism to sewing machines if there were any reason to do so!" (Stachel [28]).

Just what is quantization? Fundamentally, it is a way accounting for the effects of h, the quantum of action, on *any* process involving some system — or rather on a theoretical model of such a system — be the model regarded as "fundamental" or as "composite;" i.e. one that models the collective behavior of a set of more fundamental entities. Here, are a couple of early examples of such composite, non-fundamental quanta:

(1) quasi-particles: particle-like entities arising in the collective description of certain systems of interacting particles, such as phonons and rotons in hydrodynamics (Landau [24]).
(2) collective excitations: phenomenological photons are the result of applying standard quantization techniques to the macroscopic description of electromagnetic waves in a homogeneous, isotropic dielectric (Ginzburg [22]).

So successful quantization of some classical formalism does *not necessarily* mean that one has achieved a deeper understanding of reality — or better, an understanding of a deeper level of reality. It means that one has successfully understood the effects of the quantum of action on the phenomena (processes) described by the formalism. As Bohr put it, there is:

> an element of wholeness, so to speak, in the physical processes, a feature going far beyond the old doctrine of the restricted divisibility of matter. This element is called the universal quantum of action. It was discovered by Max Planck in the first year of this century and came to inaugurate a whole new epoch in physics and natural philosophy. We came to understand that the ordinary laws of physics, i.e. classical mechanics and electrodynamics, are idealizations that can only be applied in the analysis of phenomena in which the action involved at every stage is so large compared to the quantum that the latter can be completely disregarded (Bohr [12]).
>
> The essential lesson of the analysis of measurements in quantum theory is thus the emphasis on the necessity, in the account of the phenomena, of taking the whole experimental arrangement into consideration, in complete conformity with the fact that all unambiguous interpretation of the quantum mechanical formalism involves the fixation of the external conditions, defining the initial state of the atomic system concerned and the character of the possible predictions as regards subsequent observable properties of that system. Any measurement in quantum theory can in fact only refer either to a fixation of the initial state or to the test of such predictions, and it is first the combination of measurements of both kinds which constitutes a well-defined phenomenon (Bohr [11]).
>
> On the lines of objective description, it is indeed more appropriate to use the word phenomenon to refer only to observations obtained under circumstances whose description includes an account of the whole experimental arrangement. In such terminology, the observational problem in quantum physics is deprived of any special intricacy. On the lines of objective description, it is indeed more appropriate to use the word phenomenon to refer only to observations obtained under circumstances whose description includes an account of the whole experimental arrangement. In such terminology, the observational problem in quantum

> physics is deprived of any special intricacy and we are, moreover, directly reminded that every atomic phenomenon is closed in the sense that its observation is based on registrations obtained by means of suitable amplification devices with irreversible functioning such as, for example, permanent marks on a photographic plate, caused by the penetration of electrons into the emulsion (Bohr [4]).

Having passed beyond the quantum mystique, one is free to explore how to apply quantization techniques to various formulations of the same theory without the need to single one out as the unique, "correct" one. One might say: "In my Father's house are many mansions" (*John 14:2*) or "Let a hundred flowers blossom, let a hundred schools contend" (Mao, 1956).

If two such quantizations at *different* levels are carried out, one may then investigate the relation between them. For example, Crenshaw [16] demonstrates: "A limited equivalence between microscopic and macroscopic quantizations of the electromagnetic field in a dielectric."

If two such quantizations at the *same* level are carried out, one may also investigate the relation between them. For example: the relation between loop quantization and usual field quantization of the electromagnetic field: if you "thicken" the loops, they are equivalent (Ashtekar and Rovelli [2]).

I shall now discuss a few features that should serve as guidelines in any such search.

6. Definability and Measurability

One must always establish a qualitative and quantitative *consonance* between, the *concept* of an *entity*, for which physical significance is claimed, and an ideal *measurement procedure* for that entity. If it is a quantum concept, then h (the quantum of action) must enter *both* definition and measurement procedure.

> We share the point of view emphasized by Heisenberg and Bohr and Rosenfeld, that the limits of defineability of a quantity within any formalism should coincide with the limits of measurability of that

quantity for all conceivable (ideal) measurement procedures. For well-established theories, this criterion can be tested. For example, in spite of a serious challenge [by Landau and Peierls], source-free quantum electrodynamics was shown to pass this test (Amelino-Camelia and Stachel [1]).

Darrigol [17] discusses some aspects of the development of Bohr's viewpoint:

> The discussion of these fundamental difficulties at the 1930 Solvay Congress was dominated by Bohr's viewpoint ... [T]he scope of these problems and the nature of their solutions had to be uncovered by a critique of the basic concepts of the threatened theories, by an evaluation of the possibilities of definition and of observation within them. Bohr's main message: "One can only judge the coherence of the symbolic method by examining the limits of observability in the usual sense." ....
>
> Heisenberg had been the first to consider the problem of field measurements in his Chicago lectures of spring 1929.... [I]n his analysis of the gamma-ray microscope, he tended to privilege the corpuscular and discontinuous viewpoint above that of the wave viewpoint. Bohr had succeeded, not without difficulty, in convincing him that the evaluation of the limits of the corpuscular viewpoint necessarily involved calling upon the wave theory. But then it became important for Heisenberg to show that, reciprocally, the domain of applicability of the electromagnetic field concept must be limited by the existence of corpuscular aspects.
>
> He provided new uncertainty relations $\Delta E_x \Delta H_y \geq hc/(\delta l)^4$ for the averages of the electric field E_x and magnetic field H_y over the same domain of extension ... Heisenberg, Bohr wrote...should have taken account not only of the spatial extension of the field measurements, but also of their duration, essential for the estimation of the role of quantum fluctuations of the field.... [C]ontrary to the initial arguments of Heisenberg the Bohr-Rosenfeld article contains the rigorous proof that
>
> $$\Delta E_x \Delta H_y = 0$$
>
> if E_x and H_y are measured in the same [four-dimensional] domain (Darrigol [17], pp. 154–156, my translation).

Bergmann and Smith [6] significantly refined Bohr's concept of measurement:

> [D]etection of the state of a quantum field requires at least two stages. At the first stage, some conserved physical quantity is to be transferred from the quantum field to an intermediary device, a quantum system with but a finite number of degrees of freedom. This intermediary device is not the ultimate instrument upon which the outcome of the state determination is registered, because its observable features are subject to the indeterminacies of ordinary quantum mechanics; on the other hand, it is not a quantum field, because its state vector is not subject to second quantization. At the second stage, the conserved quantity is to be transferred from the intermediary device to a classical instrument, whose readout is classically determinate, in such a manner that the state of the quantum field is minimally altered. Having distinguished between the quantum field, the intermediary device and the classical instrument, we shall avoid ... Bohr and Rosenfeld's word "Test body," which sometimes seems to refer to the classical instrument, and sometimes to what we have called the "intermediary device" (Bergmann and Smith [6]).

7. Processes are Primary, States are Secondary

Lee Smolin [27] has expressed it beautifully:

> [R]elativity theory and quantum theory each ... tell us — no, better, they scream at us — that our world is a history of processes. Motion and change are primary. Nothing is, except in a very approximate and temporary sense. How something is, or what its state is, is an illusion. It may be a useful illusion for some purposes, but if we want to think fundamentally we must not lose sight of the essential fact that 'is' is an illusion. So to speak the language of the new physics we must learn a vocabulary in which process is more important than, and prior to, stasis (Smolin [27]).

Phrases such as "at any *moment of time*," "at any *given time*" are appropriate in Newtonian–Galileian physics, which is based on the postulation of a global, absolute time. But from SR and on to GR, this phrase involves a convention defining a global time. The only convention-invariant things are *processes*, each involving a spacetime region. This suggests — as do many other considerations — that the fundamental entities in quantum theory are the *transition amplitudes*, and that states should be taken in the c.g.s. system (cum *grano salis*). And this is true of our measurements as well: any measurement involves a finite time interval and a finite three-dimensional spatial region. Sometimes, we can get away with neglecting this, and talking, for example in NR QM, about instantaneous measurements. But sometimes we most definitely cannot, as Bohr and Rosenfeld demonstrated for QFT, where the basic quantities defined by the theory are spacetime averages. Their critique of Heisenberg shows what happens if you forget this!?

8. Processes, Classical, and Quantum

What is a quantum process? A quantum process involves three stages: preparation, interaction, and registration. The big question to ask always is: How does h enter into the preparation and registration procedures.

8.1. Preparation and prediction

My approach to quantum mechanics depends on observing that it can only be applied to open systems, and to a complete *process* (Feynman's term, which will be used hereafter) or *phenomenon* (Bohr's term) that such a system undergoes. More explicitly, in non-relativistic theories, classical or quantum, such a process consists of:

(1) *Preparation* at time t_1 of system S under conditions C_i with compatible properties P_i having the prepared values p_i;
(2) *interaction* of S with some other external system(s) during the time interval

$$t_1 < t < t_2;$$

(3) *registration* at time $t_2 > t_1$ of S under conditions C_f of the final values p_f of compatible properties P_f under conditions C_f.

Classically, if one *prepares* an initial *complete* set of compatible properties P_i of S under conditions C_i with values p_i at time t_1, one can *predict* the result of a final registration at time t_2 of the values p_f of another *complete* set of compatible properties P_f of S under conditions C_f.

Conversely, if one *registers* at time t_2 a final *complete* set of compatible properties P_f of S under conditions C_f with values p_f, one can *retrodict* the values of the initial preparation at time t_1 of a *complete* set of compatible properties P_i of S under conditions C_i with values p_i.

On the other hand, if one prepares a canonical ensemble of systems S under conditions C_i by specifying the values of only "half" of a set of canonically conjugate compatible properties P_i at time t_1, then one can predict the *probability* for *a single* system S under conditions C_f , of the result p_f of a registration of only "half" of a set of canonically conjugate compatible properties P_f at time t_2.

8.2. The difference between classical and quantum ensembles

If the process is sensitive to the existence of the quantum of action h, then one cannot carry out a preparation or registration of a system involving more than "half" of a set of classically canonically conjugate compatible properties of system S under conditions C; the other "half" are now quantum-mechanically incompatible properties.

In quantum mechanics too, only the probability of a process leading from an initial preparation to a final measurement can be defined. The big difference is that in quantum mechanics, rather than directly computing the conditional probability $0 \leq P(p_f|p_i) \leq 1$, one computes a *probability amplitude* $\langle p_f|p_i\rangle$ for each process: a complex number such that $P(p_f|p_i) = |\langle p_f|p_i\rangle|^2$.

As I explained elsewhere in greater detail:

> When introduced in classical mechanics, probabilities are always symptomatic of ignorance, i.e. renunciation of full information about a classical system that, in principle, could be obtained. Classical ensembles in configuration space (see Schiller 1962, Berry and Mount 1972) provide an example that is important

for comparison with the situation in quantum mechanics (see "Probability Amplitudes and Feynman Paths" below). Consider a *complete* solution $W(q, \alpha, t)$ to the Hamilton–Jacobi equation for a system with Hamiltonian H, i.e. a solution based on stipulation of "half" the total number of variables needed to specify an individual trajectory in phase space, i.e. each pair (q, α) specifies such a trajectory. The complete solution corresponds to a real or virtual ensemble of such trajectories, one for each value of α. The density $\rho(q, \alpha, t)$ of these trajectories in the extended configuration space with coordinates (q, t) is given by the *Van Vleck determinant* $|\partial^2 W/(\partial q \partial \alpha)|$, and hence if a trajectory is chosen at random, the probability $P(q, \alpha, t)$ of its having the values (q, α) at time t is proportional to $\rho(q, \alpha, t)$.

But, whether one *knows* their values or not, at any time each individual system in a classical always *possesses* some *definite* position and momentum, and it is only renunciation of a possible measurement of "half" this information in preparing the ensemble that leads to the probabilities of values for individual systems in the ensemble. In other words, a Hamilton–Jacobi ensemble can always be subdivided into sub-ensembles, for which more narrowly-defined probabilities can be calculated; and it is ultimately divisible into individual trajectories, with each of which definite values are associated.

For later comparison with quantum ensembles, consider a complete solution of the form $W(q_i, t_i; q, t)$ — a Hamilton's principal function — and let t be later than t_i. It follows from the previous discussion that the conditional probability for a system initially prepared with position q_i at time t_i to be found in an interval dq around q at time t is given by:

$$P(q_i, t_i; q, t)\, dq,$$

where the probability density $P(q_i, t_i; q, t)$ is proportional to the Van Vleck determinant of Hamilton's principal function. This probability may be given a propensity interpretation for a virtual ensemble associated with a single system, or a frequency interpretation for an ensemble of identically prepared systems (Stachel [29], p. 635).

8.3. Observation and retrodiction

In order to explain *observations*, scientific theories must also be able to *retrodict* the causes of the observed data. The Hanbury

Brown–Twiss effect [23] is a prime example of the scientific importance of retrodiction in explaining certain observed quantum effects. More to the point here, retrodiction is important for the discussion of the foundations of quantum mechanics because it shows that one cannot attach any fundamental ontological significance to the state function of system S at any time t between the time t_1 of preparation and t_2 of observation. If one uses the state function to help calculate the probability amplitudes for processes, then the *same* process or phenomenon will be assigned a *different* state function at time t, depending on whether it is being used for *prediction* or *retrodiction*.

9. Commutation Relations in Quantum Mechanics

One central method of taking into account the quantum of action is by means of introducing commutation relations between various particle (non-relativistic QM) or field (SR QFT) quantities ("observables") into the formalism. But these commutation relations have more than a purely formal significance.

> Within quantum mechanics, the uncertainty relations — or better, using a direct translation of the German term *Unbestimmheit*, the indeterminacy relations — assert that there is a limit to the simultaneous measurability of a pair of classical canonically conjugate variables such as the position and momentum of a system. And, as Heisenberg was at great pains to demonstrate in his little book, *Physical Principles of Quantum Theory*, the limit set by the theory on the simultaneous measurement of any pair of canonically conjugate variables agrees perfectly with the limits set by any idealized measurement procedure that takes into account the finite size of the quantum of action. If an idealized experimental arrangement could be designed that, while taking the quantum of action into account, nevertheless succeeded in violating the theoretical limits set by the indeterminacy principle on the simultaneous measurement of any pair of canonically conjugate variables, then this would be taken as a sign that non-relativistic quantum mechanics was in need of significant theoretical modification. And indeed, that is just why Heisenberg took such pains to demonstrate that this is not the case for such a wide class of conceivable experiments (Stachel [36]).

As Davies has emphasized:

> It is essential to appreciate that this uncertainty is inherent in nature and not merely the result of technological limitations in measurement. It is not that the experimenter is merely too clumsy to measure position and momentum simultaneously. The particle simply does not *possess* simultaneously precise values of these two attributes. One is used to uncertainty in many physical processes — for example, in the stock market or in thermodynamics — but in these cases the uncertainty is due to missing information rather than to any fundamental limitation in what may be known about these systems (Davies [18], pp. *ix–x*).

10. Commutation Relations in Quantum Field Theory

Why do we use *equal time* commutation relations? Tradition!!! As Bryce DeWitt has emphasized:

> When expounding the fundamentals of quantum field theory physicists almost universally fail to apply the lessons that relativity theory taught them early in the twentieth century. Although they usually carry out their calculations in a covariant way, in deriving their calculational rules they seem unable to wean themselves from canonical methods and Hamiltonians, which are holdovers from the nineteenth century, and are tied to the cumbersome $(3 + 1)$-dimensional baggage of conjugate momenta, bigger-than-physical Hilbert spaces and constraints.... One of the unfortunate results... is that physicists, over the years, have almost totally neglected the beautiful covariant replacement for the canonical Poisson bracket that Peierls invented in 1952 (DeWitt [19]).
>
> When I first realized that Bohr and Rosenfeld were dealing with Peierls brackets, I became quite excited.... [T]he Peierls bracket is the appropriate concept for analyzing the quantum mechanical limitations on measurement accuracy. This analysis says that measurements can, in principle, always be made to an accuracy equal *to but no better than* that allowed by the *a priori* uncertainties implied by the quantum mechanical formalism (DeWitt-Morette [20]).

Here is the abstract of Peierls original paper:

> A definition of Poisson brackets is given which is related to the action principle, but does not require the introduction of canonical

> variables. This permits the laws for forming both the commutators of canonical theory and the anticommutators of Fermi–Dirac particles to be stated in a manifestly covariant way. Examples of the use of this method are given. The last section discusses tentatively the extension to the case of equations which cannot be written in canonical form (Peierls [25]).

Pierre Cartier has explained the meaning of the Peierls bracket and its importance:

> Bryce DeWitt constructs the operator formalism of quantum physics from the Peierls bracket which leads to the Schwinger variational principle and to functional integral representations.
>
> The bracket invented by Peierls in 1952 is a beautiful, but often neglected, covariant replacement for the canonical Poisson bracket, or its generalizations, used in canonical quantization.... The precise definition follows from the theory of measurement (Cartier [15], p. 5).

As Bryce DeWitt emphasized:

> The remarkable thing about the Peierls' brackets is that they do not depend for their definition on the introduction of a canonical formalism. They are completely determined by the laws of propagation of Jacobi fields, and their definition emphasizes the global spacetime view of dynamics (DeWitt-Morette [20]).

This makes Peierls' brackets a potentially important tool in the search for a theory of quantum gravity: we need a theory that can somehow encompass the achievements of both Quantum Field Theory (which is background-dependent) and General Relativity (which is background-independent). But that is another story, which I have discussed elsewhere at great length (Stachel [30–32, 35]).

References

[1] Amelino-Camelia, G. and Stachel, J. (2009). Measurement of the spacetime interval between two events using the retarded and advanced times of each event with respect to a time-like world-line. *General Relativity and Gravitation*, **41**(5): 1107–1124.

[2] Ashtekar, A. and Rovelli, C. (1992). A loop representation for the quantum Maxwell field. *Classical and Quantum Gravity*, **9**: 1121–1150.

[3] Baggott, J. (2013). *Farewell to Reality: How* Modern *Physics has Betrayed the Search for Scientific Truth.* New York: Pegasus Books.

[4] Bai, T. and Stachel, J. (2015). Bohr's Diaphragms. *This Volume.*

[5] Beller, M. and Fine, A. (1994). Bohr's Response to EPR. In Faye and Folse *Niels Bohr and Contemporary Philosophy.* Dordrecht: Kluwer Academic Publishers, pp. 1–31.

[6] Bergmann, P. and Smith, G. (1982). Measurability analysis of the linearized gravitational field. *General Relativity and Gravitation*, **14**: 1131–1166.

[7] Bohr, N. (1958). *Atomic Physics and Human Knowledge.* New York: John Wiley & Sons.

[8] Bohr, N. (1928). The quantum postulate and the recent developments of atomic theory. *Nature*, **121**: 580–590.

[9] Bohr, N. (1935). Can quantum-mechanical description of physical reality be considered complete? *Physical Review*, **48**: 669–702.

[10] Bohr, N. (1939). Comments on Paul Langevin, The Positivistic and the Realistic Trends in the Philosophy of Physics. In *New Theories in Physics*, pp. 11–30. Paris: International Institute of Intellectual Cooperation and [11]

[11] Bohr, N. (1939). The Causality Problem in Atomic Physics. In *New Theories in Physics*, pp. 11–30. Paris: International Institute of Intellectual Cooperation and [10].

[12] Bohr, N. (1949). Discussion with Einstein on epistemological problems in atomic physics. In *Albert Einstein: Philosopher Scientist*, pp. 199–241, Evanston, Illinois: Library of Living Philosophers, Vol. 7, 1949. Reprinted in Bohr [4], pp. 32–66.

[13] Bohr, N. (1955). Unity of Knowledge. In *The Unity of Knowledge*, New York: Doubleday; reprinted in Bohr [7], pp. 67–82.

[14] Camillieri, K. Bohr. (2007). Heisenberg and the divergent views of complementarity. *Studies in History and Philosophy of Modern Physics*, **38**: 514–528.

[15] Cartier, P. A. Legacy. Online at (http://swc.math.arizona.edu/dls/DLSCartierCh1.pdf).

[16] Crenshaw, M. (2003). Microscopic foundation of macroscopic quantum optics. *Physical Review A*, **67**: 033805.

[17] Darrigol, O. (1991). The problem of the measurability of quantum fields, pp. 153–175 of Cohérence et complétude de la mécanique quantique: l'exemple de Bohr–Rosenfeld, *Revue d'histoire des sciences*, **44**: 137–179.

[18] Davies, P. (1989). *Introduction* to Heisenberg W, *Physics and Philosophy*, pp. vii–xvii. London: Penguin.

[19] DeWitt, B. (1999). The Peierls Bracket, in DeWitt-Morette and Zuber J.-B. (eds.), *Quantum Field Theory: Perspective and Prospective*, pp. 111–136. Dordrecht: Kluwer Academic.
[20] DeWitt-Morette, C. *The Pursuit of Quantum Gravity: Memoirs of Bryce DeWitt from 1946 to 2004*. Springer 2011.
[21] Fine, A. (2004). The Einstein–Podolsky–Rosen argument in quantum theory, in *The Stanford Encyclopedia of Philosophy*. Current edition, Edward N. Zalta (ed.), (http://plato.stanford.edu/archives/win2014/entries/qt-epr/).
[22] Ginzburg, V. L. (1940). Quantum theory of luminous radiation from an electron traveling uniformly in a medium, *J. Phys. U.S.S.R.*, **2**: 441–452.
[23] Hanbury, B. R. and Twiss, R. Q. (1956). A test of a new type of Stellar interferometer on sirius. *Nature*, **178**: 1046–1048.
[24] Landau, L. D. (1941). The Theory of Superfluidity of Helium II. *J. Phys. U.S.S.R.*, **5**: 71. Reprinted in ter Haar, (ed.), *Collected Papers of L. D. Landau*, pp. 301–330. Oxford: Pergamon Press (1965).
[25] Peierls, R. E. (1952). The commutation laws of relativistic field theory. *Proc. Royal Soc. A*, **214**: 143–157.
[26] Shaw, T. (2014). Review of *Divine Fury: A History of Genius*, *New York Review of Books*, pp. 4, 6, 8.
[27] Smolin, L. (2001). *Three Roads to Quantum Gravity*. New York: Basic Books.
[28] Stachel, J. (1986). Do quanta need a new logic? in Colodny (ed.), *From Quarks to Quasars: Philosophical Problems of Modern Physics*, pp. 229–348. Pittsburgh: University of Pittsburgh Press.
[29] Stachel, J. (2006). Quantum Logic. in Jessica Pfeifer and Sahotra Sarkar (eds.), *The Philosophy of Science: An Encyclopedia*, Vol. 2, pp. 633–644. NewYork/London: Routledge Press.
[30] Stachel, J. (2006). Structure, individuality and quantum gravity. In Don Rickles, Steven French & Juha Saatsi (eds.), *Structural Foundations of Quantum Gravity*, pp. 53–82. Oxford: Clarendon Press.
[31] Stachel, J. (2009). Prolegomena to any future quantum gravity. In Oriti, D. (ed.) *Approaches to Quantum Gravity: Toward a New Understanding of Space, Time and Matter*, pp. 44–67. Cambridge: Cambridge University Press.
[32] Stachel, J. and Bradonjić, K. (2013). Quantum Gravity: Meaning and Measurement, Preprint submitted Feb 7, 2013 to *Studies in the History and Philosophy of Modern Physics* [arXiv:1302.2285].
[33] Stachel, J. (2009) Bohr and the Photon. In Myrvold and Christian (eds.), *Quantum Reality, Relativistic Causality, and Closing the*

Epistemic Circle: Essays in Honour of Abner Shimony. The Western Ontario Series in Philosophy of Science, Vol. 73, pp. 69–83. Heidelberg: Springer.

[34] Stachel, J. F. (1997). Paths and quantum entanglement: Is there any more to the mystery? in Cohen, Horne, Stachel (eds.), *Potentiality, Entanglement, and Passion-at-a-Distance: Quantum Mechanical Studies for Abner Shimony*, pp. 245–256. Dordrecht: Kluwer Academic Publishers.

[35] Stachel, J. (2014). Quantum gravity: A heretical vision. In Sidharth, B. G. *et al.* (eds.), *Frontiers of Fundamental Physics and Physics Education, Springer Proceedings in Physics 145.*, pp. 165–174. Heidelberg: Springer.

[36] Stachel, J. Some Measurement Problems in Quantum Gravity. Unpublished.

[37] Stone, A. D. (2013). *Einstein and the Quantum.* Princeton: Princeton University Press.

[38] Styer. D. F. *et al.* (2002). Nine formulations of quantum mechanics. *American Journal of Physics*, **70**: 288–297.

[39] Żurek, W. (2014). Quantum Darwinism, classical reality, and the randomness of quantum jumps. *Physics Today*, **67**: 44–50.

Philosophical Aspects

Chapter 4

Beyond Complementarity

R. E. Kastner

Foundations of Physics Group
University of Maryland
College Park, USA
rkastner@umd.edu

It is argued that Niels Bohr ultimately arrived at positivistic and antirealist-flavored statements because of weaknesses in his initial objective of accounting for measurement in physical terms. Bohr's investigative approach faced a dilemma, the choices being (i) conceptual inconsistency or (ii) taking the classical realm as primitive. In either case, Bohr's "Complementarity" does not adequately explain or account for the emergence of a macroscopic, classical domain from a microscopic domain described by quantum mechanics. A diagnosis of the basic problem is offered, and an alternative way forward is indicated.

1. Introduction

In this volume,[a] Bai and Stachel [4] offer a rebuttal of arguments by Beller and Fine [6] that Bohr's philosophy of quantum mechanics was positivist. That discussion addresses Bohr's reply [8] to the Einstein, Podolsky, and Rosen ("EPR") paper [17]. The purpose of this chapter is not to enter into the specific debate concerning whether Bohr's basic approach was positivist or not (although this

---

[a] *Quantum Structural Studies*, (eds.) R. E. Kastner, J. Jeknić-Dugić, G. Jarosz-kiewicz, World Scientific Publishers, forthcoming.

author tends to agree with Bai and Stachel that Bohr's interpretive intentions were not antirealist). Rather, the intent is to argue that Bohr inevitably lapsed into antirealist-flavored statements about quantum systems because his notion of "Complementarity" cannot consistently account for the emergence of classicality from the quantum level. It is argued that ultimately this problem arises from Bohr's implicit assumption that all quantum evolution is unitary; i.e. that there is no real, physical non-unitary collapse.

It should be noted that Bohr's ideas changed and evolved over several decades and this chapter does not attempt to trace the intricate development of this evolution. Rather, attention is focused on Bohr's initial reply to EPR and on certain methodological and metaphysical constraints that, it is argued, led inexorably to a final antirealist position toward quantum level, as evidenced in his famous statement "There is no quantum world. There is only an abstract quantum mechanical description [28]." While a reader might disagree with whether Bohr was instrumentalist or antirealist at any particular stage of the development of his thought, the point of this chapter is to argue that the end result of Bohr's investigations into the problem was a form of antirealism about the quantum level that is not in fact forced on us but arises from certain unacknowledged metaphysical, theoretical, and methodological assumptions which acted as unnecessary constraints on his interpretive investigation (and which continue to constrain such investigations today).

2. Bohr's Initial Arguments

It should first be noted that the original EPR experiment involving position and momentum has some significant differences from the more commonly discussed later version due to Bohm, the latter based on a spin- $1/2$ singlet state. In the former case, measuring one observable involves a coupling with its complementary quantity, while that is not the case with the latter spin experiment. In the spin case, however, it can still be argued that the measurement conditions suitable for one spin observable are incompatible with measurements of a non-commuting spin observable.

With that in mind, let us attempt to distill Bohr's much-analyzed reply to EPR down to its essence. First, consider his discussion of measurement of a single quantum system S's position or momentum using a diaphragm D (screen with a single slit in it). The basic thought experiment can be described as follows:

1. Assume that S has an initial well-defined longitudinal momentum p, with zero transverse component (corresponding to the plane of the diaphragm), as it approaches the diaphragm D with slit.
2. Note that upon exiting D, S's state is one with greatly decreased transverse position uncertainty Δq and correspondingly increased transverse momentum uncertainty Δp.
3. Ask whether one could "foil" the uncertainty relation by taking into account any exchange of momentum between S and D in order to reduce the uncertainty Δp.
4. Assert that this is impossible because the exchange of momentum is "uncontrollable."

Regarding (4), Dickson [16] notes that the characterization of the exchange of momentum as "uncontrollable" is basically "an article of faith" on Bohr's part, and suggests that one should more conservatively call the momentum exchange "unpredictable."

What remains ill-defined in Bohr's account is whether the uncertainties and unpredictabilities in the measurement processes are to be understood as genuine ontological indeterminacies or merely epistemic ignorance of determinate values. This, I suggest, is a crucial equivocation in Bohr's treatment of the problem. When dealing with objects that are decidedly quantum systems (such as the particle S going through the slit), he seems to allow (at least implicitly, at this stage) these incompatible properties to be fundamentally indeterminate. On the other hand, when dealing with macroscopic systems, he uses epistemic language, referring to the relevant interactions and properties as "uncontrollable," "inaccessible," "unpredictable," etc. This is so even when he argues that under certain circumstances even a macroscopic object such as D should be considered one of the quantum "objects of study." Such a circumstance would apply, for example, to his proposal to delay the final measurement of D's displacement after passage of S and leave it as a matter of "free choice" — thus treating S and D as quantum-entangled in the EPR sense.

Of course, this equivocation concerning the nature of uncertainty (ontological versus epistemic) serves to evade the undesirable result that a macroscopic object like D could have a genuinely indeterminate position; if one pursues that line avenue of inquiry, we are led immediately to the Schrödinger Cat paradox (more on that in the next section). One might argue that, even if taken as ontologically indeterminate, under the discussed thought-experiment the indeterminacy of D's displacement would be so tiny as to be effectively microscopic and therefore not observable. But one could, at least in principle, reversibly amplify the displacement of D to macroscopic proportions, in which case D would be in many places at once. Bohr clearly does not accept this idea; thus he must take position uncertainty pertaining to D as epistemic.

With regards to statement (3) above, Bohr notes that what makes step (2) a position measurement is that D is anchored immovably to the lab frame, which establishes a spacetime frame of reference. Without that spacetime frame, the notion of a position value would be ill-defined even in a classical sense. This however is yet another form of equivocation on Bohr's part. As Dickson further points out, "there is nothing that, quantum mechanically, can really serve to define a reference frame, because reference frames are (by definition!) well-defined both in position and momentum. Quantum theory tells us that there is no such thing, but for the sake of making our notions of position and momentum meaningful, we voluntarily choose to accept a given physical object (the apparatus, or whatever) to serve as a reference frame." ([16] p. 14, my emphasis).

Here we encounter a form of the dilemma faced by Bohr and, I hope to persuade the reader, ultimately not resolved by his notion of Complementarity. Our world of experience is clearly classical in that we can legitimately consider our lab and macroscopic measuring instruments as inhabiting a well-defined inertial frame. *But these are the very phenomena that cry out for explanation in view of the fact that the microscopic quantum objects upon which we experiment, according to the theory describing them, do not inhabit well-defined reference frames.* (This seemingly paradoxical situation actually is amenable to resolution, which is the subject of Section 4.) Bohr deals with these apparently disparate realms by equivocation concerning their physical nature, and that equivocation is aided by his use of qualitative description rather than quantitative application of the relevant theoretical formalism. The next section aims to remedy this reliance on qualitative description in order to more clearly identify the underlying weaknesses in Bohr's account.

3. Analysis of Bohr's Second Thought Experiment

Suppose we apply the quantum formalism to Bohr's thought experiment of the second case considered; i.e. the case in which D is allowed a transverse degree of freedom in order to have the possibility of measuring the momentum of S. (This is termed experiment A-2 by Bai and Stachel.) This is to be done via momentum conservation by measuring D's momentum before and after passage of S. But according to Bohr, this leaves us with the "free choice" whether or not to measure position instead of momentum, by choosing whether or not to make a final position or momentum measurement of D. Thus, Bohr seems to be describing the interaction between S and D as a non-disturbing "measurement of the first kind," sometimes termed a "pre-measurement". The initial state of S is $|p\rangle$, a state of well-defined momentum with zero transverse component, and D is in a ready state of well-defined position $|Q\rangle$. After their interaction, Bohr seems to assume that they can be represented by an entangled state $|\Psi\rangle$ (much like the original EPR state, as noted by Bai and Stachel). As observed by EPR, such a state has an inherent basis ambiguity and can be written in terms of any orthogonal set of states. For reference, Eqs. (7) and (8) of the original EPR paper are reproduced here:

$$\Psi(x_1, x_2) = \Sigma_n \psi_n(x_2) u_n(x_1), \tag{1}$$

where $\Psi(x_1, x_2)$ is a two-particle wave function expressed in terms of eigenfunctions $u_n(x_1)$, of some observable A; and the coefficients $\psi_n(x_2)$ are viewed as amplitudes for the expansion in this basis.

On the other hand, as EPR note, the same two-particle state can be expressed in terms of a different set of eigenfunctions v_n corresponding to a different observable B, with different expansion coefficients:

$$\Psi(x_1, x_2) = \Sigma_n \phi_n(x_2) v_n(x_1). \tag{2}$$

Let us define EPR's first observable A as applying to relevant aspects of the position of D. For convenience, take the eigenstates to be a discrete set of small transverse position ranges $|Q_i\rangle$ (one of which would act as a pointer to the localized wave packet $|q_k\rangle$ emerging from the slit). The second observable B will apply to the transverse momentum of D and its eigenstates will be a discrete set of small transverse momentum ranges $|P_j\rangle$ (which would act as a pointer to the transverse momentum state $|p_j\rangle$ of the emerging particle). The corresponding discrete states for S will be $|q_i\rangle$ and $|p_j\rangle$, respectively. So $|\Psi\rangle$ will look like:

$$|\Psi\rangle \sim \Sigma_i \alpha_i |q_i\rangle |Q_i\rangle = \Sigma_j \beta_j |p_j\rangle |P_j\rangle, \tag{3}$$

where α_i and β_i are amplitudes. Thus, if we choose to measure the final position Q of D and find it within the range Q_k, then the "entire experimental arrangement" allows us to attribute to S the corresponding state $|q_k\rangle$; or if we measure the final momentum P of D and find it within the range P_n, then similarly in virtue of Bohr's "wholeness" criterion, we can attribute to S the corresponding state $|p_n\rangle$.

Now, presumably the designated unitary evolution of the initially independent systems to the above entangled state would have to be treated as a correlation arising via scattering of the particle from the edges of the slit in D. To get some feel for the magnitudes involved, assume an incoming electron energy of roughly 20 Mev, and a mass for D of as little as 1 gram (small but still macroscopic), in an elastic scattering process. The maximum possible outgoing velocity for D would be negligible: of the order of 10^{-17} m/s. This is good and bad news for Bohr. The good news is that such a microscopic effect might make it seem reasonable to consider S and D as two entangled quantum systems (D being on the same footing as S as an "object of study"). But the bad news is that D could not serve as a credible measuring instrument for the momentum of S, and therefore we would not really have a "free choice" at this point to make that measurement given the putative entangled system as described. In order to accomplish the latter, and still provide the "free choice" that Bohr asserts, D would need to be entangled with some sort of amplifying degree of freedom. But in that case, we have a Schrödinger's Cat situation: any indeterminacy in either D's position or momentum would be visible at the macroscopic level, but it never is.

Thus, we can see that this is just the usual problem in which macroscopic objects, when assumed to be described by quantum states entangled with quantum systems, become "infected" with any indeterminacy pertaining to the quantum system. That is, it is the measurement problem. As Dickson notes: "Presumably, to consider the interaction between [the particle] and the apparatus a genuine measurement we must ignore the subsequent entanglement between them and take the apparatus to be in a definite state of indication, even if in fact it is not." ([16] p. 28, preprint version.) This inconsistency problem is not addressed by the notion of Complementarity. That is, it is fine to note that certain observables are incompatible and cannot be simultaneously measured, and that it may be inappropriate to regard values of such sets of observables as all well-defined under specified circumstances. But since such an observation does not resolve the above consistency issue, it would appear to amount to little more than just restating the uncertainty principle. "Complementarity" is not enough.

4. Bohr's Epistemological and Methodological Assumptions as Unnecessary Restrictions on His Investigation

At this point, we consider some methodological and epistemological pronouncements by Bohr, which represent the constraints under which his investigation took place, but which can in fact be questioned. One important example is this rather lengthy categorical assertion:

> "The essential lesson of the analysis of measurements in quantum theory is thus the emphasis on the necessity, in the account of the phenomena, of taking the whole experimental arrangement into consideration, in complete conformity with the fact that all unambiguous interpretation of the quantum mechanical formalism involves the fixation of the external conditions, defining the initial state of the atomic system concerned and the character of the possible predictions as regards subsequent observable properties of that system. Any measurement in quantum theory can in fact only refer either to a fixation of the initial state or to the test of such predictions, and it is first the combination of measurements of both kinds which constitutes a well-defined phenomenon." [9]

One can make the above assertion considerably less lengthy. Omitting some of the categorical and emphatic aspects, the basic claims are found to be:

1. Measurement in quantum theory can only be physically defined by reference to a macroscopic experimental arrangement.
2. A well-defined phenomenon, taken as defining "measurement," requires an initial preparation and final (macroscopic) observation.
3. There is no unambiguous interpretation of the quantum formalism as applied to any system without reference to externally fixed conditions defining the initial and final states of that system, where "externally fixed conditions" means macroscopic phenomena accessible to an observer.

In what follows, I critique these claims. A refutation of all three is presented in the final section, through a counterexample: a formulation that unambiguously specifies how the determinacy inherent in measurement arises without necessary reference to macroscopic phenomena.

Firstly, while Bohr's insistence on the "necessity. . . of taking the whole experimental arrangement into consideration" is well known, and is often taken as a benign statement of "quantum wholeness," it is actually a very strong (and, I will argue, unnecessary) prohibition on taking any degree of freedom as physically specifiable independently of macroscopic phenomena. This prohibition is sharpened in claim 3 which effectively asserts that one is not allowed to say that the quantum formalism, as applied to any subsystem of an "entire experimental arrangement," has an unambiguous physical referent, even if one cannot describe that referent in "ordinary" — meaning classical — terms. Note that this is a stronger claim than merely saying "an unmeasured subsystem does not have classically observable properties"; rather, it says that one should not try to understand the physical nature of any degrees of freedom that are *correctly* assigned a quantum theoretical description.

Overall, Bohr's quoted statement assumes that unambiguous physics only obtains in the context of a "measurement," where that term is considered to be definable only in terms of a macroscopic experimental arrangement leading to an "observation" or "phenomenon". This use of the term "measurement" is a conflation, ongoing in much of the literature, of two distinct ideas: (i) the intervention of an observer whose intent is to gain determinate knowledge about something under study; and (ii) the existence of a fact of the matter — or determinate a value of some property — whether or not anyone has intent to discover it (or whether or not it results from a macroscopic "phenomenon"). The preceding two different notions of the determinacy obtaining in measurement (but not necessarily confined to a knowledge-gathering measuring operation) can be labeled as (i) epistemic and (ii) ontological, respectively. Bohr's pronouncement of course denies (ii) by asserting that it is only through an in-principle macroscopic "phenomenon" that any physical quantity is well-defined, and that the quantum formalism is not even interpretable outside that condition. But this denial can and will be questioned.

Besides the above conflation, Bohr's insistence that one must take "the whole experimental arrangement into account" does not remedy the consistency problem concerning S and D in their purported entanglement that he describes in his reply to EPR. One supposedly has a "free choice" whether to measure the momentum of D and thereby gain knowledge of the momentum of S on passing through the slit, or to measure the position of D and thereby gain knowledge of the position of S. In this case, S and D are in an entangled pure state and D and S are described by improper mixed states. There is no basis from within the theory to say why, at the time when the choice is supposedly available, any uncertainty pertaining to D should be of a different sort than that pertaining to S. Yet clearly Bohr needs D's uncertainty to be epistemic rather than ontic in nature to avoid a Schrödinger's Cat situation; while on the other hand, since he views any attributes of a quantum system such as S in need of (at least) irreversible amplification [10] in order to be considered determinate, the uncertainty pertaining to S cannot be considered epistemic. However, the theoretical description provides no justification for attributing different sorts of uncertainties to S and D.

Ultimately, Bohr's response to this conundrum is to deny reality to quantum objects, and to assert by fiat that at some point in the (assumed as linear) evolution, a determinate world of experience occurs and classical "reality" begins — since we routinely see objects like D with determinate position and momentum. This is not an explanation of classical emergence, but rather an equivocation concerning the application of quantum theory. A crude analogy is that the unitary quantum evolution is like a car engine engaged via the clutch with the gear shaft (which carries the entanglement of the relevant degrees of freedom); but at the point in which we find ourselves empirically describing objects that are classically determinate (or, in which the dimensions of the experiment are much larger than Planck's constant), we disengage the clutch. This is an *ad hoc* move; there is no consistent theoretical account for suspension of the unitary evolution. (It will not do to reply, in Bohrian fashion, that "the lesson of quantum theory is that there can be no consistent theoretical account," since one is provided in the final section.)

However, could we see this sort of move as justified by seeing it as form of pragmatism? I think the answer is negative, and arguably does a disservice to pragmatism. Pragmatism primarily concerned itself with reforming the concept of truth from an abstract and absolute notion to a concrete and functional one. It is one thing to say that our criteria for truth must require that truth claims pass some test of functionality; it is quite another to suspend the quantum formalism to force the theoretical description to correspond to our empirical experience and/or to classical mechanics at a specified limit, even though it apparently does not. That is the essence of equivocation, and pragmatism was not equivocal.

It is worth mentioning in this context that Bub [14] has given an interesting formal account of Bohr's "Complementarity." Bub has shown that the Hilbert space structure of quantum states allows for a generalization of the "Bohmian" theory in which the position of a quantum system is taken as an always-determinate "beable" (Bell's term, [5]). It turns out that one can always choose one particular observable as having preferred status, such that its eigenvalues attain "beable" status; and any other observable commuting with that preferred observable will have determinate values as well. Meanwhile, properties corresponding to observables not commuting with the preferred observable have indeterminate status (there are no yes/no answers to questions about those properties, where the questions are represented by projection operators on the Hilbert Space). According to Bub's observation, Complementarity consists of conferring "preferred" status on the observable selected as being determinate by the "entire experimental arrangement," such that its eigenvalues become "beables."

Does this allow Bohr to escape from the above inconsistency problem? I believe the answer is "no". Recall that Bohr says we have a free choice whether to measure position or momentum of the diaphragm in the case in which S and D are assumed to be entangled; he asserts that D is to be viewed as a quantum system at this stage of the experiment. Clearly, the availability of this "free choice" means that we have not yet completed the "entire experimental arrangement" that would bring about a preferred observable according to Bub's formulation. But this means that (at this stage of the experiment, prior to the choice), there is no fact of the matter about either D's position or its momentum, since neither is a preferred observable. Thus, invoking a preferred observable-based beable does not rescue Bohr from the inconsistency, since his "entire experimental arrangement" criterion for the preferred observable implies the undesirable conclusion that at certain preliminary stages of an experiment, a macroscopic object has no determinate physical property. It should be kept in mind that the tiny displacement of D does not help here: according to Bohr's assumptions, in principle, one could reversibly entangle another degree of freedom with D that would amplify the tiny displacement to macroscopic proportions and yet still be described, according to Bohr, as a quantum system (since there has been no "irreversible amplification" such as a change in the chemical properties of photographic plate emulsion that Bohr takes as heralding a "measurement").

As noted in the Introduction, Bohr's views evolved over time. For example, as Stachel points out, "Bohr's later approach places primary emphasis on four-dimensional processes; from this point of view, a "state" is just a particular spatial cross-section of a process, of secondary importance: all such cross-sections are equally valid, and any such sequence of states merely represents a different "perspective" on the same process." ([32], p. 1, preprint version.) It should however be noted that such an approach — dissolving the measurement problem by noting that some outcome always in fact obtains at the phenomenal, classical, spacetime level — amounts to an epistemic interpretation of the quantum state. That is, the quantum state and its unitary evolution are taken as describing only our limited perspective on a process that is assumed to be complete as an element of a classically determinate block world. In this approach, the classical world of phenomenal experience does not emerge from the quantum level. It is taken as ontologically given and primary, with quantum theory relegated to a partial and perspectival description of that classical reality.[b]

[b]Stachel (private communication) gives another argument for denying ontological reality to the quantum state. This consists in the observation that a time-symmetric approach to the Born Rule will attribute a different state to the same system depending on whether it is considered a pre-selected or post-selected. In

In addition to the pronouncement which opened this section, Bohr made many other emphatic, categorical statements concerning the interpretation of quantum theory that are nevertheless subject to challenge as being based on (a) unacknowledged metaphysical and conceptual premises, or (b) even on an ill-defined ontology. An example of (a) is the following:

> "It must not be forgotten that only the classical ideas of material particles and electromagnetic waves have a field of unambiguous application, whereas the concepts of photons and electron waves have not. Their applicability is essentially limited to cases in which, on account of the existence of the quantum of action, it is not possible to consider the phenomena observed as independent of the apparatus utilized for their observation. I would like to mention, as an example, the most conspicuous application of Maxwell's ideas, namely, the electromagnetic waves in wireless transmission. It is a purely formal matter to say that these waves consist of photons, since the conditions under which we control the emission and the reception of the radio waves preclude the possibility of determining the number of photons they should contain. In such a case we may say that all trace of the photon idea, which is essentially one of enumeration of elementary processes, has completely disappeared." [11, pp. 691–692].

The phrase "electromagnetic waves in wireless transmission" means the classical electromagnetic field. Such a field is instantiated by the quantum coherent state, which is a superposition of photon

terms of the Aharonov–Bergmann–Lebowitz rule [2], this is seen in the fact that the ABL rule gives a probability of unity for an intermediate measurement of either the pre- or post-selected state. But what this implies for interpretation of the quantum state depends crucially on one's presumed ontology. If one presumes that there is a block world (i.e. no ontological difference between past, present, and future), then the foregoing results simply restate that ontology, since in a block world each system is both prepared and fated at any intermediate time during its lifetime. On the other hand, in a growing universe ontology with indeterminate future, the foregoing results do not indicate any inconsistency for an ontological quantum state. The prepared state can be understood as describing the system prior to its detection, while the attribution of the post-selected state is only applicable *a posteriori*. (This is essentially the case for the interpretation to be discussed in the final section.)

number. To obtain a detectable classical field, one needs a very large average photon number.[c] Note that Bohr has slid from the fact that the coherent state is a quantum superposition of photon number to the conclusion that "the photon idea has disappeared". But it has not: the coherent state can be understood as a well-defined physical quantity, whether or not that it is visualizable "in the ordinary (classical) sense" [11, p. 21]. He thus simply disallows an ontology in which there could be a physically real state of the field, involving an indeterminate number of photons, that is not visualizable in a classical way. But, as Ernest McMullin has noted, "[I]maginability must not be made the test for ontology. The realist claim is that the scientist is discovering the structures of the world; it is not required in addition that these structures be imaginable in the categories of the macroworld" [26, p. 14]. In contrast, Bohr routinely insisted on the latter condition as a basic methodological requirement for doing physics. Moreover, that condition is precisely his criterion for what is to be regarded as physically real: according to Bohr, what is not "visualizable in the usual (classical) way" is deemed "abstract" and even "undefined," as we will see further below in considering an example of (b) (an ill-defined ontology).

Thus, Bohr's assertion peremptorily rules out *even the possibility* of an unambiguous physical referent for the key theoretical objects of quantum theory — discrete quanta and de Broglie waves. Yet it is dependent on the implicit and unnecessary assumption that all real physical processes must be classically visualizable spacetime processes, and on the accompanying assumption that quantum discreteness can only mean spacetime localizability as a "corpuscle."

The statement was made in the context of Bohr's inability to reconcile the idea of a wavelike frequency with the presumed corpuscular idea of a "photon," and the inverse problem of specifying within spacetime any wavelike (extended) nature of a "material particle" such as an electron. But Bohr's negative conclusion is not

[c]Sakurai notes that "The classical limit of the quantum theory of radiation is achieved when the number of photons becomes so large that the occupation number may as well be regarded as a continuous variable." [29, p. 36]

forced on us: a quantum of electromagnetic radiation or "photon", as the singular entity heralding a quantum discontinuity, need not be considered as a spatially localized object. The quantized, indivisible aspect of the photon can be reinterpreted as a component of the process of emergence of spacetime events and their discrete connections, the photon being the latter. Thus the discrete photon can be understood as emerging under certain suitable physical conditions, and the coherent state discussed above can be understood as a pre-emergent form of the underlying field. Meanwhile, the wavelike character of the photon and other material quanta (i.e. the de Broglie oscillation) can be retained on a sub-empirical, pre-spacetime level.

Such an approach, in which quantum processes are precursors to the emergence of localized spacetime events and their connections, is briefly reviewed in the final section. (It should also be noted that the present author is not the only one currently exploring spacetime emergence; cf. Sorkin [31], Oriti [27].) Thus, with a suitable relaxing of conceptual barriers and unnecessary metaphysical presumptions, one can indeed gain an unambiguous application for the basic physical concepts of quantum theory, contrary to Bohr's categorical negative claim.

An example of (b), a statement from Bohr exhibiting an ill-defined ontology is:

> "Isolated material particles are abstractions, their properties being definable and observable only through their interaction with other systems." [7]

This statement is problematic in several ways. First, many abstractions are perfectly well defined (such as mathematical concepts); so lack of definition has nothing to do with whether or not something is abstract. But more importantly, how does a non-physical, allegedly undefined abstraction undergo physical interactions? And if the interactions are not physical, how does a process that *could* be deemed concrete and physical come out of any of that? This is essentially the same "remove the clutch" inconsistency encountered above, where Bohr describes the initial degrees of freedom (S and D) by an entangled state and its unitary evolution, but then assumes that something real and determinate (somehow) occurs so that at least one of the same degrees of freedom (D) is no longer described by the entangled quantum state and its unitary evolution. There is a gap between the allegedly "abstract and ill-defined" and the allegedly "non-abstract and well-defined" that is not bridged by any amount of "amplification." This problem can be seen as the same type of metaphysical inconsistency facing Cartesian mind-matter dualism in that one has two fundamentally different substances that have no way to "interact."

In an epistemic approach to the quantum state, Bohr could finesse the inconsistencies described above by saying that we can suspend unitary evolution when it is no longer useful because we now have access to information that we lacked previously. Thus, neither the quantum state nor its unitary evolution ever directly described objects that physically existed. All that exists is the phenomenal, classical level of experience. But again, this leads Bohr to his ultimately antirealist view of quantum entities; i.e. to his utterance that "There is no quantum world. There is only an abstract quantum mechanical description." If there is no quantum world, then we need not give any account of classical emergence from such a world, since all that exists is the classical world of experience.

Bohr can thus retain a kind of consistency, but only (at least it seems to this author) at a rather high cost. Bohr spent the bulk of his career developing a detailed and revolutionary theory of the hydrogen atom in terms of its applicable quantum states. In order to retain consistency in the face of reconciling quantum mechanics with the classical realm of experience under the assumption of unitary-only evolution, Bohr ultimately felt forced to deny that hydrogen atoms could count as real physical referents for the very quantum states that he helped to formulate for them. Perhaps this turn to antirealism about the constructs of his pioneering theory was not really necessary. We consider an alternative in the next and final section.

Before turning to that alternative, it should be noted that Bohr's formulation legitimately takes measurement and determinacy as contextual; but it goes further than that by presuming that the contextuality is necessarily always a macroscopic one, dependent on a "phenomenon." The latter term essentially means "appearance," and thus is an intrinsically observer-dependent notion (since any appearance is always relative to a perceiving subject or subjects). This is a symptom of the fact that Bohr is unable to say why only one outcome occurs if one applies linear evolution to a quantum system and all its correlates; that of course always leads to a Schrödinger's Cat situation. So Bohr instead assumes that one must start with the observer's experience, where only one outcome is perceived; then one at least apparently has a well-defined physical situation.

But it is not in fact the case that this is the only way to obtain a well-defined physical quantity, and therefore it is not necessary to appeal to macroscopic "phenomena" as an ostensibly necessary starting point. The fundamental unnecessary constraint on Bohr's thinking is the presumption that the condition giving rise to a determinate value of a quantum mechanical operator cannot be defined from within the quantum formalism alone. But in fact it can.

5. Beyond Complementarity

The above-discussed apparent discrepancy between theory and observation, to which Bohr's Complementarity and its attendant antirealism about quantum objects is sometimes taken as a perplexing but inescapable response, is not a necessary one. The problem arises from demanding that all interactions between physical degrees of freedom are unitary ones. This is the key assumption that leads to the measurement problem and the "shifty split" between the quantum and classical realms, expressed in the *ad hoc* suspension of the unitary evolution and quantum-entangled state when it obviously no longer correctly describes the situation at hand. If nature in fact involves real non-unitary processes of a well-defined sort — including the circumstances that give rise to them — then the chain of unitary correlations is broken, and real physical collapse occurs, resulting in determinacy. Thus, the present author suggests that what Bohr needs to avoid the dilemma of theoretical inconsistency on the one hand,

and antirealism about quanta on the other, is genuine, non-unitary physical collapse.

What is also needed is an expansion of our metaphysical notions concerning what qualifies as 'physically real' — specifically, the acknowledgment that there may be real entities, referred to by the theoretical constructs such as quantum states, that are not be confined to 3 + 1 spacetime. Thus, the present proposal differs with the "primitive ontology" (PO) approach discussed by Allori [3]: the starting point for the PO is the assumption that any fundamental ontology referred to by a theoretical construct must be an element of the spacetime manifold. This restriction under PO of the "primitive variables" to 3 + 1 spacetime is prompted by the following consideration:

> "Roughly, the three-dimensionality of the primitive variables allows for a direct contact between the variables in the theory and the objects in the world we want them to describe. In fact, a PO represented by an object in a space of dimension d, different than 3, would imply that matter lives in a d-dimensional space. Thus, our fundamental physical theory would have to be able to provide an additional explanation of why we think we live in three-dimensional world while we actually do not." [3]

The proposed solution to this challenge is that quantum states refer to sub-empirical, pre-spacetime entities that can constitute precursors to observable spacetime events.[d] That is, the ontology has distinct levels: (i) actuality (observable, element of the spacetime manifold) versus (ii) physical possibility (still real but unobservable, pre-spatiotemporal). Level (ii) is essentially the Heisenbergian

[d]The question of why the observable spacetime manifold is 3 + 1 dimensions is a deep one with many different proposed answers. One relevant fact is that photons, which create observability, have four polarization directions. But for our purposes, it is sufficient to note that observable processes are always spacetime phenomena, while intrinsically unobservable quantum processes need not be required to inhabit the same manifold as the observable ones, as long as an account can be given of the transition from one manifold to the other. This is indicated in Kastner [21] and later in this section; the transition is precisely the collapse process.

"potentiae" [19]. Such an ontology, to be described in more detail below, is consistent with the reasonable view that real entities should be capable of leading to observable results, even if they themselves are not observable. In fact, the latter view is attributed to Bohr by Bai and Stachel, who say: "[Bohr's] (and Einstein's) view is that what exists must be measurable, or more accurately, must have measurable consequences." [4] However, despite this apparent initial openness to allowing physical existence to non-classical, unobservable entities, Bohr steadily evolved toward a form of antirealism that denied reality to objects not in-principle capable of a classical description, i.e. "which cannot be visualized in the ordinary sense", as his above-quoted assertions clearly demonstrate.

Returning now to the need for real collapse: there are "spontaneous collapse" models out there, the best known being that of Ghirardi, Rimini, and Weber [18]; but these involve changing the Schrödinger equation (by adding nonlinear terms designed to bring about dynamical collapse). The model that does not modify the basic quantum evolution (although it incorporates an additional step resulting in collapse) is based on the direct-action theory of fields, called the Transactional Interpretation (TI) [15, 21]. TI defines the usual retarded solution to the Schrödinger equation as an "offer wave" (OW). But it also includes an additional process beyond the unitary evolution of the offer wave, namely an advanced response from absorbers. The advanced response, called a "confirmation wave" (CW), is a solution to the complex conjugate Schrödinger Equation. This response is what precipitates collapse by breaking the linearity of the evolution of the quantum state (OW).

In general, one OW will elicit responses from many absorbers, where each such absorber receives only a component of the original OW. (A typical example of this is an interferometer experiment in which a beam splitter directs OW components to different detectors.) The process of CW response to OW components corresponds to the von Neumann "Process 1" measurement transition from a pure state (the OW) to a mixed state (weighted projection operators corresponding to the different OW components and their respective CW responses). As discussed in [21], Section 3.2.3, this mixed state represents a set of incipient transactions, only one of which can be actualized. It is proposed that the "collapse" to one outcome among the many (now in a well-defined basis due to the inclusion of absorber response) occurs through an analog of symmetry breaking, which is ubiquitous in physics (cf. [12]) The actualization of the transaction constitutes a transfer of measurable conserved quantities (energy, momentum, spin, etc.) from the emitter to the "winning" absorber. In the transactional picture, a photon is just this transfer of in-principle detectable electromagnetic energy, momentum, and angular momentum; and it is a discrete quantity where the energy $E = h\nu$. Thus, there is a real physical, non-unitary collapse in this model. There is also a clear physical referent for the "photon" concept independently of whether any macroscopic, observable "phenomenon" (involving an observer) results from it.

TI has been extended by this author to the relativistic domain, together with an ontological reinterpretation of the OW and CW as pre-spatiotemporal physical possibilities (reminiscent of Heisenbergian "potentiae" as noted above). This version is called the "Possibilist Transactional Interpretation" (PTI) [21]. In this picture, the collapse is not a spacetime process (which is already known to be problematic [1]); rather, it is a discontinuous process by which spacetime events (actualities) emerge from a quantum level of potentiality. The current paper focuses on Bohr's views, and will not present a detailed case for TI or PTI (that has been presented in [21], and also in [25]). The point is just to note that Bohr's conclusions are not inevitable, since they are based on certain methodological and metaphysical assumptions and constraints that need not be accepted; and that they do contain gaps and equivocations, which can in principle be remedied in an appropriate non-unitary collapse model of measurement.

However, in view of Bohr's rejection of the quantum coherent state as a purely "formal" construct in which the "idea of the photon is lost," it should be pointed out that [21], Chapter 6 discusses the physical relationship between the coherent state and the classical electromagnetic field that emerges from it through sustained actualized transactions. In this context, the term "photon" can also refer to the offer wave capable of transferring one quantum of electromagnetic energy, and a coherent state is just an offer wave that is capable of transferring a varying number n of detectable photons where n is characterized by a well-defined probability. It is the fact that the coherent state is an eigenstate of the field destruction operator that allows it to function in this way; the repeated absorption of photon(s) from the field does not change the field state, which is what allows a detectable classical field to be sustained. So the photon as a physical entity remains quite meaningful — even crucial — in the quantum coherent state. A detailed account of the well-developed theory of coherent states, including experimental verification of the theoretical predictions for photon detections, is found in [13]. To say that the "photon idea disappears" just because there is an indeterminate number of photons in the pre-detection field is at variance with both theory and experiment on coherent states.

Another aspect of PTI should be mentioned here: recall the point made in Section 2 concerning the inconsistency of defining an inertial frame of reference at the quantum level. This problem is remedied under PTI by taking spacetime as an emergent structure, supervenient on actualized transactions between quantum level emitters and absorbers. In order to describe this emergence, PTI takes literally the idea that energy and momentum are the generators of temporal and spatial displacements, respectively. Thus an actualized transaction resulting in the transfer from emitter x to absorber y of a quantum with energy E and momentum p defines a spacetime displacement $(y^{\mu} - x^{\mu})$ that is characterized by an invariant interval. (In the rest frame of a transferred material quantum, $p = 0$ and there is zero spatial displacement; this defines the temporal axis for the particle.)

In addition, macroscopic (classical) objects are distinguished from quantum systems in a well-defined (although inherently probabilistic way): they are overwhelmingly likely to bring about collapse, since they are huge collections of potential emitters and/or absorbers ([22], Section 5 and for a non-technical presentation, see [24], pp. 96–106). This makes it virtually impossible to coherently entangle an object like D with a quantum system S such that unitary evolution is preserved. This is a form of decoherence, but one based on a physically irreversible process. As such, it avoids the circularity problem of the traditional decoherence program (cf. [23]). Moreover, in this picture, irreversibility arises naturally as a previously unsuspected law of nature; thus the second law of thermodynamics is explained without having to assume special low-entropy conditions. For example, thermal interactions are irreversible transactions, thus legitimizing Boltzmann's assumption of "molecular chaos" in his derivation of his H-theorem.

Since a macroscopic object is a nexus of frequent and persistent transactions giving rise to well-defined spacetime intervals, macroscopic objects can be described by simultaneous spacetime (x, t) and dynamical (E, P) descriptions, and as such are clearly distinguished from quantum systems described by quantum states, which are elements of an underlying substratum. Thus, we have classical phenomena in PTI as well; they are simply a naturally emergent result rather than a necessary starting point in interpreting the theory.

Concerning the matter of contextuality, Bohr was of course correct that one cannot simultaneously define incompatible quantities when dealing with quantum systems. In terms of PTI, that is because determinate physical quantities only obtain as a result of actualized transactions. The latter occur by way of specific interactions between an OW and its responding CW. Confirmations define the basis for the measurement, by setting up the applicable mixed state (for example, two weighted projectors corresponding to each of two detectors in an interferometer experiment). Only the projectors in that mixed state are eligible for spacetime existence (i.e. as transfers of detectable energy, momentum, etc.); so quantities corresponding to non-commuting observables are simply not in play at that point. The CW thus constitute the physically well-defined "contextuality" that Bohr felt forced to define only with appeal to final, external observations — "phenomena".

To emphasize the fact that such contextuality has nothing to do with macroscopic "phenomena," an example of a well-defined physical quantity under PTI is the energy/momentum of a photon emitted from an excited state atom and absorbed by a ground state atom, regardless of whether that single photon is ever amplified to the level at which it could in principle be perceived by a scientist in a laboratory. All the objects involved are quantum systems, all described by quantum mechanics, and Planck's constant plays a crucial role in the interaction. Yet there is an unambiguous interpretation of the quantum formalism, applying to the degrees of freedom described by the formalism. No appeal to "the entire experimental arrangement" or necessarily observable "phenomenon" is required for this interpretation. The context consists of any forces acting on the photon offer wave (i.e. the applicable Hamiltonian) and the set of advanced absorber responses to the photon offer (the latter being described by the usual forward-propagating quantum state). The context is entirely physical. The transactional process, which heralds the advent of classicality (because it confers determinate properties on the degrees of freedom involved) occurs at a microscopic level, independently of whether any particular scientist is able to identify any macroscopic phenomenon arising from it.

6. Conclusion

Complementary cannot help us to explain measurement or the nature of physical reality in a consistent fashion unless we can explain why the quantum formalism applies correctly to quantum degrees of freedom (such as the "quantum particle" S in Bohr's thought experiments with S and D) but not to macroscopic objects; that is, why the ontic uncertainty of quantum objects does not "infect" macroscopic objects such as Bohr's diaphragm D, and why we can view the latter's uncertainty as being epistemic. If we include absorber response, we have a way forward to make this distinction in physical terms. Bohr was unable to do this through Complementarity alone, and he lapsed into instrumentalist and antirealist utterances as a result.

Recall Bohr's famous statement that "It is wrong to think that the task of physics is to find out how nature is. Physics concerns what we can say about nature."[e] Clearly, there is an implicit assumption here: "physics cannot say how nature is." But in fact, quantum theory certainly can be telling us "how nature is." Why should we presume that nature has to be determinate and classical at all levels, just because we cannot visualize it "in the ordinary way"?

Elsewhere in this volume, George Jaroszkiewicz [20] notes that the reductionistic assumptions behind classical physics need to be re-examined, and that physics is an empirical science. I certainly agree with both points. However, the fact that physical theory begins by engaging with empirical phenomena, and must be rigorously tested by experiment, does not negate the longstanding tradition in physics of theoretical description in terms of unobservables. Boltzmann's atomic hypothesis is a prominent example.[f] It is well known that the idea of unobservable atoms was highly controversial, and that Ernst Mach strongly objected to it on the basis that physics is an empirical science. Yet the atomic hypothesis was clearly the fruitful path, and it is reasonable to take that theoretical success as evidence for the existence of atoms, especially now that we can (indirectly) image atoms.

Similarly, it is reasonable to take the success of quantum theory as evidence for the existence of additional structure in nature that gives rise to the kinds of phenomena predicted by the theory, even if it is difficult (or even impossible) to visualize this structure "in the ordinary (classical) way." This is "inference to the best explanation" for the empirical success of a theory. The new challenge from quantum theory is that such referents cannot be classical (i.e. not Einsteinian "elements of reality"). But that in itself does not mean there can be no physical referent for the theory. In contrast, an instrumentalist, observer-dependent interpretation of quantum theory can provide no explanation for the success of the theory in predicting (at the statistical level) our observations. It essentially says that we have a very good instruction manual for predicting

---

[e]These peremptory sentences followed Bohr's antirealist statement "There is no quantum world. There is only an abstract quantum mechanical description."
[f]Faraday's "lines of force" is another.

the experiences of an observer, but there is nothing in the world corresponding to the manual, and/or it is wrong to think there should be a reason or explanation for its predictive power. Such an attitude would appear to be based on the assumption that if the explanation is not classical in nature (i.e. not in terms of determinate spacetime objects), there can be no explanation. But why should we demand that the explanation behind the success of quantum theory be classical? That expectation, I suggest, is what needs to be given up.[g]

Finally, the proposed PTI picture of an intrinsically unobservable, pre-spacetime quantum substratum giving rise to an empirical, classically determinate realm of experience may seem startling, even farfetched. But it does provide a clear physical referent for the quantum formalism (at least in a structural sense, [33, 34]), and a well-defined basis for the emergence of classical determinacy — describable by classical physics — from that formalism. In that regard, I have noted elsewhere ([21], Chapter 7) that the PTI ontology provides a natural correspondence for Kantian "noumenon" as describing the quantum level and "phenomenon" as describing the classical level. Here it is advisable to recall again McMullin's observation that the structures of the microworld are not required to be "imaginable in the categories of the macroworld." And as Bohr himself commented in a remark to Pauli, it might just be "crazy enough to be true."[h]

[g]Of course, the disagreement between instrumentalists and realists can also be understood as a disagreement about the nature of scientific inquiry and explanation. Mach argued for a limited descriptive role for physical theory, and considered matters of ontology as strictly outside the domain of physics. However, such a methodological limitation on the discipline of physics does not *preclude* reasonable ontological inferences based on the success of physical theory, whether or not one considers such inferences as within the proper purview of physics. And such ontological inferences may even prove fruitful in constructing new theories or in resolving anomalies or other remaining challenges in physics. This situation illustrates the ongoing fundamental dependence of physics on philosophy.

[h]Bohr's famous remark concerning a theory by Pauli, as quoted in [30].

Acknowledgments

I would like to thank John Stachel, Miroljub Dugić, Jasmina Jeknić-Dugić, Christian de Ronde, Bernice Kastner, and George Jarozskiewicz for valuable discussions.

References

[1] Aharonov, Y. and Albert, D. (1981). Can we make sense of the measurement process in relativistic quantum mechanics? *Physics Review D*, **24**: 359–370.

[2] Aharonov, Y., Bergmann, P. and Lebowitz, J. (1964). Time symmetry in the quantum process of measurement. *Physical Review B*, **134**: 1410–1416.

[3] Allori, V. (2016). Primitive Ontology and the Classical World. This Volume.

[4] Bai, D. and Stachel, J. (2016). Bohr's Diaphragms. Forthcoming in this Volume.

[5] Bell, J. S. (1975). A theory of local beables. CERN Ref. 2052, Presented at the 6th GIFT Seminar.

[6] Beller, M. and Fine, A. (1994). Bohr's response to EPR. in: Faye J., Folse, H., (eds.), *Niels Bohr and Contemporary Philosophy*. pp. 1–31. Dordrecht: Kluwer Academic Publishers.

[7] Bohr, N. (1934). *Atomic Theory and the Description of Nature*. Cambridge: Cambridge University Press.

[8] Bohr, N. (1935). Can quantum-mechanical description of reality be considered complete? *Physical Review*, **48**: 696–702.

[9] Bohr, N. (1939). The Causality Problem in Atomic Physics. In *International Institute of Intellectual Cooperation*, pp. 11–30.

[10] Bohr, N. (1958). *Atomic Physics and Human Knowledge*. New York: John Wiley Sons, p. 88 (originally published 1955).

[11] Bohr, N. (1985). In Kalckar, J., (ed.) *Collected Writings. Vol. 6. Foundations of Quantum Physics I.* Amsterdam, North-Holland/Elsevier.

[12] Brading, K. and Castellani, E. (2013). Symmetry and symmetry breaking, The Stanford Encyclopedia of Philosophy (Spring 2013 Edition), Edward N. Zalta (ed.), (http://plato.stanford.edu/archives/spr2013/entries/symmetry-breaking.)

[13] Breitenbach, G., Schiller, S. and Mlynek, J. (1997). Measurement of the quantum states of squeezed light. *Nature*, **387**: 471.

[14] Bub, J. (1997). *Interpreting the Quantum World*. Cambridge: Cambridge University Press.

[15] Cramer, J. G. (1986). The transactional interpretation of quantum mechanics. *Reviews of Modern Physics*, **58**: 647–688.

[16] Dickson, M. Bohr. (2002). Bell: A proposed reading of Bohr and its implications for Bell's Theorem, in: *Proceedings of the NATO Advanced Research Workshop on Modality, Probability and Bell's Theorem*, Butterfield, J. and Placek, T. (eds.), Amsterdam: IOS Press.

[17] Einstein, A., Podolsky, B. and Rosen, N. (1935). Can quantum-mechanical description of reality be considered complete? *Physical Review*, **47**: 777.

[18] Ghirardi, G. C., Rimini, A. and Weber, T. (1986). Unified dynamics for microscopic and macroscopic systems. *Physical Review D*, **34**: 470.

[19] Heisenberg, W. (1962). Physics and Philosophy. New York: Harper Collins, p. 41.

[20] Jaroszkiewicz, G. (2016). Principles of Empirical Science and the interpretation of quantum mechanics. Forthcoming in This Volume.

[21] Kastner, R. E. (2012). *The Transactional Interpretation of Quantum Mechanics: The Reality of Possibility*. Cambridge: Cambridge University Press.

[22] Kastner, R. E. (2012). The possibilist transactional interpretation and relativity. *Foundations of Physics*, **42**: 1094–1113.

[23] Kastner, R. E. (2014). Einselection of pointer observables: The new H-Theorem? *Studies in History and Philosophy of Modern Physics*, **48**: 56–58.

[24] Kastner, R. E. (2015). *Understanding Our Unseen Reality: Solving Quantum Riddles*. London: Imperial College Press.

[25] Kastner, R. E. (2015). Haag's theorem as a reason to reconsider direct-action theories. *International Journal of Quantum Foundations*, **1**(2): 56–64.

[26] McMullin, E. (1984). A Case for Scientific Realism. In J. Leplin (ed.) *Scientific Realism*. Berkeley: UCLA Press, pp. 8–40.

[27] Oriti, D. (2006). A quantum field theory picture of simplicial geometry and the emergence of spacetime, in the *Proceedings of the DICE 2006 workshop*, Piombino, Italy, Journal of Physics: Conference series.

[28] Petersen, A. (1963). The Philosophy of Niels Bohr. In: *Bulletin of the Atomic Scientists*, vol. 19, No. 7.

[29] Sakurai, J. J. (1973). *Advanced Quantum Mechanics*. Reading: Addison-Wesley.

[30] Scoular, S. (2007). *First Philosophy: The Theory of Everything*. Universal Publishers.

[31] Sorkin, R. (2007). Relativity does not imply that the future already exists: a counterexample. Vesselin Petkov (ed.), *Relativity and the Dimensionality of the World*. Springer (arXiv:gr-qc/0703098).

[32] Stachel, J. (2016). It Ain't necessarily so: Interpretations and misinterpretations of quantum theory. Forthcoming in this Volume.

[33] Worrall, J. (1989). Structural realism: The best of both worlds? *Dialectica*, **43**: 99–124. Reprinted in D. Papineau (ed.), *The Philosophy of Science*, Oxford: Oxford University Press, pp. 139–165.

[34] Worrall, J. (2007). Miracles and models: why reports of the death of structural realism may be exaggerated. *Royal Institute of Philosophy Supplements*, **82**(61): 125–154.

Chapter 5

Representational Realism, Closed Theories and the Quantum to Classical Limit

Christian de Ronde*

CONICET, Buenos Aires University, Argentina
Center Leo Apostel and Foundations of the Exact Sciences
Brussels Free University, Belgium

In this chapter, we discuss the representational realist stance as a pluralist ontic approach to inter-theoretic relationships. Our stance stresses the fact that physical theories require the necessary consideration of a conceptual level of discourse which determines and configures the specific field of phenomena discussed by each particular theory. We will criticize the orthodox line of research which has grounded the analysis about QM in two (Bohrian) metaphysical presuppositions — accepted in the present as dogmas that all interpretations must follow. We will also examine how the orthodox project of "bridging the gap" between the quantum and the classical domains has constrained the possibilities of research, producing only a limited set of interpretational problems which only focus in the justification of "classical reality" and exclude the possibility of analyzing the possibilities of non-classical conceptual representations of QM. The representational realist stance introduces two new problems, namely, the superposition problem and the contextuality problem, which consider explicitly the conceptual representation of orthodox QM beyond the mere reference to mathematical structures and measurement outcomes. In the final part of the chapter, we revisit, from a representational realist perspective, the quantum to classical limit and the orthodox claim that

*Fellow Researcher of the Consejo Nacional de Investigaciones Científicas y Técnicas.

this inter-theoretic relation can be explained through the principle of decoherence.

Keywords: Representational Realism; Closed Theories; Quantum Limit.

1. The Classical Representation of Physical Reality and QM

The general characterization and representation of the idea of "classical reality", which encompasses the whole of classical physics (including relativity theory), can be condensed in the notion of *Actual State of Affairs* (ASA).[a] This particular representation was developed by physics since Newton's mechanics and can be formulated in terms of *systems constituted by a set of actual (definite valued) preexistent properties.* Actual classical properties are defined in terms of three logical and ontological principles proposed by Aristotle. Indeed, the Principle of Existence (PE), the Principle of Non-Contradiction (PNC), and the Principle of Identity (PI) founded not only classical logic itself but also the basis of our classical metaphysical understanding of the world. In fact, during the 18th century, these principles were also used implicitly by Newton in order to provide a metaphysical definition of the notion of actual entity in classical mechanics.[b] Through these principles, Newton was able to conceive a Universe constituted by bodies which existed, possessed non-contradictory properties and remained always identical to themselves.[c] This physical representation of reality also allowed Laplace to imagine a demon who, given the complete and exact knowledge

[a]See for discussion and definition of this notion in the context of classical physics [17].

[b]For a detailed analysis of the relation between these metaphysical principles and Newton's representation of physical reality, see [18].

[c]As noticed by Verelst and Coecke [36, p. 167], these principles are "exemplified in the three possible usages of the verb 'to be': existential, predicative, and identical. The Aristotelian syllogism always starts with the affirmation of existence: something is. The principle of contradiction then concerns the way one can speak (predicate) validly about this existing object, i.e. about the true and falsehood of its having properties, not about its being in existence. The principle of identity states that the entity is identical to itself at any moment (a=a), thus granting the stability necessary to name (identify) it."

of all particles in the Universe (i.e. the set of all actual properties), would have immediate access, through the equations of motion, to the future and the past of the whole Universe.[d]

However, it is well known that in the case of QM, we have serious difficulties to interpret the formalism in terms of "classical reality". An evidence of the deep crisis of physical representation within QM is the fact that more than one century after its creation the physics community has reached no consensus about what the theory is really talking about. A possibility with many adepts today, is to simply deny the need of conceptual representation within physics and understand the discipline as a mathematical scheme restricted to the prediction of measurement outcomes. This idea goes back to logical positivism and the Machian understanding of physical theories as an "economy of experience" with no metaphysical background. Taking a radical epistemic viewpoint Fuchs and Peres [22, p. 70] have argued that "[...] quantum theory does not describe physical reality. What it does is provide an algorithm for computing probabilities for the macroscopic events ('detector clicks') that are the consequences of experimental interventions. This strict definition of the scope of quantum theory is the only interpretation ever needed, whether by experimenters or theorists." This instrumentalist perspective is satisfied with having a empirically adequate "algorithmic recipe". There is no need for an interpretation because the mathematical structure already provides the correct predictions for measurement outcomes. And that is what physics is all about.

In contraposition to that radical epistemic perspective, in this chapter we discuss a representational realist stance according to which the task of both physics and philosophy of physics is to

[d]According to the French physicist and mathematician Pierre Simon Laplace [33]: "We may regard the present state of the universe as the effect of its past and the cause of its future. An intellect which at any given moment knew all of the forces that animate nature and the mutual positions of the beings that compose it, if this intellect were vast enough to submit the data to analysis, could condense into a single formula the movement of the greatest bodies of the universe and that of the lightest atom; for such an intellect nothing could be uncertain and the future just like the past would be present before its eyes."

produce conceptual representations which allows us to understand and explain the features of the world and reality beyond mathematical structures and measurement outcomes. According to our stance, physical representation requires the coherent interrelation between mathematical formalisms and conceptual schemes. Thus, representational realism imposes the need of introducing adequate physical concepts in order to understand QM beyond the mere reference to mathematical structures and measurement outcomes. If we accept this challenge, there seems to be two main lines of research to consider. The first one is to investigate the possibility that QM makes reference to the same physical representation provided by classical physics; i.e. that it talks about "classical reality". This is the main idea presupposed, for example, by the hidden variables program which, as noticed by Bacciagaluppi [3, p. 74], attempts to "restore a classical way of thinking about *what there is*." The second line would be to consider the possibility that QM might describe physical reality in a different — maybe even incommensurable — way to that of classical physics. Apart from some few unorthodox attempts (e.g. [1, 30]), this second line of research has not been truly addressed within orthodoxy. The reason, we believe, is related to the introduction by Niels Bohr of two metaphysical presuppositions which assume, as a general standpoint, the idea that the only possible representation of experience and physical reality is necessarily provided by the language of classical physics. We will analyze these strong metaphysical presuppositions in the following section. We remark that our analysis is only concerned with the philosophy of QM and does not attempt to derive from it a general conclusion regarding the general debate already present in philosophy of science.

2. The Two Bohrian Dogmas of QM and the Attempt to Restore "Classical Reality"

In [14], we argued that Bohr is the main one responsible for producing an epistemological interpretation of QM that does not only limit physical representation in terms of classical language and classical phenomena but also precludes the very possibility of introducing and developing new (non-classical) concepts. Bohr is responsible for having introduced in the debate about the interpretation of QM two strong metaphysical presuppositions which have constrained the possibilities of analysis and development of quantum theory. Since the mid 20th century what became orthodoxy, took these ideas as necessary standpoints to think about any possible interpretation of QM. Accepted as unquestionable, these metaphysical presuppositions were turned into dogmas that all interpretations of QM would have to respect.

The first metaphysical dogma is the idea that there must exist a "quantum to classical limit", implying what Alisa Bokulich calls an "open theory approach" [5]. This reductionistic inter-theoretic relation between QM and classical mechanics was introduced by Bohr through his *correspondence principle.*[e]

Dogma I. Quantum to Classical Limit: The principle that there must exist a continuous "bridge" or "limit" between classical mechanics and QM.

However, this is certainly not the only possible way to approach the problem of inter-theory relation. In fact, this reductionistic understanding of the relation between classical mechanics and QM has been severely questioned by many pluralist accounts in the last decades. Pluralists approaches open different possibilities of analysis which due to the single-viewed orthodox perspective have not been discussed nor developed within the literature.

The second metaphysical dogma which has guided orthodoxy can also be traced back to Bohr's claim that physical experience needs to be expressed exclusively in terms of classical physical language [8]. Bohr [37, p. 7] stated that: "[...] the unambiguous interpretation of any measurement must be essentially framed in terms of classical physical theories, and we may say that in this sense the language of Newton and Maxwell will remain the language of physicists for all time." In this respect, he added [*Op. cit.*, p. 7] that, "it would be a misconception to believe that the difficulties of the atomic theory

[e]For a detailed analysis of Bohr's correspondence principle, see [7].

may be evaded by eventually replacing the concepts of classical physics by new conceptual forms."

Dogma II. Classical Representation of Physics: The principle that one needs to necessarily presuppose the classical representation of physics in order to interpret QM and the phenomena it talks about.

Both principles go clearly against a non-classical conceptual understanding of QM. Bohr's insistence in the necessity of only using classical physical notions can be understood in relation to his interpretation of QM as a "rational generalization of classical mechanics" [8].

A direct consequence of such commitment against new non-classical notions has been his analysis of the double-slit experiment in terms of the classical notions of "wave" and "particle". However, today we know that a quantum state Ψ cannot be physically interpreted either in terms of "waves" or "particles" for different formal and empirical reasons. Let us briefly recall some of them. Firstly, Ψ is a mathematical entity that lives in configuration space, not in classical three-dimensional space — which in turn would allow an interpretation in terms of the Newtonian physical notions of space and time. Consequently, Ψ can represent neither a "particle" nor a "wave" which are physical notions that require a classical three-dimensional space.[f] Secondly, according to the orthodox Born interpretation of Ψ, the quantum wave function describes a probability distribution that is non-Kolmogorovian, and thus, cannot be interpreted in terms of ignorance about an ASA [34]. Thirdly, a "click" in a typical quantum experimental set up does not behave as if that which is producing the "click" is a "particle" or a "wave". At the empirical level, Bell's inequalities have proven explicitly that "quantum clicks" cannot be represented in terms of a classical local-realistic theory — to which "particles" and "waves" obviously pertain. Ever since the first experiment of Aspect, the weirdness of such non-classical "clicks" has been repeatedly confirmed. Fourthly,

[f] In fact, it was this reason which led Bohm to abandon, at least for a while, his own hidden variable version of QM.

the Kochen–Specker theorem [32] proves that Ψ does not possess definite valued properties independently of the context. On the contrary, "waves" and "particles" are non-contextual entities which, as described by classical physics in phase space, do possess definite valued properties. In short, regardless of the many references made to these classical concepts within the literature, it simply makes no sense to use the notions of "wave" and "particle" in order to interpret the orthodox formalism of QM.

It is quite clear that the knowledge we have acquired of QM today is more detailed and accurate than the one Bohr and Einstein had at the beginning of the 20th century when they discussed the interpretation of QM in terms of *Gedeankenexperiments*. As a matter of fact, in the meantime, many of these imagined experiments have become testable! With the knowledge we have today, it would also seem wise to recall the old logical positivist lesson that the use of inappropriate notions within a language can only create pseudoproblems.

3. The Bohrian Project Today: Bridging the Quantum Gap

Following Bohr's footsteps, one of the present philosophy of physics's major concerns remains to try to explain QM in terms of our "common sense" understanding of the world. As Mauro Dorato remarks [21, p. 369], the project amounts to try to look for "the best candidate to bridge the gap between the manifest and the physical image of the world emerging from quantum mechanics." Sellars characterizes the manifest image as "the framework in terms of which man came to be aware of himself as man-in-the-world",[g] which, as remarked by de Vries [*Op. cit.*], means more broadly "the framework in terms of which we ordinarily observe and explain our world." According to orthodoxy — in close relation to Bohr's dogmas — rather than to imagine new physical experiences or new ways of representing reality, the philosopher should focus in "bridging

[g]See [19].

the gap" between the new technical scientific developments and our common sense-manifest image of the world.

The focus in trying to explain QM in terms of "classical reality" and our "common sense" observability, is shared not only by most interpretations of QM, but also by most philosophical perspectives within the field — both realists and antirealists. The general scheme which allows to state this particular problem is grounded — implicitly or explicitly — in the distinction between *observable* and *non-observable* put forward by the logical positivists. Contrary to our definition of realism which considers representation as a construct of physical theories, both realist and antirealist orthodox positions accept as a standpoint the idea that physical observation provides a direct access to reality *as it is*. This idea was already implicit in the logical positivist distinction between *theoretical terms* and *empirical terms*. But even though the philosophy of science community itself has characterized this distinction as "naive",[h] the problems discussed in the literature today still presuppose implicitly such distinction. As remarked by Curd and Cover:

> "Logical positivism is dead and logical empiricism is no longer an avowed school of philosophical thought. But despite our historical and philosophical distance from logical positivism and empiricism, their influence can be felt. An important part of their legacy is observational–theoretical distinction itself, which continues to play a central role in debates about scientific realism." [12, p. 1228]

We must remark that even scientific realism accepts the empiricist standpoint of common sense observability. Indeed, as Musgrave [12, p. 1221] makes the point: "In traditional discussions of scientific realism, common sense realism regarding tables and chairs (or the moon) is accepted as unproblematic by both sides. Attention is focused on the difficulties of scientific realism regarding 'unobservables' like electrons." In line with both Bohr, the logical positivist "observable non-observable distinction" also closes the door to the

[h]This distinction was strongly criticized since the 60s by many, including Kuhn and Feyerabend.

development of new physical representations since it assumes that we already know what reality *is* in terms of the (naive) observation of tables and chairs — also known as "classical reality". From this conservative perspective, the task of the philosopher should be to restore a classical understanding about *what there is*. Cao's criticism is in this respect very strong:

> "The old-fashioned (positivist or constructive empiricist) tradition to the distinction between observable and unobservable entities is obsolete. In the context of modern physics, the distinction that really matters is whether or not an entity is cognitively accessible by means of experimental equipment as well as conceptual, theoretical, and mathematical apparatus. If a microscopic entity, such as a W-boson, is cognitively accessible, then it is not that different from a table or a chair. It is clear that the old constructive empiricist distinction between observables and nonobservables is simply impotent in addressing contemporary scientific endeavor, and thus carries no weight at all. If, however, some metaphysical category of microscopic entities is cognitively inaccessible in modern physics, then, no matter how basic it was in traditional metaphysics, it is irrelevant for modern metaphysics." [10, pp. 64–65]

4. Representational Realism: A Pluralist-Ontict Approach

Within philosophy of physics, realism has been characterized as a stance which assumes the existence of a reality independent of the actions of any human subject or conscious being. In short, realism is committed to the belief of an independent reality. However, this account falls short when attempting to grasp the *praxis* of realist physicists themselves. Representational realism attempts to capture exactly this aspect; i.e. the specific way through which realist physicists produce a representational (meta-physical) account of reality. In this respect, the main presupposition of representational realism is that physical theories relate to reality, not only through their mathematical formalisms, but also through a network of physical concepts. The coherent interrelation between mathematical and conceptual structures allows physical theories to represent (in different ways) physical reality. In turn, every new representation allows the physicist to imagine and explore new physical phenomena.

For the representational realist, the task of philosophy of physics is to produce physical representations which would allow her to grasp the features of the world beyond mathematical schemes and measurement outcomes. In order to provide such representation, we must necessarily complement mathematical formalisms with networks of physical concepts. In this sense, for the representational realist, it is not enough to claim that "according to QM the structure of the world is like Hilbert space" or that "quantum particles are like vectors in Hilbert space". Mixing improperly classical notions with the quantum formalism is simply not doing the job of providing a conceptual physical representation in the sense discussed above.

Our representational realist stance is intimately linked to Heisenberg's closed theory approach [5] according to which there is no need of a reductionistic understanding of inter-theory relation.[i] Every physical theory needs to develop its own conceptual scheme — independent of those produced by different (closed) theories. The multiplicity of physical representations of reality does not invalidate by itself a realist stance. In fact, we have argued elsewhere that a Spinozist scheme would allow us to provide the missing *univocity condition* required to make sense of such *multiple representations* as expressions of the same *one reality.*

Our pluralist scheme avoids the reductionistic requirement imposed by orthodoxy. Theories are created discontinuously through

---

[i]An important example to understand this non-reductionistic characterization is how Newtonian mechanics has to be understood in relation, for example, to relativity theory. According to Heisenberg [24, pp. 97–98]: "New phenomena that had been observed could only be understood by new concepts which were adapted to the new phenomena. [...] These new concepts again could be connected in a closed system. [...] This problem arose at once when the theory of special relativity had been discovered. The concepts of space and time belonged to both Newtonian mechanics and to the theory of relativity. But space and time in Newtonian mechanics were independent; in the theory of relativity they were connected."

jumps. As remarked by Heisenberg in an interview by Thomas Kuhn [6, p. 98]: "The decisive step is always a rather discontinuous step. You can never hope to go by small steps nearer and nearer to the real theory; at one point you are bound to jump, you must really leave the old concepts and try something new... in any case you can't keep the old concepts." The only important aspect to consider a physical theory as "closed" is the internal *coherency* between the formal mathematical elements, the conceptual structure and the physical experience involved.[j] The coherency reflects a wholeness present in every "closed theory" which Heisenberg beautifully expresses in the following manner:

> "One finds [in closed theories] structures so linked and entangled with each other that it is really impossible to make further changes at any point without calling all the connections into question [...] We are reminded here of the artistic ribbon decorations of an Arab mosque, in which so many symmetries are realized all at once that it would be impossible to alter a single leaf without crucially disturbing the connection of the whole." [6, p. 95]

Physical representation allows us to think about experience and predict phenomena without the need of actually performing any measurement. It allows us to imagine physical reality beyond the here and now.[k] Following both Einstein and Heisenberg, representational realism assumes a radical metaphysical stance regarding the possibility of observability in physics. Every new physical theory creates a new experience and thus, also new possibilities of observability. Observability is not "a given", observability in physics is always

[j]It is important to remark that the coherency to which we relate in this case is not the one discussed in "coherence theory of truths" in philosophy of science (see for example [38]). While coherence theory of truths discuss about the relation between *propositions* within a physical theory, we discuss about the relation between *concepts*, *mathematical expressions*, and *physical experience*.

[k]This is of course the opposite standpoint from empiricists who argue instead that the fundament of physics is "actual experimental data". As remarked by van Fraassen [35, pp. 202–203]: "To develop an empiricist account of science is to depict it as involving a search for truth only about the empirical world, about what is actual and observable."

constructed, always grounded on metaphysical principles. Classical physical observation — restricted by the (metaphysical) PE, PNC, and PI (see Section 1) — is just a particular way of observing the world. A particular viewpoint grounded on classical physics and metaphysics.

According to our stance, physical theories contain a conceptual level from which physical notions allow us to *represent* what the theory is talking about. This basic assumption implies that mathematical structures are not enough to produce by themselves a meaningful physical representation of reality. Physics cannot be exclusively reduced to mathematical structures which predict measurement outcomes. As Heisenberg makes the point: "The history of physics is not only a sequence of experimental discoveries and observations, followed by their mathematical description; *it is also a history of concepts.* For an understanding of the phenomena, the first condition is the introduction of adequate concepts. *Only with the help of correct concepts can we really know what has been observed.*"[1] As said before, in a slogan, to explain is to represent.

Representational Realism: *A representational realist account of a physical theory must be capable of providing a physical representation of reality in terms of a network of physical concepts which coherently relates to the mathematical formalism of the theory and allows to articulate and make predictions of definite phenomena. Observability in physics is always theoretically and metaphysically laden, and thus must be regarded as a consequence of each particular physical representation.*

5. Formalism, Concepts, and Observation

The distance between orthodoxy and our representational realist project, can be understood from the radical difference in which these general schemes place the constitutive elements of a physical theory, namely, the mathematical formalism, the network of physical

[1]See [26, p. 264] emphasis added.

concepts, and the field of phenomena it is able to describe. This distance is somewhat analogous to the distance between Bohr's and Einstein's philosophy of physics. While Bohr takes as a standpoint of his analysis "classical phenomena" — or "observational terms" in philosophy of physics — both Einstein and the representational realist consider observations and phenomena always as metaphysically laden. As Einstein remarked in a letter to Schrödinger in the summer of 1935: "The problem is that physics is a kind of metaphysics; physics describes "reality". But we do not know what "reality" is. We know it only through physical description..." [12, p. 1196] According to the representational realist perspective, in order to truly understand a phenomenon we need to have, beforehand, an adequate set of physical notions. Only with the help of correct concepts and metaphysical principles can we really know what has been observed. This implies that there is in principle no exclusive commitment to classical phenomena. Observation is always both theory and metaphysically laden.

Even though classical observation — which is in fact constrained by the metaphysical PE, PNC, and PI — is regarded by orthodoxy as the "self evident" standpoint for the development and analysis of theories, it is in principle possible to imagine — specially after the creation of non-classical logics — that different principles could be considered in order to produce a new understanding of physical reality. It is only the theory, and the metaphysical principles in which it is grounded, which can tell you what can be observed. According to the representational realist stance, empirical adequacy is only a way to decide whether the theory is adequate or not, but the main goal of physics is to provide always new representations of experience and reality.

Physical observability is always dependent on the physical representation provided by each closed theory. From our constructive metaphysical line of research, the task of both physicists and philosophers is to jointly construct new mathematical formalisms and networks of physical concepts which allows us to imagine new physical phenomena. These three elements interrelate in such a way that one can help the other to correct itself, to become more adequate. It is the interrelation between the mathematical structure, the physical concepts and the phenomena which allows a theory to provide a particular representation of physical reality. As Einstein [20, p. 175] made the point: "[...] it is the purpose of theoretical physics to achieve understanding of physical reality which exists independently of the observer, and for which the distinction between 'direct observable' and 'not directly observable' has no ontological significance". Observability is secondary even though "the only decisive factor for the question whether or not to accept a particular physical theory is its empirical success." For the representational realist, empirical adequacy is part of a verification procedure, not that which "needs to be saved".

We believe our representational realist stance is close in many ways to Einstein's realism who, as noticed by Howard: "was not the friend of any simple realism" [27, p. 206]. Einstein did not consider observation as 'a given'[m] and stressed — like the representational realist — the importance of the metaphysical level of physical representation of reality.[n] We also find in Einstein a focus of attention regarding the importance of physical concepts, as well as a recognition of the threat of remaining captive of "common sense realism":

> "Concepts that have proven useful in ordering things easily achieve such an authority over us that we forget their earthly origins and accept them as unalterable givens. Thus they come to be stamped as 'necessities of thought,' 'a priori givens,' etc. The path of scientific advance is often made impossible for a long time through such errors. For that reason, it is by no means an idle game if we become practiced in analyzing the long common place concepts and exhibiting those circumstances upon which their justification and

[m]This can be seen from the very interesting discussion between Heisenberg and Einstein [25, p. 66] were the latter explains: "I have no wish to appear as an advocate of a naive form of realism; I know that these are very difficult questions, but then I consider Mach's concept of observation also much too naive. He pretends that we know perfectly well what the word "observe" means, and thinks this exempts him from having to discriminate between 'objective' and 'subjective' phenomena."

[n]See for a detailed analysis of the importance of metaphysics for Einstein [27].

> usefulness depend, how they have grown up, individually, out of the givens of experience. By this means, their all-too-great authority will be broken. They will be removed if they cannot be properly legitimated, corrected if their correlation with given things be far too superfluous, replaced by others if a new system can be established that we prefer for whatever reason." [28]

The representational realist stance argues in favor of a conceptual-metaphysical understanding of physical theories. Indeed, to provide a coherent metaphysical picture is the only way to truly understand what a physical theory is talking about. As Cao makes the point:

> "An important point here is that metaphysics, as reflections on physics rather than as a prescription for physics, cannot be detached from physics. It can help us to make physics intelligible by providing well-entrenched categories distilled from everyday life. But with the advancement of physics, it has to move forward and revise itself for new situations: old categories have to be discarded or transformed, new categories have to be introduced to accommodate new facts and new situations." [10, p. 65]

Going back to QM, since the standpoint of the representational realist stance is different from many empiricist approaches, it also confronts a different set of problems an limiting conditions. In the following section, we attempt to discuss two new problems, introduced by the representational realist project, which consider explicitly the representation of QM beyond mathematical structures and the prediction of measurement outcomes.

6. The New Representational Problems of QM

The main idea behind the representational realist stance is that when a mathematical formalism is coherently related to a network of physical concepts, it is possible to produce a physical representation of reality which allows us to describe particular physical phenomena. It is not enough to say that QM cannot be explained in terms of "classical reality". Neither is enough to argue that the world can be understood as a mathematical structure. This is not doing the job of producing a conceptual representation of QM. QM has proven already to be empirically adequate and the orthodox formalism is mathematically rigorous. However, instead of changing the formalism or adding *ad hoc* rules in order to restore a classical discourse and representation, there is a different strategy that could be considered. According to this new strategy, what needs to be done is to construct a new net of (non-classical) physical concepts capable of interpreting the formalism *as it is*. There are some few attempts in this direction which we mentioned above.

Contrary to our stance, the orthodox line of research deals with a specific set of problems which analyze QM from a classical perspective. This means that all problems assume as a standpoint "classical reality" and only reflect about the formalism in "negative terms"; i.e. in terms of the failure of QM to account for the classical representation of reality and the use of its concepts: separability, space, time, locality, individuality, identity, actuality, etc. The "negative" problems are thus: *non*-separability, *non*-locality, *non*-individuality, *non*-identity, etc.[o] These "no-problems" begin their analysis considering the notions of classical physics, assuming implicitly as a standpoint the strong metaphysical presupposition that QM should be able to represent physical reality in terms of such classical notions.

Two of the main problems discussed in the literature which attempt to provide an answer to the emergence of our manifest image from the quantum formalism are the famous measurement and basis problems. While the basis problem attempts to justify the choice of a particular basis among the many possible incompatible ones escaping the contextual character of QM, the measurement problem attempts to justify actual outcomes arising from strange superposed states. Contrary to orthodoxy, the representational realist project attempts to consider the formal features behind these problems, namely, quantum contextuality and the superposition principle, as positive elements which must guide us in the creation of a conceptual representation of QM. The replacing of these main problems and their

[o]I am thankful to Bob Coecke for this linguistic insight. Cagliari, July 2014.

consequences for the interpretation of QM will be discussed in more detail in the subsequent subsections.

6.1. The superposition problem: Representation beyond actual outcomes

Maybe the most famous of all interpretational problems of QM is the so-called "measurement problem". Let us briefly recall it.

Measurement Problem: *Given a specific basis (or context), QM describes mathematically a quantum state in terms of a superposition of, in general, multiple states. Since the evolution described by QM allows us to predict that the quantum system will get entangled with the apparatus and thus its pointer positions will also become a superposition,*[p] *the question is why do we observe a single outcome instead of a superposition of them?*

The measurement problem is also a way of discussing the quantum formalism in "negative terms". In this case, the problem concentrates in the justification of actual measurement outcomes. It should be remarked that the measurement problem presupposes that the basis (or context) — directly related to a measurement set up — has been already determined (or fixed). Thus it should be clear that there is no question regarding the contextual character of the theory within this specific problem. As we have argued extensively in [15], the measurement problem has nothing to do with contextuality. The problem arises when the actualization process is considered. There is what could be described as a mix of subjective and objective elements when the recording of the experiment takes place — as

[p]Given a quantum system represented by a superposition of more than one term, $\sum c_i|\alpha_i\rangle$, when in contact with an apparatus ready to measure, $|R_0\rangle$, QM predicts that system and apparatus will become "entangled" in such a way that the final "system + apparatus" will be described by $\sum c_i|\alpha_i\rangle|R_i\rangle$. Thus, as a consequence of the quantum evolution, the pointers have also become — like the original quantum system — a superposition of pointers $\sum c_i|R_i\rangle$. This is why the MP can be stated as a problem only in the case the original quantum state is described by a superposition of more than one term.

Wigner clearly exposed with his famous friend [37, pp. 324–341]. The problem here is the shift from the physical representation provided when the measurement was not yet performed (and the system is described in terms of a quantum superposition), to when we claim that "we have observed a single measurement outcome" which is not described by the theory. Since there is no physical representation of "the collapse", the subject (or his friend) seems to define it explicitly. The mixture of objective and subjective is due to an incomplete description of the state of affairs within the quantum measurement process (or "collapse").

The focus of the measurement problem is on the actual realm of experience. In this sense, the measurement problem is an empiricist problem which presupposes the controversial idea that actual observation is perfectly well defined. However, from a representational realist, stance things must be analyzed in a radically different perspective. Indeed, for the representational realist, it is only the theory which can tell you what can be observed. This means that if we are willing to truly investigate the physical representation of quantum superpositions, then we will need to "invert" the measurement problem and focus on the formal–conceptual level — instead of trying to justify what we observe in classical terms. Thus, attention should be focused on the physical representation of the mathematical expression instead of attempting to somehow "save" the measurement outcomes, justifying through *ad hoc* rules the "collapse" of the quantum superposition to one of its terms.

The new technological era we are witnessing through quantum information processing requires that we, philosophers of QM, pay attention to the developments that are taking place today. We believe that an important help could be provided by philosophers of physics, in case we were able to develop a conceptual representation of quantum superpositions (see also [16]).

Superposition Problem: *Given a situation in which there is a quantum superposition of more than one term,* $\sum c_i \; |\alpha_i\rangle$, *and given the fact that each one of the terms relates through the Born rule to a meaningful physical statement, the problem is how do we conceptually* *represent this mathematical expression? Which is the physical notion that relates to each one of the terms in a quantum superposition?*

The superposition problem opens the possibility to truly discuss a physical representation of reality which goes beyond the classical representation of physics. Instead of continuously trying — as we have done for almost a century — to restore dogmatically our manifest image of the world, this new problem allows us to reflect about possible non-classical solutions to the problem of interpretation of QM.

6.2. The contextuality problem: Representation beyond classical contexts

Following the analysis of the measurement problem as a negative one, the basis problem is also interesting to analyze in these terms. In fact, the basis problem is also a way of discussing quantum contextuality in "negative terms". The problem already sets the solution through the specificity of its questioning. It presupposes that *there exists a path* from the "weird" contextual quantum formalism to a classical non-contextual experimental set up in which classical discourse holds.

Basis Problem: *Given the fact that* Ψ *can be expressed by multiple incompatible bases (given by the choice of a Complete Set of Commuting Observables) and that due to the KS theorem the observables arising from such bases cannot be interpreted as simultaneously preexistent, the question is: how does Nature make a choice between the different bases? Which is the objective physical process that leads to a particular basis instead of a different one?*

If one could explain that path through an objective physical process, then the choice of the experimenter could be regarded as well, as being part of such an objective process — and not one that determines reality explicitly. Unfortunately, the problem remains with no solution within the limits of the orthodox formalism. Today, many interpretations attempt to "explain" the process with the addition of strange *ad hoc* rules, unjustified mathematical jumps and the like. These rules "added by hand", not only lack any physical justification but, more importantly, also limit the counterfactual discourse of the meaningful physical statements provided by the theory.

From the representational realist viewpoint, instead of trying to escape the contextual character of the theory, quantum contextuality should be understood as a consequence of the mathematical structure of the theory — a formal scheme which has been allowing us for more than one century to produce the most outstanding physical predictions. The feature of contextuality emerges from the orthodox formalism of QM itself, it is not something external to it. But instead of regarding quantum contextuality as a new interesting feature of a theory, many orthodox interpretations discussed in the literature focus their efforts in trying to escape or bypass quantum contextuality so as to restore a classical non-contextual representation of physical reality. We believe that to deny contextuality just because it obstructs an interpretation of the theory in terms of actual (definite valued) properties would be tantamount to trying to deny the Lorentz transformations in special relativity simply because of its implications to the contraction of rigid rods. Indeed, this was the attempt of most conservative physicists until Einstein made the strong interpretational move of taking seriously the formalism of the theory and its phenomena, and derived a new net of physical notions in order to coherently understand the new theory.

Instead of trying to escape quantum contextuality, our representational realist stance proposes to discuss the following contextuality problem:

Contextuality Problem: *Given the fact that Hilbert space QM is a contextual theory, the question is: which are the concepts that would allow us to coherently interpret the formalism and provide a representation of physical reality that accounts for this main feature of the theory?*

If we accept the orthodox formalism, then contextuality is the crux of QM. It is contextuality that needs to be physically interpreted, instead of something that needs to be bypassed because of its non-classical consequences. Like the superposition problem, the contextuality problem opens the possibility to truly discuss a physical representation of reality which goes beyond the classical representation of physics in terms of an ASA.

While the measurement and basis problems attempt to justify "classical reality", independently of the quantum formalism, the superposition and contextuality leave open the possibility to the development of new non-classical schemes. Orthodox problems escape right from the start the possibility to analyze and discuss non-classical representations. This is the reason why, without a replacement of the problems addressed in the literature there is no true possibility of discussing an interpretation of QM which provides an objective non-classical physical representation of reality. We know of no reasons to believe that this is not doable.

The failure of orthodoxy to solve the measurement and basis problems is also the failure to solve the orthodox (reductionistic) problem which attempts to bridge the gap between our common sense-classical description of reality and the quantum formalism. We will now turn our attention to this more general problem.

7. The Quantum to Classical Limit Revisited

From an orthodox perspective, the most important problem that needs to be solved still remains — after almost one century of attempts — to explain the inter-theoretic relationship between QM and classical physics. Indeed, reductionistic approaches must explain necessarily the emergence of the classical world from QM. As John Hawthorne remarks:

> "[A] natural question to ask is how the familiar truths about the macroscopic world that we know and love ('the manifest image') emerge from the ground floor described by the fundamental book of the world. Assuming that we don't wish to concede that most of our ordinary beliefs about the physical world are false, we seem obliged to make the emergence of the familiar world from the ground floor intelligible to ourselves." [23, p. 144]

In this respect, it becomes of outmost importance to recognize the fact that the problem of "the quantum to classical limit" is a conceptual representational realist problem. It is obviously a problem which goes beyond our observable classical world. It attempts to provide a physical explanation and representation of what *is* the relation between the quantum and the classical realms.

The quantum to classical limit cannot be understood from an exclusive epistemic perspective, for if there is no reference of the theory to "something happening within physical reality", beyond our conscious recognition of measurement outcomes, the question cannot be even posed. In fact, in the explanation of the quantum to classical limit, the subject is simply out of the picture. If we assumed, from an epistemic viewpoint, that physical theories provide "an economical account of experience" with no metaphysical referent whatsoever, then there seems to be no interesting "limit" or "relation" to be discussed between these two theories. The problem simply disappears. As a matter of fact, both classical mechanics and QM (within their specific limit of applicability) already accomplish their means with respect to their empirical findings. Let us be explicit about this point. The problem of the quantum limit addresses the description and representation of a real physical process, it seeks to explain the conceptual representation of what is exactly going on within physical reality. Any realist description must be provided necessarily beyond mathematical structures and measurement outcomes, for if there is no conceptual metaphysical level of analysis the "solution" seems to differ in no way from instrumentalism itself.

The reductionistic perspective requires an explanation of the limit from QM to "classical reality". Without a clear explanation of such limit, orthodoxy seems to run into trouble. The limit cannot be regarded exclusively in terms of a formal or empirical reduction. It is also a problem which requires necessarily a conceptual level of analysis capable of providing a physical explanation of such a seemingly incompatible relation. And the explanation of a physical relation between theories seems to necessarily imply the understanding of the relation. In order to explain the path from QM to classical mechanics, it seems one should understand, first of all, what is QM really talking about.

Bohr himself made a special emphasis on the fact that the measurement process was the key to recover a "rational account of physical phenomena" [8]. In line with Bohr's concerns, the principle of decoherence introduced by Zurek in 1981, attempted to provide the missing physical explanation of the "the quantum to classical limit" [40, 41]. Indeed, would there be an objective explanation of such physical process that would turn "quantum particles" into "stable macroscopic objects", it would also imply a solution to the basis and measurement problems of QM. Such objective explanation would then provide the key to a unified (reductionistic) representation of physical reality which would contain both quantum and classical physics.

As remarked by Jeffrey Bub, in the last decades decoherence has become "the new orthodoxy" [9, p. 212]. The popularity of the principle came from the repeated claims which argued that decoherence provided a suitable solution to the problem of the quantum to classical limit. As explained by Zurek [42, p. 20]: "[Classical] reality emerges from the substrate of quantum physics: Open quantum systems are forced into states described by localized wave packets. They obey classical equations of motion, although with damping terms and fluctuations that have a quantum origin. What else is there to explain?" However, when decoherence theorists reflect about the physical meaning of such isolated quantum states, things become quite bizarre:

> "If the unknown state cannot be found out — as is indeed the case for isolated quantum systems — then one can make a persuasive case that such states are subjective, and that quantum state vectors are merely records of the observer's knowledge about the state of a fragment of the Universe (Fuchs and Peres 2000). However, einselection is capable of converting such malleable and 'unreal' quantum states into solid elements of reality." [*Op. cit.*, p. 22]

Indeed, it is claimed by Zurek that decoherence is capable of creating the "real" from the "unreal", objective states from subjective choices.

> "Quantum state vectors can be real, but only when the superposition principle — a cornerstone of quantum behavior — is 'turned off' by einselection. Yet einselection is caused by the transfer of information about selected observables. Hence, *the ontological features of the state vectors — objective existence of the einselected states — is acquired through the epistemological 'information transfer'*." [*Op. cit.*, p. 22] (emphasis added)

Unfortunately, regardless of Zurek's strong claims, it was soon recognized that the promise of the new orthodoxy to account for the quantum to classical limit was neither physically nor philosophically justified. In fact, the principle of decoherence has been found to have many problems, *ad hoc* moves as well as unjustified shortcuts. There are several important technical problems which cast severe doubts on the validity of the original project of decoherence to explain the limit. Let us mention some of them:

I. The fact that the diagonalization of quantum states is not complete, since "very small" is obviously not "equal to zero". It is true that decoherence seems to produce approximately diagonal mixtures in very "short times", and it is also true that from an epistemic viewpoint "very small" might be considered as superfluous when compared to "very big"; however, this is clearly not the case when considering an ontological problem such as the one addressed by the limit. From an ontological perspective, there is no essential difference between "very big" and "very small", they both have exactly the same importance. The problem is not if a subject is capable of observing something "big" or "small", the problem is to provide a coherent representation — both formal and conceptual — of *what there is* — even if it is very, very small!
II. The fact that the diagonalization can recompose itself into un-diagonalized mixtures if enough time is considered [2, 11]. The restitution of non-diagonal terms comprises the whole program since the fact that time is "long" or "short" is neither important from an ontological perspective. Once again, the problem is not epistemic, it is not about *what a subject might be*

able to observe; the problem of the limit is ontic problem, it is a question about *what there is in physical reality* — independently of conscious beings.

III. The fact that the principle only allows to turn — at least for a while — non-diagonal improper mixtures into "approximately" diagonal improper mixtures does not solve anything. Even if decoherence was able to accomplish the complete diagonalization — something it does not — of the improper mixture, the problem remains that all improper mixtures simply cannot be interpreted in terms of ignorance. As remarked by Zeh:

> "[...] by decoherence I mean the practically irreversible and practically unavoidable (in general approximate) disappearance of certain phase relations from the states of local systems by interaction with their environment according to the Schrödinger equation. Since phase relations cannot absolutely disappear in a unitary evolution, this disappearance can only represent a delocalization, which means that the phases 'are not there' any more, neither in the system nor in the environment, although they still exist in the total state that describes both of them in accordance with quantum nonlocality [...] Decoherence is thus a normal consequence of interacting quantum mechanical systems. It can hardly be denied to occur — but it cannot explain anything that could not have been explained before." [39]

IV. The late recognition by Zurek of the fact that the principle of decoherence only leads to (approximately diagonal) improper mixtures has forced him to venture himself into Everett's many-worlds interpretation and quantum bayesianism (see [42]). However, also here, as it has been discussed by Dawin and Thébault, serious inconsistencies threaten the project [13].

V. Even more worrying is the analysis provided by Kastner, who has pointed out that — even if all these just mentioned problems would be left aside — the main reasoning of the whole decoherence program is circular [31]. There has been, up to date, no reply to the analysis provided by Kastner.

Due to all these well-known facts about decoherence, today it is becoming to be more and more recognized, mainly within the specialized literature, that the principle of decoherence has failed to provide a convincing physical explanation of the quantum to classical limit (see [29]). As remarked by Guido Bacciagaluppi:

> "[some physicists and philosophers] still believe decoherence would provide a solution to the measurement problem of quantum mechanics. As pointed out by many authors, however (e.g. Adler 2003; Zeh 1995, pp. 14–15), this claim is not tenable. [...] Unfortunately, naive claims of the kind that decoherence gives a complete answer to the measurement problem are still somewhat part of the 'folklore' of decoherence, and deservedly attract the wrath of physicists (e.g. Pearle 1997) and philosophers (e.g. Bub 1997, Chapter 8) alike." [4]

In fact, today, there seems to be more questions than answers when considering the solution provided by decoherence to the quantum to classical limit. So it seems, the principle of decoherence might be regarded, in this specific respect, at best, as a proto-principle or proto-model of the quantum process of measurement, but never — at least at this stage of its development — as a coherent physical representation of the path from the quantum to the classical. This is not to undermine the importance of decoherence. One might recall that proto-models have been of great importance in the development of physical theories. Even in QM, we find a very good example of the importance of proto-models when recalling the history behind Bohr's atomic model. Quite regardless of its empirical success, this model was abandoned due to the incoherent physical description it provided. The critical recognition by Heisenberg and Pauli of the failure of Bohr's atomic model to produce a coherent formal and conceptual account might be regarded as the very precondition of possibility for the development of QM itself. We could say that matrix mechanics and Schrödinger's formulation were developed because it was recognized that the Bohrian model was simply not a closed physical theory.

8. Is the 'FAPP Solution' a 'Solution'?

Unfortunately, after more or less having recognized that decoherence does not solve that which it had promised to solve originally, something very strange happened within the new community. Instead of reconsidering the problem and the set of presuppositions involved in order to re-develop the principle of decoherence, a new justification was advanced by the new orthodoxy. Even though it was accepted that decoherence did not "really" solve the quantum to classical limit, it was argued that "the principle of decoherence solves the problem For All Practical Purposes (FAPP)!" This has been a way of the new orthodoxy to argue — more or less in disguise — that "we cannot really explain the path in physical terms, but don't worry, it works anyhow!" This instrumentalist justification escapes any physical explanation grounding itself, once again, in the predictive power of the theory — a predictive power no physicist had ever doubted in the first place.

Following an epistemic viewpoint, it might then be argued that decoherence is an "economy of experience", that it helps us to work in the lab, that it is in fact an "epistemic solution". But then the original ontological problem has been completely abandoned, the original question completely forgotten. This new "FAPP solution" — which is in itself a revival of instrumentalism placed in the heart of realist discussions — has been imported in order to discuss the ontological question of the quantum to classical limit, not only without a proper recognition of the failure of the original project but also turning the original problem into a pseudoproblem. This re-cooked quantum omelette produced by decoherence — which like many interpretations today create physical reality from epistemic choices — has penetrated physics so deeply that today many physicist seem to uncritically accept there actually exists a physical process called "decoherence" that *really* takes place in the lab. But when asked to explain what is exactly this physical process about, the new orthodoxy rapidly shifts the debate and using this new instrumentalist justification argue: "decoherence works FAPP!"

The main problem of the "FAPP solution" is that it solves nothing, it just sweeps the (quantum) dirt under the (classical) carpet. By repeating that "it works!", many physicist and philosophers believe the problem has been actually solved. And there is nothing less interesting in physics than engaging in a problem that already has a solution. There is nothing there to be done, nothing to be thought or in need of development. But problems are the true gas of science, they are that which scientist work on, concentrate with passion, allow us to produce new physical theories. Problems in science should not be regarded as ghosts or monsters that we need to destroy. In fact, there is nothing more interesting, more encouraging than a good difficult scientific problem.

In QM, it was Heisenberg and Pauli's unsatisfaction with Bohr's "magical" model of the atom — as Sommerfeld used to call it — which led them to develop matrix mechanics and the exclusion principle. Instead of confronting the problem, orthodoxy has advanced the most weird type of justifications: "more or less solved", "approximately solved", "almost solved", and "solved FAPP"? These solutions simply escape the question at stake. But in fact, it is quite simple, ontological problems cannot be solved epistemically. A representational realist problem, such as the quantum to classical limit, cannot be solved FAPP.

Acknowledgments

I want to thank R. Kastner and G. Domenech for the private discussions regarding these subjects. This work was partially supported by the following grants: FWO project G.0405.08 and FWO-research community W0.030.06. CONICET RES. 4541-12.

References

[1] Aerts, D. (2009). Quantum particles as conceptual entities: A possible explanatory framework for quantum theory. *Foundations of Science*, **14**: 361–411.

[2] Anglin, J. R., Laflamme, R. and Zurek, W. H. (1995). Decoherence, re-coherence, and the black hole information paradox. *Physical Review D*, **52**: 2221.

[3] Bacciagaluppi, G. (1996). *Topics in the Modal Interpretation of Quantum Mechanics*, PhD dissertation, University of Cambridge: Cambridge.

[4] Bacciagaluppi, G. (2012). The role of decoherence in quantum mechanics, *The Stanford Encyclopedia of Philosophy (Winter 2012 Edition)*, Edward N. Zalta (ed.), (http://plato.stanford.edu/archives/win2012/entries/qm-decoherence/).

[5] Bokulich, A. (2004). Open or Closed? Dirac, Heisenberg, and the Relation between classical and quantum mechanics", *Studies in History and Philosophy of Modern Physics*, **35**: 377–396.

[6] Bokulich, A. (2006). Heisenberg meets Kuhn: Closed theories and paradigms. *Philosophy of Science*, **73**: 90–107.

[7] Bokulich, A. (2014). Bohr's Correspondence Principle, *The Stanford Encyclopedia of Philosophy (Spring 2014 Edition)*, E. N. Zalta (ed.), (http://plato.stanford.edu/archives/spr2014/entries/bohr-correspondence/).

[8] Bokulich, P. and Bokulich, A. (2005). Niels Bohr's Generalization of Classical Mechanics. *Foundations of Physics*, **35**: 347–371.

[9] Bub, J. (1997). *Interpreting the Quantum World.* Cambridge University Press: Cambridge.

[10] Cao, T. Y. (2003). Can we dissolve physical entities into mathematical structures? *Synthese*, **136**: 57–71.

[11] Cormick, C. and Paz, J. P. (2008). Decoherence of Bell states by local interactions with a dynamic spin environment. *Physical Review A*, **78**: 012357.

[12] Curd, M. and Cover, J. A. (1998). *Philosophy of Science. The central issues*, Norton and Company (eds.), Cambridge University Press: Cambridge.

[13] Dawin, R. and Thébault, K. (2015). Many worlds: incoherent or decoherent? *Synthese*, **192**: 1559–1580.

[14] de Ronde, C. (2015). Epistemological and Ontological Paraconsistency in Quantum Mechanics: For and Against Bohrian Philosophy, in *The Road to Universal Logic (Volume II)*, pp. 589–604, A. Koslow and A. Buchsbaum (eds.), Springer.

[15] de Ronde, C. (2015). Hilbert space quantum mechanics is contextual. (Reply to R. B. Griffiths), Preprint. (quant-ph/arXiv:1502.05311).

[16] de Ronde, C. (2016). Quantum Superpositions and the Representation of Physical Reality Beyond Measurement Outcomes and Mathematical Structures. Preprint.

[17] de Ronde, C., Freytes, H. and Domenech, G. (2014). Interpreting the modal Kochen–Specker Theorem: Possibility and many worlds

in quantum mechanics. *Studies in History and Philosophy of Modern Physics*, **45**: 11–18.

[18] de Ronde, C., Freytes, H. and Domenech, G. (2014). Quantum Mechanics and the Interpretation of the Orthomodular Square of Opposition, in *New dimensions of the square of opposition*, J.-Y. Béziau and K. Gan-Krzywoszynska (ed.), pp. 223–242, Philosophia Verlag, Munich.

[19] de Vries, W. (2015). Wilfrid Sellars, *The Stanford Encyclopedia of Philosophy* (Fall 2015 Edition), Edward N. Zalta (ed.), (http://plato.stanford.edu/archives/fall2015/entries/sellars/).

[20] Dieks, D. (1988). The formalism of quantum theory: an objective description of reality. *Annalen der Physik*, **7**: 174–190.

[21] Dorato, M. (2015). Events and the ontology of quantum mechanics. *Topoi*, **34**: 369–378.

[22] Fuchs, C. and Peres, A. (2000). Quantum theory needs no interpretation. *Physics Today*, **53**: 70.

[23] Hawthorne, J. (2009). A metaphysician looks at the everett interpretation, in *Many Worlds*, pp. 144–153, S. Saunders, J. Barrett, A. Kent, and D. Wallace (eds.), Oxford University Press: Oxford.

[24] Heisenberg, W. (1958). *Physics and Philosophy*, World perspectives, George Allen and Unwin Ltd., London.

[25] Heisenberg, W. (1971). *Physics and Beyond*, Harper & Row.

[26] Heisenberg, W. (1973). Development of Concepts in the History of Quantum Theory, in *The Physicist's Conception of Nature*, pp. 264–275, J. Mehra (ed.), Reidel, Dordrecht.

[27] Howard, D. (1993). Was Einstein Really a Realist? *Perspectives on Science*, **1**: 204–251.

[28] Howard, D. (2010). Einstein's Philosophy of Science, *The Stanford Encyclopedia of Philosophy (Summer 2010 Edition)*, E. N. Zalta (ed.), (http://plato.stanford.edu/archives/sum2010/entries/einstein-philscience/).

[29] Joos, E. (1999). Elements of Environmental Decoherence, in *Decoherence: Theoretical, Experimental, and Conceptual Problems*, P. Blanchard, D. Giulini, E. Joos, C. Kiefer, and I.-O. Stamatescu (eds.), Springer: Berlin.

[30] Kastner, R. (2012). *The Transactional Interpretation of Quantum Mechanics: The Reality of Possibility*, Cambridge University Press: Cambridge.

[31] Kastner, R. (2014). 'Einselection' of pointer observables: The new H-theorem? *Studies in History and Philosophy of Modern Physics*, **48**: 56–58.

[32] Kochen, S. and Specker, E. (1967). On the problem of hidden variables in quantum mechanics. *Journal of Mathematics and Mechanics*, **17**: 59–87. Reprinted in Hooker, (1975), pp. 293–328.

[33] Laplace, P. S. (1902) 1951. *A Philosophical Essay on Probabilities*, Dover, New York.

[34] Rédei, M. (2012). Some historical and philosophical aspects of quantum probability theory and its interpretation, in *Probabilities, Laws, and Structures*, pp. 497–506, D. Dieks *et al.* (eds.), Springer.

[35] Van Fraassen, B. C. (1980). *The Scientific Image*, Clarendon, Oxford.

[36] Verelst, K. and Coecke, B. (1999). Early greek thought and perspectives for the interpretation of quantum mechanics: Preliminaries to an ontological approach, in *The Blue Book of Einstein Meets Magritte*, pp. 163–196, D. Aerts (ed.), Kluwer Academic Publishers: Dordrecht.

[37] Wheeler, J. A. and Zurek, W. H. (eds.) 1983. *Theory and Measurement*, Princeton University Press: Princeton.

[38] Young, J. O. (2008). The coherence theory of truth, *The Stanford Encyclopedia of Philosophy (Fall 2008 Edition)*, E. N. Zalta (ed.), (http://plato.stanford.edu/archives/fall2008/entries/truth-coherence/).

[39] Zeh, H. D. (1997). What is achieved by decoherence? in *New Developments on Fundamental Problems in Quantum Physics* M. Ferrero and A. van der Merwe (eds.), Springer: Berlin.

[40] Zurek, W. (1981). Pointer basis of quantum apparatus: into what mixture does the wave packet collapse?, *Physical Review D*, **24**: 1516.

[41] Zurek, W. (1982). Environment-induced superselection rules. *Physical Review D*, **26**: 1862.

[42] Zurek, W. (2002). Decoherence and the transition from quantum to classical — revisited. *Los Alamos Science Number*, **27**: 2–25.

Chapter 6

Principles of Empiricism and the Interpretation of Quantum Mechanics

George Jaroszkiewicz

School of Mathematical Sciences, University of Nottingham, University Park, Nottingham, NG7 2RD, UK

The interpretation of quantum mechanics (QM) is discussed in terms of the principles and logic of empiricism. First, we list a set of issues that should be settled before any consistent interpretation is attempted. This includes questions such as whether we can use an exophysical perspective or an endophysical perspective, and whether a completely reductionist approach makes sense or are we forced to incorporate emergent laws of physics. We then list the scientific principles that should be strictly adhered to in any debate on QM. We follow this with a list of cautions and warnings about misleading concepts that should be avoided, such as ignoring contextuality and the meaning of scientific truth values. These principles and warning are then used to decide on the issues we first identified, giving us a basis for an interpretation of QM from the perspective of observers and quantum signal states of apparatus, rather than in terms of quantum states of systems under observation. Finally, we review a proposed mathematical formalism that encodes this interpretation in terms of quantum registers.

1. Introduction

The ongoing debate (OD) about the interpretation of quantum mechanics (QM) has gone on for over a century now. Seeking to look at the issue in a novel light, our objective in this chapter is to discuss QM from the perspective of the observer and their apparatus, rather than as a description of systems under observation (SUOs). We base our arguments on principles of empirical science that should be generally agreeable.

OD is a nexus, a confluence of several key points of critical significance, each of which contributes to the overall complexity of the problem. Each point involves two conflicting, incompatible views of the same issue, so we call them *binary conflicts* (BCs). In Section 2, we list the most important BCs. In Section 3, we list the principles of science that should apply in OD. Then in Section 4 we give a number of cautions about inappropriate concepts that are frequently used in OD. In Section 5, we use the principles given in Section 3 to support logical reasons for either **Alternative 1**: accepting one side of a BC and certainly not the other, or **Alternative 2**: accepting an intermediate position from which either side of a BC can be contextually relevant. Alternative 1 is generally favored but Alternative 2 cannot always be ruled out. In Section 6, we give resolutions of the above BCs. In Section 7, we review our concept of generalized proposition and a classification scheme for such propositions that allows us to identify and filter out vacuous and contextually incomplete propositions in OD. Finally, in Section 8 we discuss a mathematical implementation of our ideas that emphasizes the observer's processes of acquisition of information.

2. Binary Conflicts

BC 1: Endo versus exo

Exophysics is the description of the physics of an SUO from the perspective of an observer standing outside of that SUO and looking in on it, rather like a God overseeing the Universe. For example, Newton's absolute time is the time of a God-like exophysical observer in classical mechanics (CM) [30].

In contrast, *endophysics* is the physics of SUOs that takes into account the fact that real observers are inside the Universe and subject to the same laws of physics as SUOs. For example, *the Unruh effect* requires a description of an accelerating detector as well as the SUO it is observing [38].

BC 2: Reduction versus emergence

Reductionism is the belief that all SUOs can be explained in terms of a relatively small number of simple principles or laws of physics. For example, chemistry is based on the reductionist atomic theory of matter. Another example is the idea that a single Lagrangian, such as that of the Standard Model, can be used to discuss all elementary particle processes.

In contrast, *emergence* (or *emergentism*, or *non-reductionism* [29]) is the view that there are situations that cannot be so reduced. For example, *life* is a process and not just a collection of atoms.

BC 3: Classical versus quantum

CM asserts that SUOs have absolute properties that can be determined completely by observers at no cost. For example, the CM state of an SUO at any given time is specified completely by a single point (the system point) in the corresponding phase space at that time.

In contrast, the Kochen–Specker theorem in QM supports the view that SUOs do not "have" properties in a classical sense [23]. According to QM, whatever "properties" an SUO is deemed to have are contextually dependent on experimental protocol (i.e. processes of observation). For example, according to the Heisenberg–Kennard uncertainty relation in QM, a sharp measurement of total momentum of an SUO precludes us from knowing where its center-of-mass is [15, 22].

BC 4: Vacuous propositions versus logical propositions

A *vacuous proposition* has no truth value, meaning that there is no way of deciding whether such a proposition is true or false. An example is the Multiverse hypothesis [6].

In contrast, a *logical proposition* is a proposition that has a truth value, either true or false, relative to some explicit contextual test. For example, Pythagoras' theorem is a false logical theorem relative to the axioms of hyperbolic geometry. The point is here that this theorem can be *proved* to be false relative to the axioms of hyperbolic geometry.

BC 5: Absolute truth versus contextual truth

An *absolute truth* is a proposition asserted to be true for all possible contexts, and is therefore invariably stated without any context. It is a logical error to believe that an absolute truth can be proved by any countable number of experiments.

In contrast, a *contextual truth* is a logical proposition that has a truth value of one (i.e. is deemed *true*) relative to a specific context, i.e. true provided certain conditions are met.

BC 6: Upwards causation versus downwards causation

Upwards causation is the reductionist view of time, the idea that complex systems behave the way they do because of the evolution and interaction of the component elements and their laws of such systems. For example, humans age because of molecular changes in their bodies.

In contrast, *downwards causation* is the view that complex (emergent) systems can influence upwards causation. For example, humans have the capacity to slowdown ageing processes by improving their environment.

BC 7: Epistemic uncertainty versus aleatoric uncertainty

Epistemic uncertainty is uncertainty due to an observer's ignorance of information that exists in some meaningful, classical sense and could in principle be acquired. For example, the result of a fair election exists in a classical sense on paper immediately after the polling stations have closed but is an epistemic uncertainty until the votes have been counted and announced.

In contrast, *aleatoric uncertainty* is uncertainty that arises from factors that cannot be mitigated, in a given context. For example, there is no way of predicting the trajectory of a Brownian particle in suspension, so such a trajectory is an aleatoric uncertainty.

BC 8: The nature of the wave function

Hidden variables (HV) theorists believe that the wave function in Schrödinger mechanics is a "thing" with objective existence in one form or another. For HV theorists, the concept of wave function collapse is generally anathema.

In contrast, QM theorists of the Heisenberg/Copenhagen School variety believe that wave functions are contextual representations of the processes of observation [31]. For such theorists, wave function collapse is not only logically necessary but also quantifiable [27, 36].

BC 9: Classical locality versus quantum non-locality

Special relativity (SR) is a theory based on classical principles of observation. SR invokes a light-cone structure that dictates the laws of classical causality. For example, the Einstein locality principle requires events that are spacelike separated to be causally disconnected, so that there is no "spooky action at a distance".

In contrast, QM invariably represents states of SUO in terms of nonlocal theoretical constructs, such as extended wave packets. In particular, the non-locality attributes of entangled states have generated much discussion, as in the EPR debate [4, 7].

BC 10: Open and closed systems versus isolated systems

An *open* system is one that can exchange both energy and matter with its environment, whilst a *closed* system is one that exchanges energy but not matter.

On the other hand, an *isolated* system exchanges nothing with its environment, including information.

3. Principles of Science

Science operates according to certain principles that have withstood the test of time and should not be dispensed with. These principles should be used to determine which side of any BC should be rejected as unphysical. Some of these principles are well known, others are implicit in the way science operates.

P1: *Nullius in verba*

Nullius in verba is the motto of the Royal Society of London. It means "*take no one's word for it*" and is a reminder that science is not a democracy but the dictatorship of empirically validated propositions.

P2: Dialogue architecture is significant

This principle has been given little regard over the last two and a half millennia but it is perhaps the most import one in science and any other subject. It concerns the psychology (that is, thinking patterns) of humans. If this should appear as a misplaced topic in the interpretation of QM, we need only reflect on what we are doing in this book to see its relevance: an interpretation without either speaker, interpreter, or audience, is pointless. Dialogue architecture concerns the assertion that scientific debate, discourse, and understanding is not absolute and does not take place in a vacuum: it is an interaction between parties that should be identified. For instance, what do the laws of physics mean? It is a vacuous assertion to say that science has meaning outside of human context. How could this proposition be tested?

Here the term *architecture* refers to a verbal and/or diagrammatic description of the parties involved in any discussion. This means identifying *who* is making *what* sort of proposition to *whom* and *why*. Science recognizes this principle in several unstated but important respects. For example, high energy particle physics experiment reports generally start with a list of authors, sometimes many hundreds of them. Such reports have the following implicit architecture: that people other than the authors of those reports will read them and act on them. There is no point in writing reports otherwise.

When we introduce our notion of *generalized proposition* (GP) in Section 5, the dialogue architecture will be of an observer talking to themselves as part of the process of observation. Such a dialogue must take place at some point in any process of science, for observation is more than just the accumulation of data. It involves planning, construction of equipment, implementation, and interpretation. These comments are not metaphysical but commentary on what actually happens in observation.

The significance of how observers think and decide on what they do was commented on by Feynman, who must have thought such issues significant, despite his reputation for avoiding metaphysics. He wrote [10]:

> "*...we have an illusion that we can do any experiment that we want. We all, however, come from the same universe, have evolved with it, and don't really have any 'real' freedom. For we obey certain laws and have come from a certain past. Is it somehow that we are correlated to the experiments that we do, so that apparent probabilities don't look like they ought to look if you assume that they are all random.*" (Feynman)

P3: Wheeler's participatory principle

The physicist John Wheeler stated the *participatory principle*, which asserts that the only statements that mean anything are our observations [40]:

> "*...we could not even imagine a universe that did not somewhere and for some stretch of time contain observers because the very building materials of the universe are these acts of observer-participancy. You wouldn't have the stuff out of which to build the universe otherwise.*" (Wheeler)

This is a direct appeal to empiricism.

P4: Contextuality

This is the principle that in science, vacuous propositions should be avoided. A more pointed version of this principle is that **there are no absolute truths in science.**[a] An important corollary is that **the laws of physics are empirical truths, not absolute truths**. Therefore, science requires the laws of physics to be constantly tested in order to determine any limitations.

A refinement of this principle is *contextual completeness*, which is discussed in Section 6, where we introduce our notion of GP. The principle of contextual completeness states that **all GPs in quantum physics should be contextually complete with a generalized proposition classification (GPC) of 3.**

[a]This principle is not self-contradictory: it is about *how* to do science.

P5: There are no empirical infinities in science

All observations are based on the discrete accumulation of discrete information taking finite time. We should not confuse actuality (or reality if you prefer) with *models* of observation that invoke continuity: the notion that there are continuous-in-time observations is such a model. Similarly, all theoretical quantities implying an infinity in one form or another (such as continuous space and time,[b] divergent renormalization constants in quantum field theory, and black hole singularities) are conceptual devices used to encode theoretical belief patterns economically. In a private communication, the eminent computer scientist Marvin Minsky wrote that in long conversations with Feynman about computers and computation, Feynman had said to the effect that "*he (Feynman) did not believe that a finite volume of space could contain an infinite amount of information*".

By the same logic, the vacuous concept of *infinite regress* has to be ruled out of order in science by this principle.

P6: The laws of reductionist physics cannot explain the laws of emergent physics

This principle is a recognition that some known phenomena cannot be rationalized but have to be treated on a different level to reductionist physics (r-physics). In other words, there are scenarios for which there is no known reductionist theory that gives a satisfactory account. There might not be a "Theory of Everything" (ToE) after all. Branches of physics that fall into this category involve the laws of thermodynamics, cosmology, life, consciousness, space and time, and (we suggest) quantum physics.

The use of reductionist concepts in an emergent scenario, such as the use of atomic theory to explain the workings of drugs in medicine or the explanation of genetics by the structure of DNA, is not a pure reductionist account of what is going on: there is invariably an implied emergent contextuality that is necessary for

[b] A continuous set has an uncountable number of elements.

these explanations to mean what they are intended to mean. For example, in cosmology, the reductionist account of the Big Bang has led to a great understanding of the successive phases in Early Universe Evolution, but can give no "explanation" of what generated the Big Bang itself. That stands outside the reductionist domain of validity.

4. Cautions and Warnings

A lack of progress in some theoretical disciplines such as string theory and Multiverse has led to some unscientific ideas that have impinged on OD in one way or another. We discuss some of these now in the form of cautions or warnings.

W1: Mathematics is not physics

The notion that mathematics alone can be used to develop theories of the physical universe crops up on occasion. Anyone reading for the first time the Newtonian gravitation explanation of planetary orbits, Maxwell's electromagnetic wave explanation of light, Einstein's general relativistic account of the precession of the orbit of Mercury, or Dirac's theory of relativistic electron spin, stands a good chance of experiencing the equivalent of a religious conversion to this notion. These are all excellent and inspiring theories. What is really significant about them however is not their elegance, which is remarkable indeed, but that each has been tested empirically and passed with flying colors in significantly large domains of validity that makes them very useful.

In contrast, various recent theories developed by mathematical physicists have no current empirical support or prospect of it, but continue to be favored only on account of their mathematical qualities. This has led to a serious non-scientific confidence bordering on the religious that should be avoided at all cost. Woit has discussed the development, within the string theory community, of the scientifically unacceptable belief by some string theorists that the theory must be right because of its elegance, and that a lack of empirical evidence for string theory is therefore of no consequence [41]. Another example is the *Mathematical Universe hypothesis* [37], which violates every one of the principles stated in Section 3.

Applying this warning to the situation at hand, we should regard QM **not** as a logical mathematical discipline that explains everything but a branch of e-physics (the physics of emergence) that to date has used the mathematics of quantum theory primarily to discuss only one side of the observer-SUO coin in any detail and on that account looks like an r-physics (physics based on reductionism) theory. A notable and very early exception to this was Mott's explanation in 1929 of the straight-line trajectories observed in the ionization trails of alpha particles emitted from decaying nuclei [28]: his approach was to consider the detectors of the ionized tracks and not just the emitted particle.

It is our thesis that QM is not a reductionist, mathematical explanation of the structure and behavior of SUOs, in the sense that Newtonian mechanics explains the orbits of planets, but the e-physics based rules of observation that necessarily requires the emergent concepts of observer and apparatus for the subject to make sense. Moreover, it is not enough to identify Hermitian operators with 'observables' if a deeper understanding of QM is desired: we have to give an empirically motivated and based account of what those Hermitian operators mean *in the laboratory*. It is vacuous to believe that every Hermitian operator represents a physically possible experiment or measurement.

W2: The laws of physics might be ugly

The principles of symmetry that have been so successful in the twentieth century may have reached their natural sell-by-date, in the same way that the theory of epicycles, based on the perfection of spheres, reached its limit of usefulness by Renaissance times. The earth is *not* a sphere: it is *roundish*. Protons and neutrons are *not* perfect isospin partners: they have very different electric charges and slightly different masses. The existence of the cosmic background radiation field is an empirical contradiction of the SR principle that all inertial frames are "as good as each other". The principle of general covariance in general relativity (GR) is a classical concept that has not been incorporated into quantum physics to date. The extremely symmetrical theory known as superstring theory makes no unique-to-itself currently observed or observable empirical predictions whatsoever, and its further development, M-theory, has not yet even been defined. The apparent geometrical symmetry of Minkowski spacetime has been broken by the discovery of parity non-conservation in 1957 [42], not to mention the laws of thermodynamics and cosmological expansion. The list seems endless.

Our conclusion is that there are no absolute symmetries in empirical physics, and those that are incorporated into theories are r-physics approximations, it being hard to think of an empirically validated example of a perfect symmetry, apart from the CPT theorem (that continues to be empirically tested vigorously). One candidate is color symmetry in quantum chromodynamics, but that is a gauge symmetry, and such are generally considered mathematical artefacts requiring the techniques of constraint mechanics to extract the physics [1]. Another example that comes to mind involves the wave-function symmetries associated with bosons and fermions, but in the long run that has everything to do with signal detection, which is an endophysical activity.

To reinforce our concerns about fermionic symmetry in QM, we make two comments. First, Jordan and Wigner gave an early account of where such antisymmetry might come from [3, 21]. Their approach was to set up a theoretical collection of qubits over a region of space and construct what is essentially an emergent type of quantum field that had the required anticommutation properties of fermions. And as for space and time themselves, Schwinger encapsulated precisely our own views on the subject [33]:

> "*The mathematical machinery of quantum mechanics is a symbolic expression of the laws of atomic measurement, abstracted from the specific properties of individual techniques of measurement. In particular, the space-time manifold that is the background of any quantum-mechanical description is an idealization of the function of a measurement apparatus to define a macroscopic frame of reference.*" (Schwinger)

These two comments taken together suggest that the concept of fermions and their anticommutation properties is a contextual, emergent one. Does it really make sense to believe that a quantum field "knows" that it is fermionic rather than bosonic? What makes sense is the view that in the context of an observer-centric arena of space and time defined contextually as suggested by Schwinger above, the observer can efficiently incorporate some of that contextuality in terms of theoretical quantum fields with anticommutation properties, thereby giving an impression that they come from r-physics. Jordan and Wigner's description of fermions is manifestly an emergent one.

W3: Ockham's razor is just a useful guide

The idea that *simple* beats *complex* is a faulty principle when evaluating explanations. The real world is complex beyond measure, and the only way our brains can handle that is to filter the flood of data that we receive from our environment and simplify it so as to fit various templates of expectation held in memory. That is hardly a recipe to understand complexity. The principle that the simplest model is the best, otherwise known as Ockham's razor, is no more than a reasonable guide, but not an invariant principle of science.

We should be prepared, therefore, to find that the laws of physics that incorporate the processes of observation as they are actually carried out, may not be simple or elegant. Indeed, that seems precisely the reason why the interpretation of QM has been so arduous: QM as a theory of unitary Schrödinger evolution of pure wavefunctions appears elegant and simple, stimulating various contextually incomplete interpretations such as HV, Many Worlds, decoherence, and consistent histories. Whenever theorists have attempted to see QM as a theory of observation, however, then they have inevitably found that the technical description becomes very much harder, precisely because it involves the emergent phenomenon of the observer and their apparatus. That should come as no surprise, as emergence is wrapped up with *complexity*, a term that speaks for itself when mathematics is being applied to it. The oft-quoted criticism of state reduction (wave function collapse) as a blemish on Schrödinger evolution is another example of a category error. Lüder's reduction hypothesis [27] is as much about the observer of an SUO as it is about states of that SUO. When properly formulated as GPs, it is our view that such hypotheses will always be contextually complete with a GPC of 3,[c] as required of good QM propositions.

W4: The concept of a reductionist 'ToE' is vacuous

A common theme in particle physics is that there is a universal Lagrangian that will be the key to unlock all the doors to an understanding of the universe, i.e. a ToE. This is a reductionist view of physics that does not take into account emergence and is vacuous on that account.

Our view is that there *may* be such a theory, but it would have to include the laws of e-physics, that is the laws of physics as they should apply to emergent phenomena. These e-laws are currently not understood in any significant way and are most likely totally unlike the laws of reductionist physics (r-physics). For instance, GR in its original form does not preclude the concept of closed timelike curves (CTCs). This should not come as a surprise, given that the differential equations of GR are manifestly local and reductionist in detail, whereas a CTC is necessarily a global, emergent concept not included, excluded, or referred to in any way in those differential equations. QM, when properly interpreted as a theory of observation, is not a reductionist theory of matter but a manifestation of nonlocal e-physics. Therefore, any attempt to 'quantize GR' is a category error, trying to combine a r-physics based theory (GR) with an e-physics based theory (QM). The quantum gravity program has failed not just because of technical difficulties, which appear insurmountable, but because of its basic conceptual error of neglecting or marginalizing the processes of observation.

W5: SUOs are conceptual objects

Humans are conditioned to model the world around them in terms of mental models involving SUOs. Generally, CM is a good reductionist

[c]Discussed in Section 6.

template in which to set these mental models, but not invariably so, as evinced by QM. Because CM models are generally extremely good in non-technical settings, humans have become accustomed to think of SUOs as real, rather than conceptual. This is particularly the case with the concept of *particle* [5]. The history of science shows that today's reductionist elementary particle is tomorrow's emergent bound state. Even the photon does not look elementary when observed in high energy physics, its quantum field theory propagator having a complex, energy dependent dynamical structure [8].

W6: Non-sequiturs

There are places in the annals of OD where errors of logic cloud an already confused debate. Three examples come to mind. First, it is vacuous to assert that a lack of proof that a proposition is false implies that it *may be true*. By such means, we can believe in fairies. There are many propositions that may or may not be true, but to be taken seriously in science, in each case the truth value has to relate to some empirical context of validation. In the absence of any such context, the correct scientific stance is *to say nothing*. In particular, science has nothing to say for or against the *god exists* hypothesis. It is a category error to believe otherwise.

Our second example involves Heisenberg and an English translation of his famous Uncertainty Principle paper of 1927 [15]. At one point, the translation states that '*the "orbit" comes into being only when we observe it.*' Looking at Heisenberg's other remarks and his general quantum philosophy, it seems most probably to us that what he had in mind was something more like: '*the* ***concept of*** *"orbit"* ***is meaningful to us*** *only when we observe it.*' The phrase "*coming into being*" is vacuous and not characteristic of Heisenberg. If there is one hallmark of Heisenberg's interpretation of QM, particularly in the early days of QM proper, it was the avoidance of vacuous statements.

Our third example involves the use of the "*element of reality*" concept used by Einstein, Podolsky and Rosen in the EPR debate [7]. This concept is a form of counterfactual: it asks us to accept the principle that just because information of one kind "would have" been obtained if the right context had occurred (such as a choice to measure position and not momentum), then that information exists in an absolute sense outside of that context. This is a vacuous concept that is embedded in CM but is anathema in QM. According to Wheeler, continuing the quotation given in P3 above [40]:

> "*This participatory principle takes for its foundation the absolutely central point of the quantum: No elementary phenomenon is a phenomenon until it is an observed (or registered) phenomenon.*"
>
> (Wheeler)

W7: The laws of physics are empirical truths, not absolute truths

All experiments take time. Therefore, only a finite number of empirical contexts has been explored in the laboratory. Therefore, the laws of physics are contextual, not absolute. An *empirical truth* or *law of physics* is a contextual truth that has been found to be true in a sufficient number of contexts for it to be regarded as an absolute truth *for all practical purposes*. Empirical truths must be tested as much as possible. For example, the assertion that *energy is always conserved* is not an absolute truth but an empirical truth, and then only relative to the empirically vacuous concept of an isolated SUO.

W8: The necessity for balance

There are two particularly insidious dangers facing us when we seek to clarify OD: one is an undue reliance on intuition, the other is an undue emphasis on mathematical formalism. Intuition is based on a heavily filtered, classical interpretation of the sensory data that our brains are constantly flooded with, whilst our mathematical concepts are attempts to seek logic and patterns in that data. The first danger is that the relative success of CM in modelling real life experiences leads to an unfounded belief that CM can explain everything. The facts speak otherwise: the advanced technology that shows up quantum effects is outside of our normal, everyday experience.

The second danger involves the way mathematics is done. The standard approach to mathematics is either to ignore any discussion of what mathematics is and how it is done (by humans, actually), or to invoke the belief that mathematics is physically real, an example being Tegmark's mathematical universe hypothesis [37], an extreme version of Plato's theory of forms.

It is our view that a recognition of the fact (and it is an empirical fact, not just our opinion) that science is done by human observers is long overdue. Moreover, this fact should be encoded into OD in a realistic way, such as describing QM from the observer perspective, rather than that of the states of SUOs, as is standard in QM. In Section 7, we outline our thoughts on such an encoding.

5. Resolution of the BCs

We now return to our list of BCs and use the above principles to filter out the logical choice in each case.

BC 1: Exo versus endo

By P4, the concept of an absolute exophysical observer has to be logical, not vacuous if it is to mean anything in physics. Therefore, by P1 and P3, it must be possible for the existence of such an observer to be empirically validated. Assuming this is the case, then by P2, we should state who is doing that validation. There are two cases: (i) validation is done by another exophysical observer, or else (ii) by some endophysical observer. Case (i) is the beginning of an infinite regress, which we rule out by P6. The remaining logical possibility, case (ii), is logically equivalent to the assertion that an absolute truth can be validated contextually, which is contradictory by P5. The conclusion therefore is that logically, there can be no empirical evidence for the existence of any absolute exophysical observer.

BC 2: Reductionism versus emergence

According to Laughlin in his Nobel Lecture on fractional quantization [24]:

> "... *the world is full of things for which one's understanding, i.e. one's ability to predict what will happen in an experiment, is degraded by taking the system apart, including most delightfully the*

standard model of elementary particles itself. I myself have come to suspect most of the important outstanding problems in physics are emergent in nature, including particularly quantum gravity."

BC 2 is perhaps the most critical item on our list of BCs. There is no doubting the success of reductionism in CM, and a great deal of QM. But we can argue logically that reductionism in QM without emergence is inconsistent with the above principles. Our argument goes as follows.

Suppose we have a reductionist theory of some SUO in QM, such as a hydrogen atom. Applying P1 and P2, we should ask *who is describing the atom* and *how are they doing it*? If no statement is forthcoming, then P3 and P4 tell us that the description is vacuous. Therefore, we must recognize that our reductionist theory needs an endophysical observer for that theory to make sense. But endophysical observers are emergent concepts (they are not included in the standard reductionist laws of physics, although of course they are subject to those laws). Therefore, reductionism requires emergence, else it is vacuous. Another quote from Laughlin reinforces the point [24]:

"*... those students who stay in physics long enough to seriously confront the experimental record eventually come to understand that the reductionist idea is wrong a great deal of the time, and perhaps always.*"

Two examples illustrate the point that emergent phenomena in QM can have particle-like attributes but no actual elementary 'particles' exist in a classical sense: (i) *phonons* are particle-like manifestations carrying energy and momentum in bulk crystals, but have no identity when those crystal are taken apart; and (ii) in high energy hadronic collisions, there is an exchange process of energy and momentum called the *pomeron*, which has been objectivized as if it were a particle, but there is no quantum field associated with it [20]. There are many other examples of such collective phenomena in science that have invoked objectification, such as cells, solitons and black holes.

Our resolution of BC 2 is certainly *not* to abandon reductionism but to recognize that its role in science is always subordinate to that of emergence. Without emergent observers, reductionism is vacuous. P3, Wheeler's participatory principle, is a direct assertion of that resolution.

BC 3: CM versus QM

The apparent conflict between CM and QM is in our view due to a false comparison, a type of category error. The inherent philosophy underpinning CM is a reductionist one, describing the absolute properties of SUOs on the assumption that such properties exist.[d] In contrast, QM is theory of observation: it started off as such with Planck and its subsequent history amply confirms this. QM is emphatically **not** a theory of *things* but a manifestation of e-physics, the laws of which we know very little about. Because the laws of e-physics are inexplicable in terms of the laws of r-physics, that explains why we do not "understand" QM. The numerous failed attempts to "explain" QM remind us of the scenario in Lem's novel *Solaris*, where decades of attempts by Earth scientists to "understand" the Ocean on Solaris have come to nothing [25].

Our resolution of BC 3 involves recognizing two sorts of context: *relative external context* (REC) and *relative internal context* (RIC). REC is the information that describes an observer's relationship with their immediate environment and the wider universe. REC is always a non-empty set, because as we have argued above, all observers are endophysical.

Moreover, REC is generally classical, in the sense that the parameters describing it are known to the observer involved, or at worse, they are epistemically uncertain. Some or all of these parameters will be statistical in nature, such as the temperature of the laboratory environment.

Turning to RIC, we have to accept that context dictates whether it involves CM alone or a combination of CM and QM. RIC involves

[d]Used non-contextually, terms such as *existence* and *reality* are vacuous.

CM alone if the observer is prepared to step down in terms of their requirements and collect data that does not carry any quantum attributes such as phase information. On the other hand, modern technology allows a fine degree of control over phase loss and under such circumstances, QM comes into play in RIC. Quantum optics illustrates this point well [26].

CM will also be involved in this scenario in the description of the *parameters* associated with the apparatus being used, such as directions of main magnetic fields in Stern–Gerlach experiments.

It is our view that the notion of *Heisenberg cut* marks the boundary between REC and RIC in any experiment, and is contextual.

Our resolution of BC 3 is that CM and QM have separate, complementary roles in a physics that properly relates observers and SUOs.

BC 4: Vacuous propositions versus logical propositions

Vacuous propositions are ruled out of order in science by P4. Our resolution of BC 4 is to ignore any theory based on such propositions.

BC 5: Absolute truths versus contextual truths

By P4, the only propositions we should deal with are logical propositions, and truth values for these can only be obtained in a finite number of contexts. Therefore, there can be no absolute truths in science.[e]

BC 6: Upwards causation versus downwards causation

We have resolved BC 2 by asserting that reductionism requires emergence to make any scientific sense. Therefore, we have to accept that we cannot begin to discuss science and its laws without recognizing the prior existence of endophysical observers (prior to any acts of observation, that is). Observers are the results of downwards causation, that is, the results of natural selection and

---

[e]This statement is a meta-scientific one, telling us *how* to do science.

the evolutionary processes that gave rise to our species. Downwards causation had to occur *before* upwards causation could be discovered.

BC 7: Epistemic uncertainty versus aleatoric uncertainty

A great debate internal to OD is the conflict between the HV adherents and the standard QM adherents such as the followers of the Copenhagen interpretation of QM. This internal debate hinges in part on the nature of randomness in the universe. Our view is that, once we accept that the e-laws of physics are different from the r-laws of physics and cannot be derived from them, then there should be no reason to assert that the reductionist laws of CM can or even should derive or duplicate all the standard predictions of QM, particularly if as we assert, QM is properly part of e-physics and not r-physics.

BC 8: The nature of the wave function

The fact that quantum wave functions and quantum state vectors can be multiplied by arbitrary, constant global phase factors without altering any of the physical predictions of the theory suggests that wave functions and state vectors should not be objectivized. Therefore, this conflict is resolved against HV by accepting the position long argued by Peres, that quantum states are contextual repositories of quantum information [31]. This is over and above any unresolved questions about the HV interpretation of the wave function, such as *what happens to "empty" wave functions once the associated particles are absorbed by detectors*, etc.

BC 9: Classical locality versus quantum non-locality

This is an example of a BC that arises only when context is ignored. By this, we mean that the concept of 'classical locality' is meaningful only relative to some observer capable of validating that concept. Therefore, classical locality cannot be a pure reductionist concept. We refer the reader now to our last paragraph in warning W4 above, on the Jordan–Wigner description of fermions. There it is argued that fermionic (anticommuting) properties require a pre-existing nonlocal structure to be established *before* any such anticommuting local framework can be discussed.

BC 10: Open and closed systems versus isolated systems

Both the open and closed system concepts make an appeal to emergence, in that they make reference to an SUO's environment, and that is not a reductionist concept. These concepts differ only to the extent of differences in relative internal context (RIC, discussed above), that is, in the details of the apparatus that the observer uses in their experiments.

On the other hand, the *isolated system* concept is a most subtle one, for it has two sides to it. On the one side, it is a vacuous concept, for if not, we should ask the question: *how does the observer even know that such a system is there?* Observers could verify the existence of an isolated system only if they attempted to interact with it in some way, thereby destroying its status as an isolated system.

On the other side, it is an absolutely essential concept in QM, in the following sense. In a standard quantum experiment, the observer first prepares an initial state. Then they leave it absolutely alone for some interval of laboratory time. Then they attempt to detect signals in their detectors. During state preparation and outcome detection, the observer may feel entitled to believe that 'there is something there' in some strange quantum limbo land, completely isolated from the rest of the universe.

Our view is that this latter view is no more than a convenient mental image reflecting the contextuality of how the state was prepared, and we may choose not to believe in it. Such a mental image does have one particularly important value, however, in that it can help us calculate the outcome probabilities of the signals seen at the end of each run. That will depend on whether we have found a reasonable model for interactions in the *information void*, our name for the realm of isolation between state preparation and outcome detection. In that realm, no information is obtainable by the observer, simply because there are no detectors there, by definition. The Schrödinger wave function is a good example of a conceptual model of the information void that generally gives good outcome predictions, particularly in the non-relativistic regime.

Precisely because we can know nothing directly about the information void, we are entitled to use any mathematical model in order to explore it. Some models will be better than others. For example, modelling the information void by Minkowski spacetime is better empirically than using Newtonian spacetime. There is scope here for perhaps spectacular novelties, such as Snyder's quantized spacetime algebra [34, 35], or even string theory, but ultimately, all we will have for sure are the signals in our detectors. There can be no absolute truths asserted about the information void.

6. Generalized Propositions

We are in a position now to specify what we believe should be a reasonable approach to the interpretation of quantum physics.

Our starting point is to accept that physics can relate meaningfully only to what endophysical observers do. Such observers are manifestations of e-physics and their laws are not currently understood. Therefore, we cannot currently explain why observers operate as they do. We have to adopt a pragmatic stance here and introduce the notion of *primary observer*. Such an observer is taken as a given, with no questions asked about them. They have all the attributes expected of any observer: memory, intent, and the ability to create and operate apparatus. They are called primary because they make GPs intended for their own use. In other words, a primary observer's dialogue architecture is internal. Lest this seems bizarre, we point out that all observation in the real world is a practical activity carried out according to *plans* of one form or another, and what is a plan but a form of a one-way conversation by a planner talking to their future self.

Primary observers will be intent on evaluating the truth values of various propositions. Given a typical proposition P, such as "the mass of the electron is of the order 9×10^{-31} kilograms", a primary observer O will need to supply RIC in the form of apparatus A, suitable for assigning a truth value of one if P is true and zero P

is false. But that is not all. The primary observer O will need to relate both their apparatus, themselves, and any relevant laboratory structure to the wider, external universe in the form of REC C.

A little reflection should confirm in the reader's mind that a complete specification of any experiment should have all of these ingredients. Indeed, in major experiments such as the search for the Higgs boson at the Large Hadron Collider (LHC), both RIC and REC required the most enormous efforts of construction and expense to ensure that both of these contexts were under suitable control. For instance, the local geological conditions (sandstone) at the LHC that had been accommodated for in the construction of its predecessor collider, the Large Electron–Positron collider (LEP) posed problems for the LHC [39]. Even the current position of the moon has to be factored in during runs of the LHC, as there are noticeable deformations of the proton beams that have to be countered when the moon is full [12].

Given all the above factors, we define a GP $\mathcal{P}$ by

$$\mathcal{P} \equiv (P, A \mid O, C), \tag{1}$$

where P is a proposition of interest to some endophysical observer O, A is the RIC such as the apparatus and the theory used by O to validate P, and C is the REC that describes O. It is our view that the only physically meaningful (i.e. non-vacuous) propositions in quantum physics are GPs or their equivalents. If either A and/or C is missing then a GP is *contextually incomplete*. For example, GPs in metaphysics are invariably of the form $\mathcal{P}_M = (P, \emptyset \mid \emptyset, \emptyset)$, where the *empty set* symbol $\emptyset$ denotes missing context.

6.1. Validation

Validation (or *testing*) is the process of establishing a truth value for a GP. Truth values are either zero for *false* or one for *true*. If a GP $\mathcal{P}$ is true, we write $\mathbb{V}\mathcal{P} = 1$ and if it is false, we write $\mathbb{V}\mathcal{P} = 0$, where $\mathbb{V}$ is the validation function that maps GPs into the truth set $\{0, 1\}$.

Not all GPs can be validated however, and these are termed *vacuous*. They are outside the domain of the validation function $\mathbb{V}$. Vacuous GPs merit the proverbial label *not even wrong* [41].

It turns out that contextual incompleteness is not sufficient to classify a GP as vacuous. GPs in mathematics are generally only partially complete, in that they do not make any reference to any observer. That is not a problem for mathematicians, because they do not claim to be operating according to the rules of empirical science. Indeed, the point of mathematics is to assign truth values to propositions using axioms, not empirical observations. A *theorem* in mathematics is a logical GP, that is, a proposition that can be proved either true or false relative to the axioms involved. It was the great achievement of Gödel to prove that there are GPs that are not completely contextually incomplete but nevertheless are vacuous, meaning that they can be neither proved nor disproved.

6.2. GPC classification

We assign a GPC to any GP by the rule

$$\mathrm{GPC}(P, A \mid O, C) = \alpha(A) + 2\alpha(C), \tag{2}$$

where $\alpha(x) = 0$ if x is the empty set and $\alpha(x) = 1$ if x is non-empty.

In this classification scheme, both A and C must be essential contexts: A must be sufficient to allow an empirical or mathematical test of proposition P, and C must be a physically realistic description of the observer. For instance, it would be empirically invalid to discuss an observer with a laboratory that had part of it outside the Schwarzschild radius of a black hole and another part inside of it. This particular scenario illustrates the two dangers listed in W6 above: a conventional coordinate patch that covers the Schwarzschild radius is an exophysical device, a mathematical contrivance, that should be used with caution: whilst the mathematics appears to allow a discussion of what goes on inside and outside the Schwarzschild radius "simultaneously", the event horizon at that radius rules out such a discussion as empirically vacuous.

The above scheme gives GPCs ranging from zero to three. A GP of the form $(P, \emptyset \mid \emptyset, \emptyset)$ is totally contextually incomplete and has a GPC of zero. Such a proposition is synonymous here with what we mean by *vacuous*. Vacuous propositions cannot be considered scientific. To illustrate the point, consider the proposition Q that "*there is a quantum wave function for the universe*", a basic premise of the Many Worlds approach to QM [9]. According to our principles, we must ask: *what apparatus A can be used to test this proposition?* and "*what sort of observer is going to test it?*" Of course, no answers can be forthcoming, so we conclude that Q is vacuous. This was the conclusion of Fink and Leschke [11].

GPs with GPC of 1, 2, or 3 that can be validated are what we term *logical*, with various degrees of contextual completeness. Mathematical GPs are of the form $(P, Axioms \,|\, \emptyset, \emptyset)$ and have a GPC of 1.[f] Here $Axioms$ is the set of rules relative to which a mathematician uses to determine the truth status of P. It is not currently considered necessary in mathematics to supply any information on the right hand side of the contextual cut in such a GP, although constructivist mathematicians might disagree (in that they might demand existence proofs in place of conjectures).

GPs of the form $(P, \emptyset \,|\, \omega, F)$ have a GPC of 2 and are typical of propositions in CM. Here ω is some observer, endophysical or exophysical, and F is some statement about their frame of reference, such as the spacetime metric, coordinate system, and so on. CM generally finds no reason to discuss the specific details of the apparatus that would be used to determine the truth status of the proposition P, simply because CM assumes that such a truth value "exists" independently of any observation.

GPs of the form $(P, A \,|\, \omega, F)$ have the greatest degree of contextual completeness, having a GPC of 3. They are characteristic of reports about actual experiments. We assert that quantum theoretical discussions should always be based on such propositions. Here the specification of the apparatus A is essential: it is not enough to say that "*A is an observable because it is represented by a Hermitian operator*". That is a mathematical statement that may have no content in the laboratory. Such content has to be proved empirically. For instance, the search for the Higgs particle at the

[f]Note that Gödel proved that GPC greater than zero does not imply logical. The converse is true, however, by definition.

LHC involved the most enormous effort to build real equipment that would validate the theory.

In line with our warnings above concerning the relationship of mathematics to physics, we should regard any GP with a GPC of one that claims to be about physics as a deception and classify it accordingly as *mathematical metaphysics*, meaning it is empirically vacuous. For instance, string theory, quantum gravity, HV, and the Multiverse paradigm, all contain mathematics but currently are empirically vacuous. They are examples of mathematical metaphysics. This does not mean theorists should not work in such areas. That is a legitimate choice that some theorists may take up, but undue claims should not be made for such choices.

6.3. Heisenberg uncertainty

Used in the right way, GPs can provide some insight into important issues in OD. An example is Heisenberg's view of the uncertainty principle and the simultaneous measurement of incompatible variables [15]. Suppose we prepare a particle state Ψ in the laboratory and determine its position expectation value $\mathbb{E}(x \,|\, \Psi)$ to be X. Then the associated GP is of the form $\mathcal{P}_x \equiv (\mathbb{E}(x \,|\, \Psi) = X, A_x \,|\, \omega, F)$, where A_x is the apparatus used to determine the truth status of the proposition $\mathbb{E}(x \,|\, \Psi) = X$. Then we may write $\mathbb{V}\mathcal{P}_x = 1$. On the other hand, a measurement of momentum expectation value would involve a GP of the form $\mathcal{P}_p \equiv (\mathbb{E}(p \,|\, \Psi) = P, A_p \,|\, \omega, F)$, where now A_p is the apparatus used to determine the expectation value of momentum $\mathbb{E}(p \,|\, \Psi)$. Then we may write $\mathbb{V}\mathcal{P}_p = 1$.

If now we wanted to determine the truth status of the combined proposition $\{\mathbb{E}(x \,|\, \Psi) = X\}\&\{\mathbb{E}(p \,|\, \Psi) = P\}$, the uncertainty principle tells us that we cannot find any apparatus that can do the job. Essentially, Heisenberg tells us that there is no apparatus A_{xp} for which

$$\mathbb{V}(\{\mathbb{E}(x \,|\, \Psi) = X\}\&\{\mathbb{E}(p \,|\, \Psi) = P\}, A_{xp} \,|\, \omega, F) = 1. \tag{3}$$

Experimentalists continue to test the truth status of proposition (3). A recent much discussed (and criticized) empirical attempt to show that (3) could hold for some apparatus A_{xp} was reported by Afshar [2].

7. Quantum Register Physics

Assume now that we have accepted the strategy of focusing on the observer's information about the quantum signals in their apparatus, rather than about possible states of an SUO (as typified by eigenvalues of observables). There are several immediate consequences.

7.1. Consequences

First, the concept of SUO itself is effectively banished from the formalism. It is now no more than an economical way of discussing contextuality, that is, what an observer knows from prior experience about their preparation and detection devices. A particle now is relegated to the status of a click in a detector (plus appropriate context).

Second, the concept of *entangled state* loses its undue aura of mystery and magic. Entanglement is, like entropy, not an r-physics property of an SUO but a manifestation of e-physics. In such cases, a correct discussion requires an implicit or explicit recognition of an observer and what they can observe. It is vacuous to discuss an entangled state without stating what the apparatus is that would identify the various entangled components. Likewise, the conceptual fog surrounding wave-particle duality is dispersed, as is the debate about wave function collapse.

Third, non-locality is no longer a mysterious "spooky" property of a quantum state vector that theorists cannot quite come to terms with, but another manifestation of contextuality. Apparatus has to be created in a laboratory *before* there is any meaning to an observation (according to Wheeler's participatory principle, (P3), and such apparatus is necessarily distributed over space, that is, non-locally. CM does not discuss processes of observation and so never requires or discusses its implied non-locality. Indeed, although action-at-a-distance is anathema to most physicists, CM based theories such as Newtonian gravitation have no natural way of avoiding it.

Fourth, the dynamical origins of apparatus and context become significant factors that should be discussed. We should always ask: where does this or that apparatus come from? Can such apparatus actually exist in the real world? Consider GR. A typical investigation is based on a discussion of some line-element, either as a solution to a pre-existing energy–mass configuration, or as a given. The question of how such energy–matter configurations were arrived at is frequently not considered. This is not a trivial point but perhaps a major stumbling block in GR. Even the spinning disk (Ehrenfest's paradox) scenario has this problem: there is a difference between a disk that starts from rest and is gradually speeded up to a final state, and a disk that is just postulated to be spinning. It is not surprising, therefore, that GR finds no natural way of eliminating or arguing away closed timelike loops whenever they occur, such as happens in Gödel's spacetime [13]. Hawking's "chronology protection conjecture", which asserts that the laws of physics are such as to prevent time travel on all but submicroscopic scales [14], is manifestly an e-physics statement, not r-physics: the concept of *submicroscopic scale* is not a reductionist one.

A fifth consequence is that the Hilbert spaces associated with apparatus are necessarily finite dimensional, if we accept principle P5. Moreover, these Hilbert spaces need not have fixed-in-time dimensions. This is because in our observer-centric approach, apparatus can be created, destroyed, merged with other apparatus, taken off-line (decommissioned), or changed with time according to the observer's requirements [18].

A sixth consequence is a conceptual simplification of the actual processes of observation as they are done in the laboratory by the observer. Instead of thinking of observables such as $\hat{x}$ as representation-dependent mathematical objects representing in some bizarre way perhaps complex processes of measurement, all that can happen now is that an observer asks a yes-or-no question at each detector site and interprets the significance of the answers using the contextual information about that apparatus that they the observer must have in order for any experiment to make any sense.

8. Quantum Register Physics

We turn finally to a review of our approach to observer-centric QM. For reasons that will become apparent, we call this approach *quantum register physics* (QRP). Its formalism has a lot in common with quantum computation but its ideological motivation is different.

8.1. Time

Time in QRP is modelled not by a continuous parameter but as a series of discrete stages predicated on real or potential processes of information acquisition. Each stage represents either one of Wheeler's 'acts of observer-participancy' that could in principle result in irreversible acquisition of information, or a change of context in one form or another. For instance, a change of context might involve a choice of *not* looking at a detector, as in the Renniger thought experiment [32], or even creating or destroying pieces of the apparatus. This latter process may result in a change in dimension (or rank) of the quantum register used by the observer to model the *lab state*, the current quantum state of the detectors in the laboratory.

In general, subscripts are used to denote successive stages in any run of an experiment.

8.2. Labstates

A primary observer O describes the labstate at stage n by either a pure state $|\Psi, n)$, an element of the current quantum register $\mathcal{R}_n$, or by a mixed states, as in standard QM.

8.3. Binary detectors

A *binary detector* is a real or conceptual device that can be found in only one of two states, denoted $|0)$ and $|1)$ respectively.[g] The former state is called the *ground state* whilst the latter state is called the

[g]Our choice of round bracket here was chosen deliberately, to highlight the conceptual difference between our labstates (states of the apparatus) and SUO state vectors, which are denoted with angular brackets in Dirac's notation.

signal state. These states are not necessarily lower and higher energy states: context alone determines the observer's interpretation of the physics of detectors.

8.4. Binary questions

Although the world around us appears unimaginably complex, Wheeler's participatory principle and our principle of finiteness allows us to confront that complexity head on with the assumption that any meaningful GP in physics is based on a finite number of binary questions. A *binary question* is one with two possible answers, such as *yes* or *no*, or equivalently, *true* or *false*, or even *one* or *zero*. Binary questions are asked of binary detectors. We represent a binary question in two ways, denoted $(0^i|$ and $(1^i|$ respectively. The former represents the binary question '*is detector i in its no-signal state*?' whilst the latter represents the conjugate question '*is detector i in its signal state*?'

8.5. Binary answers

Given a binary detector i and the corresponding binary question, then we have the rule

$$(p^i \mid q^i) = \delta_{pq}, \tag{4}$$

where p and q are zero or one and the right hand side is a *binary answer*, taking value one or zero, where *zero* represents the answer *no* and *one* represents the answer *yes*.

There is no operational meaning to $(p^i \mid q^j)$ for $i \neq j$: we cannot determine the signal state of a detector by looking at a different detector.

8.6. Quantum bits and quantum registers

The above represents a classical view of question and answer. We have to take quantum processes into account and we do this by working with *quantum bits* (qubits), the quantum versions of classical bits. A quantum bit is a two dimensional Hilbert space. Now it is possible to consider a detector in a linear superposition $|\Psi)$ of its ground state and its signal state, i.e. a state of the form

$$|\Psi) = \alpha|0) + \beta|1), \tag{5}$$

where α and β are complex and satisfy the rule $|\alpha|^2 + |\beta|^2 = 1$. If now we ask question $(p|$ when the detector is in that state, we find the *answer amplitude*

$$(p\,|\,\Psi) = \alpha\delta_{p0} + \beta\delta_{p1}. \tag{6}$$

The Born probability interpretation applies here, so the probability that the detector would be found in its ground state is just $|\alpha|^2$ whilst the probability that it would be found in its signal state is $|\beta|^2$.

Typical experiments will involve perhaps a vast number of binary detectors, and the logical way of dealing with them is to construct *quantum registers*, that is, tensor products of those qubits. Our formalism has been outlined in detail elsewhere [16–19]. For example, a pure labstate[h] at stage n involving a hundred binary detectors would be an element in the rank 100 quantum register $\mathcal{R}_n^{100}$ of the form

$$|\Psi, n) = \sum_{\varepsilon^1,\varepsilon^2,\ldots,\varepsilon^{100}=0}^{1} \Psi_n^{\varepsilon_1\varepsilon_2\cdots\varepsilon_{100}} |\varepsilon^1, n) \otimes |\varepsilon^2, n) \otimes \cdots \otimes |\varepsilon^{100}, n), \tag{7}$$

where

$$\sum_{\varepsilon^1,\varepsilon^2,\ldots,\varepsilon^{100}=0}^{1} |\Psi_n^{\varepsilon_1\varepsilon_2\cdots\varepsilon_{100}}|^2 = 1. \tag{8}$$

8.7. Projection operators and partial questions

Given a labstate $|\Psi)$ in $\mathcal{R}^r$, then we can construct projection operators $\mathbb{P}^i$ and $\overline{\mathbb{P}}^i$ that allow us to ask a question of the ith detector and not of any other detector [17]. In other words, we can ask a partial

[h] We can readily extend the formalism to cover mixed states.

question. For example, $(\Psi|\mathbb{P}^i|\Psi)$ is the probability that, if we looked, the ith detector would be found in its ground state whilst $(\Psi|\overline{\mathbb{P}}^i|\Psi)$ is the probability that it would be found in its signal state.

We may generalize this. For example, $(\Psi|\mathbb{P}^1\mathbb{P}^{31}\overline{\mathbb{P}}^{397}|\Psi)$ is the probability that, if we looked, detectors 1 and 31 would be each in their ground state whilst detector 397 was in its signal state.

9. Commentary

The quantum register approach to quantum physics has a number of features that may allow an extension of standard quantum principles into the hitherto neglected area of *observer theory*, that is, a theoretical framework that gives a comprehensive description of observers *per se* and the processes of observation. We have started to think about some aspects of that program [18], but much remains to be explored. What we have done to date may be regarded as a sketch of the architecture of observation, an emergent description of detectors and how their signals are logically related. What is currently lacking is precisely what standard QM should be good at, namely the reductionist calculation of the transition amplitudes involved in triggering outcome detector signals conditional on prepared states. That requires us to have a good model of the information void, something that is currently a major problem in standard quantum field theory: the divergent renormalization constants generally encountered are manifestations of that problem.

Another way of expressing this problem is in terms of the reductionist-emergence debate. Preparation devices and detectors live on the emergence side of the fence whilst the fundamental reductionist laws of elementary particle physics live on the other side. Bringing the two sides together is perhaps the most interesting challenge facing physics at this time. The interpretation of QM is in our view synonymous with this challenge.

Acknowledgments

I am greatly indebted to Ignazio Licata for giving me the motivation to see, after many years of reductionist thinking, that emergence is really the branch of physics that I should be working on.

References

[1] Abers, E. and Lee, B. W. (1973). Gauge theories. *Physics Reports*, **9C**(1): 1–141.

[2] Afshar, S. S. (2005). Violation of the principle of complementarity, and its implications. In *The Nature of Light: What Is a Photon?* No. 5866 in Proceedings of SPIE, pp. 229–244.

[3] Bjorken, J. D. and Drell, S. D. (1965). *Relativistic Quantum Fields*. McGraw-Hill Inc.

[4] Bohr, N. (1935). Can quantum mechanical description of physical reality be considered complete? *Physics Review*, **48**: 696–702.

[5] Colosi, D. and Rovelli, C. (2009). What is a particle? *Classical and Quantum Gravity*, **26**: 025002 (22pp).

[6] Deutsch, D. (1997). *The Fabric of Reality*. The Penguin Press.

[7] Einstein, A., Podolsky, B. and Rosen, N. (1935). Can quantum mechanical description of physical reality be considered complete? *Physics Review*, **47**: 777–780.

[8] Erdmann, M. (2002). Proton and Photon Structure. In P. Franzini J. Lee-Franzini and F. Bossi, editors, *Proceedings of the XX International Symposium on Lepton and Photon Interactions at High Energies*, pp. 259–272, Rome, Italy, July 2001. World Scientific.

[9] Everett, H. (1957). 'Relative State' formulation of quantum mechanics. *Review of Modern Physics*, **29**(3): 454–462.

[10] Feynman, R. P. (1982). Simulating physics with computers. *International Journal of Theoretical Physics*, **21**(6/7): 467–488.

[11] Fink, H. and Leschke, H. (2000). Is the universe a quantum system? *Found. Phys. Lett.*, **13**(4): 345–356.

[12] Gagnon, P. (2012). Is the moon full? just ask the LHC operators. (http://www.quantumdiaries.org/2012/06/07/).

[13] Gödel, K. (1949). An example of a new type of cosmological solutions of Einstein's field equations of gravity. *Review of Modular Physics*, **21**(3): 447–450.

[14] Hawking, S. W. (1992). The chronology protection conjecture. *Physics Review*, **D46**: 603–611.

[15] Heisenberg, W. (1927). Uber den anschaulichen inhalt der quanten theoretischen kinematik und mechanik. *Zeits. Physik*, **43**: 172–198. Reprinted English translation: The Physical Content of Quantum Kinematics and Mechanics in Quantum Theory of Measurement, (ed.) by Wheeler, J. A. and Zurek, W. H., Princeton University Press: N.J.

[16] Jaroszkiewicz, G. (2007). Multiscale quantum optical networks. In Pirich, A. R., Donkor, E. J. and Brandt, H. E. (eds.), *Quantum Information and Computation*, Proceedings of SPIE, 6573, p. 65730J. 10–12 April 2007, Orlando, Florida, USA, SPIE.

[17] Jaroszkiewicz, G. (2008). Quantized detector networks: A review of recent developments. *Int. J. Mod. Phys. B*, **22**(3): 123–188.
[18] Jaroszkiewicz, G. (2010). Towards a dynamical theory of observation. *Proc. Roy. Soc. A*, **466**(2124): 3715–3739.
[19] Jaroszkiewicz, G. and Ridgway-Taylor, J. (2006). Quantized detector representation of quantum optical networks. *International Journal of Modern Physics B [Condensed Matter Physics; Statistical Physics; Applied Physics]*, **20**(11–13): 1382–1389. 9th International Conference on Squeezed States and Uncertainty Relations (ICSSUR '05), Besancon, France, May 2–6, 2005.
[20] Jaroszkiewicz, G. A. and Landshoff, P. V. (1974). Model for diffraction excitation. *Physical Review D*, **10**(1): 170–174.
[21] Jordan, P. and Wigner, E. P. (1928). Über das Paulische äquivalenzverbot. *Zeitschrift für Physik*, **47**: 631–651.
[22] Kennard, E. H. (1927). Zur Quantenmechanik einfacher Bewegungstypen. *Zeits. f. Physik*, **44**(4–5): 326–352.
[23] Kochen, S. and Specker, E. (1967). The problem of hidden variables in quantum mechanics. *Journal of Mathematics and Mechanics*, **17**: 59–87.
[24] Laughlin, R. B. (1999). Nobel Lecture: Fractional quantization. *Review of Modern Physics*, **71**(4): 863–874.
[25] Lem, S. (1961). *Solaris*. Faber and Faber Limited.
[26] Lipfert, T., Sperling, J. and Vogel, W. (2015). Homodyne detection with on-off detector systems. *Physics Review A*, **92**: 053835.
[27] Lüders, G. (1951). über die zustandsänderung durch den meßprozeß. *Ann. Physik*, **8**(6): 322–328. Translated by Kirkpatrick, K. A., arXiv:quant-ph/0403007, 'Regarding the state-change due to the measurement process'.
[28] Mott, N. (1929). The wave mechanics of alpha-ray tracks. *Proc. Roy. Soc.*, **A126**: 79–84.
[29] Murphy, N. (2000). A non-reductive physicalist account of human nature. In *Ethics, Values and Personhood in the 21st Century*, volume (http://www.counterbalance.org/evp-mind/index-frame.html) of *Counterbalance: New Views on Complex Issues*. Seattle Pacific University.
[30] Newton, I. (1999). *The Principia (Philosophiae Naturalis Principia Mathematica)*. University of California Press, 1687. New translation by I. B. Cohen and Anne Whitman, University of California Press.
[31] Peres, A. (1993). *Quantum Theory: Concepts and Methods*. Kluwer Academic Publishers.
[32] Renniger, M. (1953). Zum Wellen–Korpuskel–Dualismus. *Zeits. für Physik*, **136**: 251–261.

[33] Schwinger, J. (1958). Spin, statistics and the TCP theorem. *Proc. N. A. S.*, **44**: 223–228.

[34] Snyder, H. S. (1947). Quantized spacetime. *Physics Review*, **71**(1): 38–41.

[35] Snyder, H. S. (1947). The electromagnetic field in quantized spacetime. *Physics Review*, **72**(1): 68–71.

[36] Svensson, B. E. Y. (2013). Pedagogical review of quantum measurement theory with an emphasis on weak measurement. *Quanta*, **2**(Issue 1): 18–49.

[37] Tegmark, M. (2014). *Our Mathematical Universe: My Quest for the Ultimate Nature of Reality*. Alfred A. Knopf, New York.

[38] Unruh, W. G. (1976). Notes on black-hole evaporation. *Physics Review D*, **14**(4): 870–892.

[39] Wallis, S. (2001). Spectacular excavations for physics research (http://www.tunneltalk.com/CERN-Aug01-Construction-of-great-excavations2.php).

[40] Wheeler, J. A. (1979). From the big bang to the big crunch. *Cosmic Search Magazine*, **1**(4): 1979. Interview with J. A. Wheeler.

[41] Woit, P. (2006). *Not Even Wrong: The Failure of String Theory and the Search for Unity in Physical Law*. Basic Books.

[42] Wu, C. S., Ambler, E., Hayward, R. W., Hoppes, D. D. and Hudson, R. P. (1957). Experimental test of parity conservation in beta decay. *Physics Review*, **105**(4): 1413–1415.

Specific Interpretive Approaches and Ontologies

Chapter 7

Primitive Ontology and the Classical World

Valia Allori

Philosophy Department, Northern Illinois University
Zulauf Hall 915, DeKalb IL 60115, USA

In this chapter, I present the common structure of quantum theories with a primitive ontology (PO), and discuss in what sense the classical world emerges from quantum theories as understood in this framework. In addition, I argue that the PO approach is better at analyzing the classical limit than the rival wave function ontology approach or any other approach in which the classical world is non-reductively "emergent:" even if the classical limit within this framework needs to be fully developed, the difficulties are technical rather than conceptual, while this is not true for the alternatives.

1. Introduction

Scientific realists believe that, in some strong sense, physical reality is accessible to us through scientific investigation. In particular, scientists can account for what exist by looking at our best scientific theories. This view has been challenged by the success of quantum mechanics: while, on the one hand, quantum theory is incredibly powerful in making new and very precise predictions, on the other hand it is extremely difficult to understand what image of the world this theory provides us. Indeed, many have thought that the real lesson of quantum mechanics is that the dream of the scientific realist is impossible, since quantum mechanics has been taken to suggest that physical objects have contradictory properties, like being in a place and not being in that place at the same time, or that properties do not exist at all independently of observation.[a] Naturally, scientific realists accepted the challenge of "making sense" of the quantum world, and today we have various proposals of quantum theories that allow for a realist reading. Among these theories, most famously we find Bohmian and Everettian mechanics, and the GRW theory: they are fundamental physical theories according to which there is an objective physical world, which can be described by non-contradictory, mind-independent properties. Most philosophers of physics recognize the legitimacy of these theories, but disagree about the metaphysical pictures these theories actually provide. Many scientific realists think that the wave function, which appears in all quantum theories, represents material objects: they defend a view that has been dubbed "wave function ontology." Others instead believe that it is not the case. The *primitive ontology* (PO) approach belongs to the latter group: the idea is that in all quantum theories (and more in general, in all fundamental physical theories), the microscopic description of reality is provided by an object which, mathematically, is defined in three-dimensional space (or four-dimensional spacetime). This has been dubbed the PO of the theory. As we will see, the wave function, being a different kind of mathematical entity, is not a suitable PO for quantum theories. Rather, it has a different role in these theories: instead of describing microscopic reality, it is a necessary ingredient in defining the law of motion for the microscopic constituents of the world described by the PO.

In this chapter, I wish to analyze in more detail than previously discussed in the literature the connection between the microscopic description of reality provided by quantum theories in the PO framework and the macroscopic, classical world of our everyday experience. In the next two sections, I describe the PO approach, first outlining the main ideas and then describing quantum theories with a PO. Then, in Section 4, I discuss the "general scheme of mechanics," namely the main ingredients and the structure that fundamental

[a]The claim that Bohr himself was a positivist has been challenged by Bai and Stachel [17]. Nonetheless, if not Bohr, many others have turned to positivism because of these considerations.

physical theories with a PO have in common and which is used to account for the behavior of matter according to the theory. In Section 5, I analyze the classical limit of quantum theories in the PO framework. Then I draw a contrast with the wave function ontology approach. Given that the wave function does not live in three-dimensional space, there is the problem of accounting for our three-dimensional perception of the macroscopic world. People defending the PO approach believe that this gap between the manifest and the scientific image is very hard to fill without too high a cost: quantum theories understood in these terms will cease to be simple and will lose explanatory power. The primitive ontologists instead maintain that, when considering quantum theories with a PO, then one can construct a simple and suitably straightforward explanatory schema which can be used to account for everyday macroscopic phenomena in terms of microscopic ones. As we will see, this schema, in contrast to what happens in the wave function ontology framework, is an extension of the one used to account for macroscopic phenomena in terms of the microscopic entities in classical physics. In this way, quantum theories understood as theories with a PO are simpler and more explanatory.

2. The Main Ideas of the PO Approach

By looking to the history of science, it seems that mathematics can describe the world around us. This is a remarkable and impressive fact, which I will not attempt to account for in this chapter. Nonetheless, if nature speaks in the language of mathematics and our aim is to study and explain the world around us, our theory will be a mathematical theory of which we will give a physical interpretation. But what does this really mean? In this chapter, I will discuss the answer provided by the PO approach, and in this section I will summarize its main ingredients.[b] Ultimately, scientific realists want fundamental physical theories to explain and account

[b]For a more detailed discussion of this approach, see e.g. [4–6, 11] and references therein.

for the objects of their experience. Thus, the theory will need to specify which mathematical objects, among the variety it contains, should be taken to represent physical objects. Once this has been established, the other mathematical entities in the theory will find their place accordingly. Concisely, the basic idea of the PO framework rests on the recognition that in a fundamental physical theory, some mathematics is representational and some is not. Some mathematical entities in the theory will represent matter (otherwise the theory will be just devoid of physical significance): these entities are the PO of the theory. Some other mathematical objects will have the role of determining how matter moves in time: these are the so-called "nomological" variables of the theory. And some other mathematical objects represent nothing in the world: they are just useful tools necessary to formalize the theory as it is. Other (realist) approaches share this idea, but the PO approach is peculiar in the fact that a suitable PO has to meet the following characteristics: (1) it needs to be defined in three-dimensional space (or four-dimensional spacetime) rather than being a more complicated mathematical object; and (2) it needs to be microscopic rather than macroscopic. Roughly, the three-dimensionality of the primitive variables allows for a direct contact between the variables in the theory and the objects in the world we want them to describe. In fact, a PO represented by an object in a space of dimension d, different from three, would imply that matter lives in a d-dimensional space. Thus, our fundamental physical theory would have to be able to provide an additional explanation of why we think we live in three-dimensional world while we actually do not. It has been argued that this is at best undesirable, in part for reasons connected to another feature that a good PO is supposed to have, namely its fundamentality.[c] Also, the primitive variables better be microscopic entities: a microscopic PO, in which the PO constitutes the building blocks of everything else, is able to ground a (bottom-up) scheme of explanation that allows determining the properties of macroscopic physical objects in

[c]In this regard, see [4, 5] and [33].

terms of the behavior of the PO, as we will explain later in this chapter.[d]

In order to provide a simple example of the PO approach that I will later use to explain how the classical world emerges from quantum theories, let us consider classical mechanics. According to this theory, matter is made of point particles. These particles are completely described by their position r, which is represented by coordinates in three-dimensional Euclidean space, taken to represent physical space. Particles' positions are the PO of the theory. The particles move in time according to a definite law of motion, given by Newton's second law. In order to specify this law, $F = m\frac{d^2r}{dt^2}$, additional mathematical objects are required: m specifies the mass of the particle, and F is the external force that produces in the body the acceleration $a = \frac{d^2r}{dt^2}$. In addition, one needs to specify the so-called laws of the force, namely what F looks like. In the case of Newton's theory of gravitation, we have the familiar $F = G\frac{m_1 m_2}{r_{12}}$, where G is the gravitational constant, m_i, $i = 1, 2$ represent the masses of the bodies involved, and´ r_{12} is their relative spatial distance. Nothing changes much in this respect if we move to classical electrodynamics: the only difference will be that one would have to specify the suitable expressions for the electromagnetic force, namely $F = q_1 v \times B + \frac{1}{4\pi\varepsilon_0}\frac{q_1 q_2}{r_{12}}$, which includes the constant ε_0, the particle velocity v, the charges $q_i, i = 1, 2$, and the magnetic field B. In contrast with the PO, these variables do not "make stuff up:" matter is not made of forces, or of masses. Matter is made of particles which are subject to forces in virtue of their masses. Nonetheless, these variables are needed to account for "how stuff moves." This is where the name "nomological variables" comes from: they help define the law of nature with which matter evolves.[e]

---

[d]See footnote n for an example of a theory with a macroscopic PO, and for an argument for why this is at best unsatisfactory.

[e]Does the electromagnetic field belong to the PO? On the one hand, one would say that it does, for it is both three-dimensional and microscopic: in this sense, the electromagnetic field is a *local beables* [19]. On the other hand, though, its role in the theory seems to be the one of generating the trajectories of matter rather than representing matter itself. If so, the electromagnetic field is a nomological

3. Quantum Theories with PO

Here is how the PO approach extends to the quantum framework. Let us focus on some quantum theories which are compatible with scientific realism, namely Bohmian mechanics [21], the GRW theory [26], and Everettian mechanics (also known as the Many-Worlds theory) [24]. As we will see in the following section, they can be taken to be the following structure in common: matter is described by the PO (which is particles, matter field, or flashes); the temporal evolution of matter is determined by the wave function, which is the main nomological variable. The wave function has its own evolution equation: it can evolve linearly (according to the "usual" Schrödinger equation), or stochastically (according to the GRW evolution discussed later).

These theories emerged as a response to the infamous measurement problem, which plagued early quantum theories and lead many people into embracing antirealism. If matter is completely described by the wave function, and it evolves according to the Schrödinger equation, then impossible macroscopic superpositions, like a dead and alive cat, promptly arise. To avoid such superpositions, many have thought that one either had to add something to the wave function, or have it evolve according to another equation [19]. Bohmian mechanics is usually taken to be an example of the first kind of theories: there is the wave function, but *in addition* there are particles. Not so in the PO approach: the world is not made of wave function, rather *only* particles are the PO of Bohmian mechanics. In fact, the wave function is a mathematical object that lives in a very abstract space, namely configuration space: the space of all the positions of all the particles in the universe, configuration space. If there are N particles in the universe, configuration space has dimension $M = 3N$. Thus by definition, the wave function is not a suitable primitive variable, given that it fails the first condition to be a PO. In addition, in Bohmain mechanics, the trajectories of a system

variable (even if it is microscopic and three-dimensional). A discussion on the metaphysical status of the electromagnetic field can be found in [8].

of particles are determined by Bohm's guidance equation, which involves the wave function that in turn evolves in time according to Schrödinger's equation.[f] Thus, the wave function has the role of generating the trajectories of the PO of Bohmian mechanics, and consequently should be regarded as a non-primitive variable.[g]

The situation in the GRW theory is more complicated: what is the PO of the GRW theory? Historically, the GRW theory was proposed as a solution of the measurement problem where the macroscopic superpositions were "cured" by having the wave function collapse into one of the terms of the superpositions as a matter of law of nature. This is the reason why sometimes the theory is called "spontaneous collapse theory." The wave function first evolves according to the Schrödinger equation until a random time. At that time, the wave function undergoes an instantaneous collapse around a point which is chosen randomly according to a given, suitably defined, probability distribution in terms of the wave function. Then the situation repeats itself: the wave function continues to evolve according to the Schrödinger equation until a second random time, after which it collapses again around a second collapse center, *et cetera, et cetera.*[h] As such, GRW has been taken

[f]For a simple but mathematically accurate description of Bohm's theory, see [15].

[g]The PO approach takes three-dimensional space (or four-dimensional space-time) as fundamental, in the sense that it provides the arena in which physical phenomena take place, and as such the PO view is in contrast with others approaches in which spacetime somehow "emerges" from the basic non-spatiotemporal framework. Examples of such approaches go from the wave function ontology approach in which the fundamental space is configuration space, to the transactional interpretation discussed in [30], which is based on "an intrinsically unobservable, pre-spacetime quantum substratum" from which spacetime emerges. In addition we have, for instance, the system developed in [18], in which spacetime is emergent in a relationist sense, and fundamental physical theories are taken to "describe how entities change," rather than "what entities are." This seems to suggest some sort of structuralism, while the PO approach seems arguably closer to a limited realism *a-la* Kitcher [29], even if recently it has been argued otherwise [23]. These issues are worthwhile investigating, and a preliminary assessment can be found in [9].

[h]The reader interested in a mathematical description of the GRW evolution in terms of the wave function as discussed here is encouraged to read [11].

to be a theory in which matter is described by the wave function,[i] for the wave function is the only object whose temporal evolution was described in the theory as originally presented. As we have already discussed, though, in the PO approach the wave function cannot represent matter. Thus, as it stands, the GRW theory is fundamentally incomplete because it misses a suitable PO. One modification of the GRW theory to include a suitable PO has first been put forward in [20] and later dubbed GRWm in [11]. In this theory, the PO is a three-dimensional matter field m defined roughly as the sum of the masses of each "particle," weighed by the square module of the wave function. In this way, just as in Bohmian mechanics, the wave function has the role of implementing the law of temporal evolution for the PO, for it defines the probability distribution that determines where the matter density will collapse. Another proposal of a GRW theory with a different PO was first suggested in [19], then adopted in [35] and called GRWf in [11]. In this theory, the PO is represented by spacetime events, dubbed "flashes." These flashes are randomly distributed in spacetime in a way determined by the GRW-evolving wave function: every flash corresponds to one of the spontaneous collapses of the wave function. As in GRWm, the wave function has the role of generating the distribution of the flashes.[j] Another possibility is a theory dubbed GRWp3 in [13]. This theory has a PO of particles, whose motion is governed by an equation that is exactly like Bohm's guiding equation. The difference is that in this theory the wave function obeys a GRW-like evolution such that the collapses occur exactly as in usual GRW theories except that, once the time for the collapse has been chosen, the collapse is centered at the actual position of the particle displaced at random. To sum up, in the PO framework, there is no

[i]See, for instance, [1].

[j]Chiatti and Licata [22] write that they "assume the only truly existent 'thing' in the physical world is events of creation and destruction." I believe that a question that is worth investigating is whether or not they have in mind a theory of flashes in which the spatiotemporal distribution of the flashes is determined by a Schrödinger evolving wave function.

single GRW theory: there are instead (at least) GRWm, GRWf, and GRWp3. Each theory describes matter differently, respectively as made of a continuous material field, as made of flashes, and as made of particles. In contrast, the GRW-evolving wave function in each of these theory has the same role: it governs the motion of matter. In this framework, then, the fundamental difference between Bohmian mechanics and the GRW theory is not that the former solves the measurement problem supplementing the description provided by the Schrödinger evolving wave function with particles, while in the latter the wave function provides the complete description of the world but it does not evolve according to the Schrödinger equation. In fact in both theories the PO needs to be specified in addition to the wave function. Since the wave function does not represent matter, the fact that in Bohmian mechanics the wave function evolves according to the Schrödinger equation, and in the GRW-like theories it does not constitute a fundamental difference. Indeed as shown in [11], one can formulate, the former in terms of a "collapsed" wave function, and one of the latter, namely GRWf, in terms of a Schrödinger evolving one. The real difference is that, while in Bohmian mechanics the PO is independent of the wave function, in GRW-like theories like GRWm, GRWf, and GRWp3 the PO is defined in terms of it: PO $= f(\psi)$. In this sense, the PO supervenes on the wave function, but naturally and not logically.[k]

Lastly, let us consider Everettian mechanics. It was originally proposed to solve the measurement problem by allowing for each macroscopic superpositions to be realized, each in a different world (hence the name "many worlds" often used to dub the theory). In virtue of this, nothing is added to the Schrödinger-evolving wave function. Accordingly, almost all the proponents of Everettian mechanics agree in considering the wave function as the object in

[k]Roughly, Y supervenes logically on X when once God has created a world with certain X, the Y comes along for free; instead Y supervenes naturally on X when after making the X, God had to do more work in order to make the Y: he had to make a law relating the X and the Y. Once the law is defined, X will automatically bring along the Y.

the theory that describes physical objects.[1] Again, this is at odds with the PO approach I am analyzing here. Nonetheless, a theory originally developed in [19] can be regarded as an Everettian theory with a PO of particles. In this theory, the wave function evolves according to the Schrödinger equation and guides the motion of the particles, like in Bohmian mechanics. But here the particles do not have a continuous trajectory in spacetime, rather, there is no temporal correlation among them. The theory was dubbed BMW (that is, Bell Many Worlds) in [11] and called Sip (S from the Schrödinger evolution of the wave function, i for independent, p for particle ontology) in [12]. AGTZ [12] also describe another Everettian theory with a three-dimensional matter field PO, whose evolution is determined by a Schrödinger evolving wave function. This theory has been dubbed Sm (S for the Schrödinger equation and m for the matter density function). It is also possible to construct a many-worlds like theory with a PO of flashes. An example of such a theory is Sf as described in [12], in which a Schrödinger evolving wave function never collapses and determines the distribution of the flashes, hence the name.

All these theory have a many-worlds character, given the linearity of the Schrödinger evolution. In fact, contrarily to Bohmian mechanics in which the positions of the particles are connected by the Bohmian trajectories, in Sip there is no connection whatsoever between where a particle is at a given instant of time and where it will be at the previous or following instant. As a consequence, many-worlds exist, not at the same time, but one after another. In Sm, the superpositions of the wave function are inherited by the matter density field and therefore there are non-interacting mass densities associated with the different terms of the superposition. Thus, they can indeed be regarded as comprising many worlds, superimposed on a single spacetime. Since the different worlds do not interact among themselves, they are, so to speak, reciprocally transparent. Similarly, in Sf, different non-interacting families of

[1]See, e.g. [36].

flashes correspond to different terms of the superposition, and hence the many-worlds character of the theory.

In addition to the theories already described, we can imagine a variety of other theories mixing up the various types of PO and the various evolution equations, with the only additional constraint that they are empirically adequate.[m] To summarize, according to the PO approach, philosophers of physics and in general scholars interested in the foundations of quantum mechanics have always focused on the wave function, but that was a mistake: the wave function should not be taken as representing material objects. For if one does, she will have to face problems like the measurement problem, which arise due to the mathematical properties of the wave function, and which disappear on the PO approach. This framework says that in quantum theories, just like in any other fundamental physical theories, matter is represented by a variable in three-dimensional space (or four-dimensional spacetime), and the wave function has the role of implementing the law of evolution for the PO. Thus, *the wave function does not describe matter, it describes how matter behaves.*[n]

4. The General Scheme of Mechanics

Let us get back to classical mechanics for a moment. As we saw above, schematically, Newtonian mechanics accounts the behavior of

[m]For more on these theories, see [11–13], and [5, 6, 10].

[n]It is interesting to notice how quantum theories with a macroscopic PO seem to have been proposed. In fact, as first discussed in [11], one could think of Bohr's quantum theory as one example of this: what might be regarded as its PO is the classical description of macroscopic objects which Bohr insisted was indispensable, with the wave function serving to determine the probability relations between the successive states of these objects. Since the notion of "macroscopic" entities is hopelessly vague, the theory is not satisfactory. In addition, as also discussed in [30], Bohr's quantum theory merely postulates rather than explains the emergence of the classical behavior from the quantum world, and because of this seems extremely *ad hoc*.

macroscopic objects as follow:

(1) It specifies what material objects are made of (the PO);
(2) It specifies how the PO moves in time (the law of motion for the PO);
(3) In order to do this, it specifies the laws of the force.

Typically, one talks about the state of the system to specify the quantities that provide the complete dynamical description of physical bodies. In Newton's theory, the state is given by the pair (r, p) of the position and momentum of the particles. As we mentioned already in Section 2, the position of particles represents the PO, and the momentum is the nomological variable: what is needed to implement the law of motion for the PO. In fact, in terms of these variables, one can rewrite Newton's second law as a set of two first-order differential equations, one for the evolution of position, and the other for the evolution of momentum as follows:

$$\frac{dr}{dt} = u(r,p), \quad \frac{dp}{dt} = F(r,p),$$

where $u = \frac{p}{m}$, in which m is the mass, and F is the force. These equations are the general equations of mechanics: the first defines the velocity, and the second its rate of change. Together, via the specification of u and F, they determine how the PO moves in time. If the PO approach is correct, then this general scheme is not a peculiarity of Newton's theory, but arguably a general trait of a fundamental physical theory.[o] The claim is that in general the state of the system is always given by the pair (χ, η), where χ represents the PO, and η the nomological variable. Typically, the PO's temporal evolution is defined in terms of some function u, and the nomological variable temporally evolves as specified by another function F.

That is:

$$\textit{temporal evolution of } \chi = u, \quad \textit{temporal evolution of } \eta = F.$$

[o]See [4–7, 10].

In Newton's theory, as we saw, the PO is $\chi = r$, and the nomological variable is $\eta = p$. In addition, $u = \frac{dr}{dt} = p/m$ and $F = m\frac{du}{dt}$, where $F = -\nabla V(r)$, V being the (gravitational) potential. The theory is defined when the spatiotemporal trajectories of the PO $\chi = \chi(t)$ are given, and this is determined once u and F are specified.

In the framework of quantum theories, if the PO evolves deterministically, the scheme described above for Newtonian mechanics is generalized rather straightforwardly. The temporal evolution of χ and η can be described by two functions, u and F, such that one can write an equation for the evolution of χ in terms of an appropriate function u and one can write an equation for η in terms of another function F. Consider, for instance, Bohmian mechanics. In this theory, the world is composed of particles, described by points in three-dimensional space, like in Newtonian mechanics. The pair (r, ψ), where ψ is the wave function and $r = (r_1, \ldots, r_N)$ represent the configuration of all the N particles under consideration, are the state of a system. To stress the formal analogies with Newtonian mechanics, let us use the same notation we used before. We have: $\chi = r$, $\eta = \psi$. The functions u and F are defined as follows:

$$u = \frac{\hbar}{m} Im \left[\frac{\eta^* \nabla \eta}{\eta^* \eta} \right], \quad F = \frac{1}{i\hbar} \left[\frac{\hbar^2}{2m} \nabla^2 + V \right].$$

The spatiotemporal histories of the PO $\chi = \chi(t)$ evolve according to $\frac{d\chi}{dt} = u$, determined by η, which in turn evolves according to the equation $\frac{\delta\eta}{\delta t} = F$.

In a quantum theory in which the PO evolves according to a stochastic law, the generalization is less evident but it is still there. The law u for the PO χ amounts to the specification of the possible probability distributions, for example, providing the generator or the transition probability of a Markov process. For example, in GRWm the PO is the mass density field, $\chi = m^{\psi}(r, t)$ and the u function is given by its definition in terms of the nomological variable $\eta = \psi$. The η variable evolves according to the function F, which the stochastic GRW-evolution expressed in terms of χ. Correspondingly, the spatiotemporal histories of the PO evolve according to $\frac{d\chi}{dt} = u$.

Let us now consider the case of quantum theories with a PO in spacetime, like GRWf, in which χ is the set of flashes $\chi = \{(X_1, T_1), \ldots, (X_k, T_k), \ldots\}^{\psi}$. Here, there is no u function because the theory is presented directly in a spacetime framework. Nonetheless, the possible histories of the PO are determined by the equation for the distribution of flashes in terms of the wave function. As in GRWm, the nomological variable η is the wave function, which evolves to the stochastic GRW-evolution.

Before moving to the emergence of the classical world in the PO framework, let me make some final remarks. Notice that any choice of χ, η, u, and F will never be logically forced or determined by experiments. Theories are in this sense underdetermined by data: once the PO χ is fixed, different mathematical choices of η, the variables to implement the law of motion of χ, can be made, since the same trajectories can be generated by different η. For instance, in Newtonian mechanics, the same particles trajectories are generated by a potential V and the potential $V + \textit{constant}$, given that the force will be the same. Similarly, as already mentioned and as it has been shown in [11], Bohmian mechanics as a theory of particles can be reformulated in terms of a GRW-evolving wave function, and that GRW as a theory of flashes can be reformulated in terms of a linearly evolving wave function. These theories are "physically equivalent:" as long as the spatiotemporal histories of the PO are the same, the details of the evolution of the nomological variable do not matter. Thus, we have different ways of freely choosing our theories and their components (that of course must be constrained by the comparison with the experimental data):

- Freedom in choosing the PO χ: changing χ we change the theory, since we are changing the mathematical objects representing matter. This is what happens, for example, in theories like string theory, in which the basic objects are one-dimensional, rather than dimensionless, as in classical mechanics.
- Freedom in choosing η: changing η does not amount to a change in the theory if the evolution of the PO stays the same. This is what happens, for example, in classical mechanics using two potentials

that differ by a constant. If instead we change η such that the evolution for χ changes, then we have a different theory. This is what happens between Bohmian mechanics and GRWp3.

- Freedom in choosing u: changing u amounts to change the law of evolution for the PO, so the theory may well change. An example of a change of u keeping fixed the PO is given by the passage from classical to Bohmian mechanics.
- Freedom in choosing F: changing F, we change the law of evolution for η and two things can happen. Either this change also affects the evolution of the PO or it does not. In the former case, the theory changes, as we can see if we compare classical mechanics with classical electrodynamics, in which the change in F amounts to the change in the potential (we add the electromagnetic potential to the gravitational one). In the latter case, the theory is the same as before. The same u for a given χ means that the theories give rise to the same trajectories for the PO even if we have a different F: they are physically equivalent. An example of this is GRWf and its linear reformulation.

5. The Emergence of the Classical World

In this section, I am going to show that within the PO approach, one can set up a framework to situate the emergence of the classical world from quantum theories, based on the fact that the general scheme of mechanics is the same in quantum and classical theories.[p]

How does the classical world emerge from the description provided by quantum mechanics? Usually, the classical limit is associated with the limit $\hbar \to 0$, meaning by this $\hbar \ll A_0$, where A_0 is some characteristic action of the corresponding classical motion [32]. This condition is often regarded as equivalent to another which involves some relevant length scales of the motion: $\lambda_{dB} \ll L$, where

[p]Note that this approach is strictly *bottom-up*: the basic idea is that we can explain why macroscopic objects behave classically assuming that they are composed by microscopic entities governed by quantum theory. In contrast, approaches like the ones developed in [25], and [16] are explicitly *top-down*.

λ_{dB} is the de Broglie wavelength of the "particle" and L is the scale on which the potential varies given by $L = \sqrt{\left|\frac{V'}{V'''}\right|}$, where V' and V''' denote respectively the first and the third derivative of the potential.[q] These two length scales allow to define the classical limit in terms of parameter $\varepsilon = \frac{\lambda_{dB}}{L}$. The basic idea is that the classical world emerges whenever $\varepsilon \to 0$. What exactly this means depends on our view on quantum theories. Within the framework of standard quantum mechanics, the classical limit is often discussed in terms of the Ehrenfest theorem: if the initial wave function is a narrow wave packet, the packet moves approximately according to Newton's second law. Thus, the emergence of the classical world is associated with the formation and preservation of narrow wave packets. While any wave packet will typically spread and consequently the classical approximation will break down, decoherence, namely the interaction of the system with its environment, helps in keeping the packet narrow [27].[r] In the PO framework, instead, we have more and different resources. In fact, the wave function is not the only ingredient of the theory. Indeed, it is not even the main ingredient: matter is described by the PO, while the wave function is a nomological variable, needed to implement the PO's temporal evolution. The main idea is therefore that classical behavior arises when the trajectories of the PO generated by the quantum theory approximate the classical trajectories as the relevant parameter ε goes to zero. Pictorially, this is equivalent to say that there is a macroscopic scale, defined by λ_{dB} and L, on which trajectories "look" classical. Consider a very big poster, like the ones one encounters in the walls of the subway: if one looks closely, they are made of small colored circles, but when one looks from a distance, the single circles blend together to form a face. Similarly, looking at quantum trajectories "from far away" (i.e. on the macroscopic scale defined by $\varepsilon \to 0$), all the quantum features fade out and the trajectories appear classical. First, I will discuss what this means in the case of quantum theories with a

[q]See [3, 14].

[r]For more on decoherence and its role on the classical limit of quantum mechanics, see [25].

PO of particles, given that the quantum PO is the same as the classical PO. Then, I will show how one can generalize this to the cases in which the quantum theory has a PO that is different from particles.

As we have seen, the paradigmatic example of a quantum theory of particles is given by Bohmian mechanics. To show the convergence of the quantum trajectories to the classical ones is not an easy task, for λ_{dB} is in general a function of the wave function, and L depends on the potential. Thus, the dimensionless parameter ε may depend in general on a combination of various quantities $x, y, z, \ldots$, which relate to the potential and to the wave function. Therefore, the limit $\varepsilon(x, y, z, \ldots) \to 0$ can be taken along different paths (different sequences) in the $x, y, z, \ldots$ space. It turns out that the limit $\hbar \to 0$ *is* only a special case of the limit $\varepsilon(x, y, z, \ldots) \to 0$: it is only one special sequence and it is not able to cover the complexity of the classical world as a whole. Nonetheless, the convergence of quantum trajectories to classical ones, as the parameter ε goes to zero, can be explicitly derived for special cases.[s] One such case is when we consider a fixed external potential acting on a particle whose motion is governed by a quasi-classical wave function, namely a wave function of the short waveform. This corresponds to a de Broglie wavelength λ_{dB} which is small when compared to L, the length of variation of the potential. Therefore, ε goes to zero as λ_{dB} goes to zero while L is fixed. Another special limit is the one of slowly varying potentials: they are potentials that vary on a scale L which is much greater than the de Broglie wavelength. Here, ε goes to zero as L goes to infinity, and λ_{dB} is fixed. It has been shown that in both these cases the quantum trajectories approximate the classical ones in the limit $\varepsilon \to 0$. Also, in both cases there is the formation of a "local plane wave:" a wave function that locally can be regarded as a plane wave having a local wavelength. That is, a local plane wave is a wave function which is slowly varying over a distance of the order of the de Broglie wavelength. The idea is that classical behavior

[s] See [2, 3, 14].

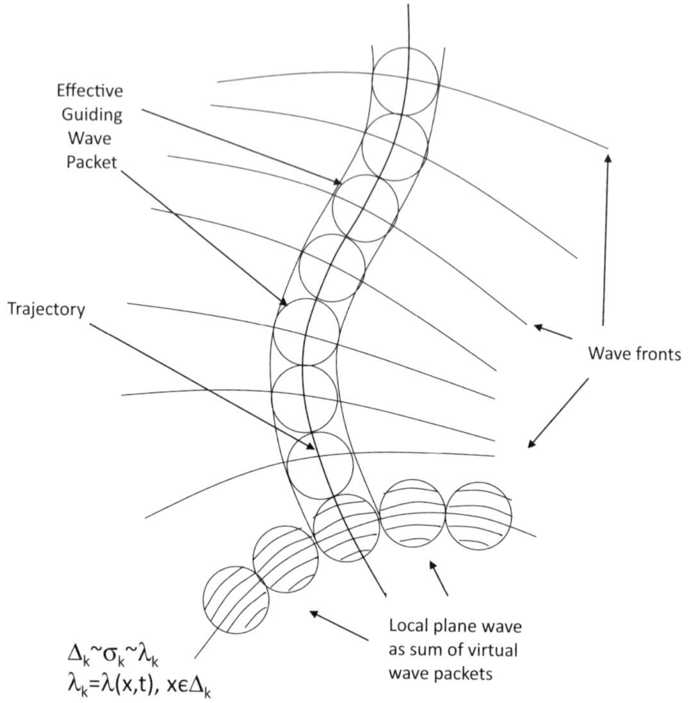

Fig. 1. Local plane wave.

arises when a local plane wave forms. Such local plane wave can be thought of a sum of non-interacting "virtual" wave packets ψ_{k}, each with an almost constant wavelength λ_{dBk} as shown in Fig. 1. Analysis shows that this local plane wave structure is preserved by the dynamics: once the local plane wave is formed, in a relevant sense, it keeps guiding the particle's motion classically. In fact, not all the wave function is relevant to determine the motion: there is an *effective* guiding wave packet which moves the particles along a classical trajectory. This effective wave packet is the part of the wave function in a local neighborhood of the trajectory at any given time. As it has been shown in [2, 3], for the model considered, the classical behavior is not preserved for long times due to interference terms coming from the different components of the local plane waves. In a more realistic model, the environmental degrees of freedom (namely decoherence) would take care of suppressing the superpositions of spatially separated wave functions.[t]

Now, let us consider the case of quantum theories whose PO is different from the one of classical mechanics. For example, consider GRWm, in which the PO is a matter density three-dimensional field. What does it mean that quantum trajectories converge to classical ones in this framework? There is practically no literature on the subject but presumably what is necessary is that there is a macroscopic scale in which: (1) the quantum PO converges to the classical PO; and (2) the quantum trajectories of the PO converge to the classical ones. The first points should be understood as saying that, on the relevant scale, certain quantum microscopic details about matter are obscured and irrelevant, and matter appears as to be made by something else: even if it is not fundamentally made of particles, it looks like it is. A sphere from far away looks like a one-dimensional point, and a hose from far away looks like a one-dimensional string. Similarly, the continuous three-dimensional matter field which constitute the PO of theories like GRWm from far away would look as if matter is concentrated only in its peaks, even if actually there is matter everywhere. The second condition, namely that the quantum trajectories of the PO, the matter density, would converge to the classical one, is in certain respects similar to the case of Bohmian mechanics: when focusing on the peaks of the matter field, their trajectories will converge to classical trajectories. But the situation is a little more complicated in GRWm because the matter density field has superpositions, generated by the superposition of the wave function. As we saw, GRWm gets rid of them as the wave function spontaneously collapses into one of the terms of the superposition with a rate that depends on N, the number of "particles" in the system. That is, the more macroscopic the body, the quicker its mass density will collapse into one of the terms of the superpositions. Therefore, in order for the classical behavior to emerge, first one needs the wave function to collapse, so that the matter density behaves accordingly: this presumably happens on a

[t]This result has also been obtained by [28].

time scale that is inversely proportional to the rate of collapse, which is proportional to the number of "particles" in the system. Then, the remaining term in the matter field has to be "looked at" on a scale in which it seems particle-like: this presumably happens on a spatial scale of the order of σ, the width of the Gaussian by which the wave function is multiplied during the collapse. The matter density will start spreading, given that the wave function will spread just as in Bohmian mechanics. In this theory, though, the local plane wave would form and be preserved by the dynamics, ensuring the classical behavior. Here, one would need the collapse of the wave function to induce the collapse of the matter density field in order to prevent matter from spreading out. Similar considerations hold true for Sm, with the difference that here one does not need the collapse since every term of the superposition is realized in a different world. No one has proven yet that in the relevant macroscopic scale, the classical world emerges in GRWm or Sm, and it is an interesting research project. In particular, it would be interesting to see the role of local plane waves and their interplay with the wave function collapse, as well as the role of decoherence. Details aside, it seems perfectly clear how to formulate the question of the classical limit in the framework of quantum theories with a PO, even if they are not particle theories: both here and in the case of Bohmian mechanics in order to obtain classical behavior one has to focus on the trajectories of the PO, not on the wave function, and one has to find the conditions under which these trajectories approximate the classical ones.

Finally, what about theories directly formulated in spacetime, like GRWf or Sf? The idea is that the classical world would emerge when the quantum distribution of flashes in spacetime approximate the spacetime classical trajectories of particles. Presumably, there is a macroscopic scale in which the set of flashes "clump together" in spacetime to form classical patterns. In GRWf, the flashes are the collapse centers of the wave function, so the more massive (i.e. classical) the object is, the more flashes there will be. There is no literature that explores the relevant macroscopic scale for the emergence of classical behavior in this context, and what mechanics or structures this emergence involves. Nonetheless, given the parameter in GRWf, if the number N of the degrees of freedom in the wave function is large, as in the case of a macroscopic object, the number of flashes is also large. Therefore, it seems reasonable to maintain that large numbers of flashes can form macroscopic shapes, such as tables and chairs.

Notice that in standard quantum mechanics, the emergence of classicality is always connected to the permanence of a narrow wave function during the motion. Instead in Bohmian mechanics, it is exactly the opposite. In fact, the crucial feature of the classical limit in Bohmian mechanics is the formation of very spread out wave function, namely the local plane wave. Thus, only in Bohmian mechanics, we can explain the emergence of the classical behavior for delocalized wave functions. In the case of GRWm, classicality arises as long as the matter density field is concentrated and evolves along a classical path, and this is seemingly ensured by the wave function collapse.

Even if there is still work to be done, the PO has an advantage over the competitor wave function ontology approach as proposed by [1], and then developed by [31] and [34]. In fact, as we just saw, in the PO approach, the classical limit of quantum mechanics is understood as the limit in which the spatiotemporal trajectories of the PO in the quantum theory converge to the classical trajectories. The difficulties one finds in this framework are merely technical, not conceptual: the route to derive the classical behavior from quantum theories is clear. In contrast, the wave function ontology approach considers quantum theories fundamentally as theories in which matter is made of the wave function. Given that, as we have previously mentioned, the wave function is not an object that is defined on three-dimensional space, it is unclear how to define the classical limit. This is obviously not an argument that the classical limit in this approach cannot be defined: to say that the road is steep does not imply it is necessarily impossible to travel. Nonetheless, it is unclear why one would want to go home struggling uphill when there is a nice route that brings us home downhill.

Similar arguments can be raised against the various top-down approaches to the classical limit: in the PO approach, like any other bottom-up approach, macroscopic properties are recovered in terms of their microscopic quantum constituents in a way that looks very similar to the derivation of these properties in terms of their microscopic classical constituents. That is, in the framework of classical mechanics, which postulated that everything is made of microscopic point-like particles, the solidity of a table or the transparency of a window are explained (exclusively) in terms of these microscopic particles and their mutual interactions. In the PO approach to quantum theories, we can do what is usually claimed that cannot be done. Namely, that one can derive the properties of macroscopic objects in terms of their microscopic constituents. The top-down approaches, instead decide to recover these properties differently, in a way that is still to be uncontroversially identified. As in the case against the wave function ontology approach, there is a question of motivation: why take the less travelled road if it is not necessary? What is the advantage? We are scientific realists: we do not need to assert our independence of thought, or to show our ability to overcome difficulties, or the like. What we need to do is explain the world around us in terms of our best fundamental physical theories. And more often than not the best explanation is the less convoluted and most boring alternative.

6. Conclusion

In the PO framework, a satisfactory fundamental physical theory represents matter using a three-dimensional entity defined on a microscopic (fundamental) scale. Given this, there is more of continuity between quantum and classical theories than one would have imagined in the first place. As it has been argued in [7], there is no need to invoke a paradigm shift, or a quantum revolution in order to explain the behavior of macroscopic objects in terms of quantum mechanics.

The three-dimensionality of the PO guarantees that we do not need to worry about the emergence of a three-dimensional space from something of higher dimensions. In addition, a microscopic PO that constitutes the building blocks of everything else is able to ground the classical limit as we have shown in the previous section: classical (macroscopic) objects are composed of quantum (microscopic) PO, and their classical behavior emerges from quantum reality on suitable macroscopic scales which are defined in terms of the evolution of the quantum PO. Even if the problem of the classical limit has been proven difficult to be rigorously treated within the PO framework, in contrast with the alternative approaches the challenges are not a matter of conceptualization. Heuristics and conjectural arguments can be provided to explain the emergence of classical behavior from classical theories, and more research just needs to be done to improve them.

References

[1] Albert, D. Z. (1996). Elementary quantum metaphysics. In: J. Cushing, A. Fine, and S. Goldstein (eds.), Bohmian Mechanics and Quantum Theory: An Appraisal: 277–284. Kluwer.

[2] Allori, V. (2001). Decoherence and the classical limit of quantum mechanics. Ph.D. Thesis, Department of Physics, University of Genova, Italy.

[3] Allori, V. (2007). Fundamental physical theories: Mathematical structures grounded on a primitive ontology. Ph.D. thesis, Department of Philosophy, Rutgers University.

[4] Allori, V. (2013). On the metaphysics of quantum mechanics, in: S. Lebihan (ed.), Precis de la Philosophie de la Physique.

[5] Allori, V. (2013). Primitive ontology and the structure of fundamental physical theories, in: D. Z. Albert, A. Ney (eds.), The Wave Function: Essays in the Metaphysics of Quantum Mechanics: 58–75. Oxford University Press.

[6] Allori, V. (2015). Primitive ontology in a nutshell. *International Journal of Quantum Foundations*, **1**(3): 107–122.

[7] Allori, V. (2015). Quantum mechanics and paradigm shift. *Topoi*, **32**(2): 313–323.

[8] Allori, V. (2015). Maxwell's paradox: classical electrodynamics and its time reversal invariance. *Analytica*, **1**: 1–19.

[9] Allori, V. (manuscript). Primitive Ontology and Scientific Realism.

[10] Allori, V. (manuscript). How to Make Sense of Quantum Mechanics (and More): Fundamental Physical Theories and Primitive Ontology.

[11] Allori, V., Goldstein, S., Tumulka, R. and Zanghi, N. (2008). On the common structure of Bohmian mechanics and the Ghirardi–Rimini–Weber theory. *The British Journal for the Philosophy of Science*, **59**(3): 353–389.

[12] Allori, V., Goldstein, S., Tumulka, R. and Zanghi, N. (2011). Many-worlds and Schrödinger's first quantum theory. *The British Journal for the Philosophy of Science*, **62**(1): 1–27.

[13] Allori, V., Goldstein, S., Tumulka, R. and Zanghi, N. (2014). Predictions and primitive ontology in quantum foundations: a study of examples. *The British Journal for the Philosophy of Science*, **65**(2): 323–352.

[14] Allori, V. and N. Zanghi (2009). On the classical limit of quantum mechanics. *Foundations of Physics*, **39**(1): 20–32.

[15] Allori, V. and Zanghi, N. (2004). What is Bohmian mechanics. *International Journal of Theoretical Physics*, **43**: 1743–1755.

[16] Arsenijevic, M., Jeknic-Dugic, J. and Dugic, M. A top-down versus a bottom-up hidden-variables description of the Stern–Gerlach Experiment. This volume.

[17] Bai, T. and Stachel, J. Bohr's Diaphragms. This volume.

[18] Baumgarten, C. Minkowski spacetime and QED from ontology of time. This volume.

[19] Bell, J. S. (1987). Speakable and Unspeakable in Quantum Mechanics. Cambridge University Press.

[20] Benatti, F., Ghirardi, G. C. and Grassi, R. (1995). Describing the macroscopic world: Closing the circle within the dynamical reduction program. *Foundations of Physics*, **25**: 5–38.

[21] Bohm, D. (1952). A suggested interpretation of the quantum theory in terms of "Hidden" Variables, I and II. *Physical Review*, **85**: 166–193.

[22] Chiatti, L. and Licata, I. Fluidodynamical representation and quantum jumps. This volume.

[23] Esfeld, M. How to account for quantum non-locality: Ontic structural realism and the primitive ontology of quantum physics. Synthese (forthcoming).

[24] Everett, H. (1957). Relative state formulation of quantum mechanics. *Review of Modern Physics*, **29**: 454–462.

[25] Fortin, S. and Lombardi, O. A top-down view of the classical limit of quantum mechanics. This volume.

[26] Ghirardi, G. C., Rimini, A. and Weber, T. (1986). Unified dynamics for microscopic and macroscopic systems. *Physical Review D*, **34**: 470–491.

[27] Giulini, D., Joos, E., Kiefer, C., Kumpsch, G. C., Stamatescu, I. and Zeh, H. D. (1996). Decoherence and the appearance of a classical world in quantum theory. Springer-Verlag: Berlin.

[28] Holland, P. The Quantum State as Spatial Displacement. This volume.

[29] Kitcher, P. (1993). The Advancement of Science: Science Without Legend, Objectivity without Illusions, Oxford University Press.

[30] Kastner, R. Beyond complementarity. This volume.

[31] Lewis, P. (2005). Interpreting spontaneous collapse theories. *Studies in History and Philosophy of Modern Physics*, **36**: 165–180.

[32] Maslov, V. P. and Fedoriuk, M. V. (1981). Semi-Classical Approximation in Quantum Mechanics, D. Reidel Publ. Co., Dordrecht, Holland.

[33] Monton, B. (2006). Quantum mechanics and 3n-dimensional space. *Philosophy of Science*, **73**(5): 778–789.

[34] Ney, A. (2012). The status of our ordinary three dimensions in a quantum universe. *Nous*, **46**(3): 525–560.

[35] Tumulka, R. (2006). A relativistic version of the Ghirardi–Rimini–Weber Model. *Journal of Statistical Physics*, **125**: 821–840.

[36] Wallace, D. (2002). Worlds in the Everett interpretation. *Studies in the History and Philosophy of Modern Physics*, **33**: 637–661.

Chapter 8

Fluidodynamical Representation and Quantum Jumps

Leonardo Chiatti

ASL VT Medical Physics Laboratory
Via Enrico Fermi 15, 01100 Viterbo, Italy
leonardo.chiatti@asl.vt.it

Ignazio Licata

ISEM, Institute for Scientific Methodology
90146 Palermo, Italy

School of Advanced International Studies on Theoretical and Non Linear Methodologies of Physics, 70121 Bari, Italy

International Institute for Applicable Mathematics and Information Sciences (IIAMIS), B.M. Birla Science Centre
Adarsh Nagar, Hyderabad 500, India
ignazio.licata@ejtp.info

"The mixture of Planck's ideas with the old mechanics makes it very difficult to form a physical picture of what is the basis of it. There appears to me one great difficulty in your hypothesis, which I have no doubt you fully realize, namely, how does an electron decide what frequency it is going to vibrate at when it passes from one stationary state to the other? It seems to me that you would have to assume that the electron knows beforehand where it is going to stop."

Letter from E. Rutherford to N.Bohr; 20 March 1913

The sophistication of recent experimental procedures has made it possible to detect experimentally the so-called Quantum Jumps (QJ), i.e. the stop and restart of the wave function as a real phenomenon.

In this work, we connect this experimental evidence to a foundational interpretation of QM, based on an a-temporal vacuum. In that picture, the QJ is a localization process, defined as the time-symmetric action of a self-conjugate projection operator. The compatibility of these discontinuous processes with the Theory of Relativity is demonstrated using the Bohm representation of quantum mechanics.

Keywords: Quantum Jumps; Bohm Potential; Bell Length; T-Symmetry; Localization; Background Vacuum Model.

1. Introduction

Since the mid-80s, the experimental evidence of the existence of quantum jumps (QJ) has been gradually accumulating. Nowadays, therefore, no one doubts the existence of QJs or concomitant "restart" of the wave function as real and genuine physical phenomena [14]. Despite this, the identification of these phenomena as the cause of the objective definition of the state of micro-entities (and consequently the state of macroscopic systems consisting of a large number of them) has not yet become commonly understood. In this chapter, we will examine QJs between stationary states of elementary quantum systems, such as the hydrogen atom or the harmonic oscillator. The transition between two of these states is known to be described by the superposition, with time-dependent coefficients, of the wave functions of the initial and final states. The localization density of the system on its configuration space, as well as the so-called "quantum potential" Q, are therefore functions of time. The finite duration of the transition corresponds, as is known, to the natural line width in the frequency domain. Although this description is correct from the statistical point of view (i.e. when applied to an *ensemble* of identical preparations), it should be noted that the single experimental run shows instead a discontinuous jump of zero duration between the initial and the final states; this jump can be observed directly.

As a result of this jump, Q undergoes a discontinuous change that cannot however constitute a real physical phenomenon, because it conflicts with relativistic invariance. This conflict can be removed by switching to a less familiar, but logically more consistent notion of quantum leap, as the action of a self-conjugate projection operator. This concept is bidirectional in time, and leads to both the transactional interpretation and that of the ABL scheme of Aharanov and collaborators [3, 25]. The search for the principle of connection between two causally connected QJs leads to an a-spatial and a-temporal background. It is argued that such a background, far from being a vague metaphysical notion, constitutes the fundamental level of physical reality and a possible preliminary model of it is provided.

The chapter is organized as follows. In Section 2, some basic notions about the meaning of Q as fluidodynamical translation of the uncertainty principle are recalled. In Section 3, we discuss QJs in this context, and we highlight problems with relativistic invariance. Section 4 describes the (transactional) projectors-based approach and the relation with the background is elucidated; the removal of troubles with relativistic invariance is the immediate consequence. In Section 5, a preliminary model of the background is formulated, and some connections with particle physics are briefly addressed.

2. The Origin of Q and its Application in Hydrogenoid Atoms

Let us consider the Schrödinger equation for the quantum analogue of a "material point":

$$i\hbar\partial_t\Psi = -\hbar^2\Delta\Psi/2m + V\Psi. \tag{1}$$

If we pose, without any loss of generality, $\Psi = \exp(R + iS/\hbar)$ [where R and S are real functions] the following pair of relations is derived after separation of the real and imaginary parts:

$$\begin{aligned} \partial_t S + (\nabla S)^2/2m + V' = 0, \\ \partial_t\rho - \nabla \bullet (\rho\boldsymbol{v}) = 0, \end{aligned} \tag{2}$$

where $\boldsymbol{v} = \nabla S/m$, $\rho = \exp(2R)$, $V' = V - (\hbar^2/2m)(\Delta(\exp R))/(\exp R)$.

These relations are completely equivalent to Eq. (1); they are the equations of a perfect (ideal) fluid of density and local speed $\boldsymbol{v}$ on the configuration space (Madelung fluid; see [24, 27]). The only difference compared to the fluidodynamical representation of classical mechanics is constituted by the appearance of the term:

$$Q = -(\hbar^2/2m)(\Delta(\exp R))/(\exp R) \tag{3}$$

in the expression of potential. This term is often considered as an additional potential (although it is not generated by any source!), and it is usually called "Bohm potential" or "quantum potential" [17]. A good model for understanding the physical meaning of (3) is the 1s ground state of the hydrogen atom, which is notoriously devoided of classical limit. The 1s orbital is expressed by:

$$\Psi(1s) = \pi^{-1/2}(Z/a_0)^{3/2}\exp(-Zr/a_0), \tag{4}$$

where Z is the nuclear charge and $a_0 = \hbar^2/e^2m$. Therefore $S \equiv 0$, $\exp(R) \equiv \Psi(1s)$, and:

$$Q = -(\hbar^2/2m)(Z/a_0)[(Z/a_0) - 2/r]. \tag{5}$$

The force associated with the quantum potential, and acting on the atomic electron, is then:

$$F_Q = -(\partial/\partial_r)Q = Ze^2/r. \tag{6}$$

It does not vanish at all for $\hbar/m \to 0$, therefore the dynamic effects of the quantum potential do not vanish in this limit. This is another way of saying that the proposed system has no classical limit.

The physical meaning of (6) is transparent: F_Q is equal and opposite to the Coulomb nuclear attraction force acting on the electron (which is associated with the "ordinary" V potential) and therefore equilibrates it. In other terms, F_Q stabilizes the electron in the 1s state preventing its fall on the nucleus (and the emission of radiation). The expression (6) forms thus the basis of the strictly quantum phenomenon of atom stability. Of course, $\hbar$ can be arbitrarily small but as long as $\hbar \neq 0$ a stable 1s orbital of size $\approx \hbar^2/e^2m$ exists, because the electron is delocalized around the nucleus. As it is well known, the size $\hbar^2/e^2m$ derives from the minimization of the electron total energy when the antagonistic effect of the uncertainty principle is taken into account. On the other hand, the atomic stability requires, in terms of classical language, the existence of a force equilibrating the nuclear attraction; this force can only be Ze^2/r, for every values of $\hbar$ or $\hbar/m$. This is precisely what is stated by (6). Therefore, the quantum potential expresses, in the classical language of forces, the physical action of the uncertainty principle.

This concept can be immediately generalized to the case of a polyelectronic atom with no entangled electrons. The time independent Schrödinger equation of such a system is:

$$-(\hbar^2/2m)\sum_{k=1,\dots,N}\Delta_k\Psi(x_1,\dots,x_N)+V\Psi(x_1,\dots,x_N)$$
$$=E\Psi(x_1,\dots,x_N). \tag{7}$$

N being the number of electrons. We search for an independent particles solution of the type:

$$\Psi(x_1,\dots,x_N)=\prod_{k=1,\dots,N}\Psi_k(x_k), \tag{8}$$

$$V(x_1,\dots,x_N)=\sum_{k=1,\dots,N}V_k(x_k), \tag{9}$$

$$E=\sum_{k=1,\dots,N}E_k. \tag{10}$$

Substituting Eqs. (8)–(10) in (7), the following N single-electron equations are obtained:

$$-\hbar^2\Delta_k\Psi_k/2m+V_k\Psi_k=E_k\Psi_k;\quad k=1,2,\dots,N. \tag{11}$$

Posing $\Psi_k=\exp[R_k(x_k)+iS_k(x_k)/\hbar]$, where R_k and S_k are real functions, each Eq. (11) is converted to a pair of relations:

$$\begin{gathered}-E_k+[\nabla S_k(x_k)]^2/2m+V_k'(x_k)=0\\ \partial_t\rho_k(x_k)-\nabla_k\bullet(\rho_k\boldsymbol{v}_k)=0,\end{gathered} \tag{12}$$

where $\boldsymbol{v}_k=\nabla_kS_k/m$, $\rho_k=\exp(2R_k)$ and:

$$V_k'=V_k-(\hbar^2/2m)(\Delta_k(\exp R_k))/(\exp R_k). \tag{13}$$

Within the limits of this approximation, each electron is represented as a cloud of charge density ρ_k in the ordinary three-dimensional space. On the generic point, x_k of this cloud acts as the potential $V'_k(x_k)$, which includes the quantum potential. The following relation holds:

$$V_k(x_k) = V_{k,\text{nuc}}(x_k) + V_{k,\text{rep}}(x_k),$$

where:

$$V_{k,\text{nuc}}(x_k) = Ze^2/\|x_k\|,$$
$$V_{k,\text{rep}}(x_k) = e^2 \sum_{k' \neq k} \langle \|x_{k'} - x_k\|^{-1} \rangle_{\Psi^{k'}}. \tag{14}$$

The complex (8)+(9)+(10)+(11) is certainly more manageable. Given an atom with N electrons, the Ψ_k orbital of k-th electron can be expanded on the basis of the hydrogenoid orbitals associated with that atom:

$$\Psi_k = \sum_{n_k,l_k,m_k} a_{k;n_k,l_k,m_k} \Psi_k(n_k, l_k, m_k). \tag{15}$$

Substituting (15) in (11) and taking advantage of the orthonormality of hydrogenoid orbitals is obtained the *secular equation*:

$$\sum_{n_k,l_k,m_k} a_{k;n_k,l_k,m_k} \big[V(n_k, l_k, m_k; n'_k, l'_k, m'_k) + (E_k^H - E_k)\delta_{n_k,n'_k}\delta_{l_k,l'_k}\delta_{m_k,m'_k}\big] = 0. \tag{16}$$

where:

$$E_k^H = -Z^2R/n_k^2; \quad R_\infty = 13.60535\,\text{eV}.$$

The (16) must be solved in order to determine the coefficients a_k and the energies E_k. The initial condition will be given by the usual filling of hydrogenoid orbitals according to the Aufbau:

$$a_{k;n_k,l_k,m_k}(\text{start}) = \delta_{n_k,N_k}\delta_{l_k,L_k}\delta_{m_k,M_k}, \tag{17}$$

after that the matrix $V(n_k, l_k, m_k; n'_k, l'_k, m'_k)$ is computed. Diagonalizing the matrix enclosed in square brackets of (16), the coefficients a_k and energies E_k are obtained. A new matrix $V(n_k, l_k, m_k; n'_k, l'_k, m'_k)$ is then calculated using new coefficients a_k. The procedure is iterated until the condition of self-consistent field is fulfilled; the total

Table 1. Light atoms.

Configuration	Total energy (a.u)	Exp. energy (a.u)	Difference (%)	Notes
He $^1S_0$ $(1s^2)$	−2.809	−2.904	+3.3	*
He $^1P_1$ (1s $2p$) parahelium, or $^3P_{2,1,0}$ $(1s, 2p)$ orthohelium	−2.089	−2.128	+1.8	—
Li $^2S_{1/2}$ $(1s^2\ 2s)$	−7.332	−7.4781	+1.9	—
Li$^+$ $^1S_0$ $(1s^2)$	−7.185	−7.2804	+1.3	—
Be $^1S_0$ $(1s^2\ 2s^2)$	−14.286	−14.657	+2.6	*

formation energy of the atom is then computed through Eq. (10). Table 1 shows the numerical results for some light atoms. As shown, the agreement with the experimentally measured formation energy is good. Calculated energies are slightly exceeding the measured energies, due to the truncation of expansion (15) at values of n lower than 4. Starting from the initial condition (17), the repulsion pushes electrons away thus partially delocalizing them on larger orbitals (higher n values). Because of the small dimension of the basis, this process stops before a complete energy optimization is reached; the final energy is thus slightly larger than real value. However, this difference is larger for electronic configurations with more electrons in the same external orbital (marked with an asterisk). This effect occurs because we have not used entangled wave functions (antisymmetrized orbitals), thus neglecting the identity of electrons. This leads us directly to the last question of this introductory section.

We have to take the last step, removing the condition of independence of the particles. The entanglement of a pair of particles 1, 2 can be easily understood. Let us suppose the particle 1 can be localized in the regions A, B and the same holds for particle 2. We say these particles are delocalized in the region A ∪ B. Let us assign now a constraint on this delocalization, stating that if the particle 1 is localized in A then the particle 2 must be localized in B and *vice versa*. We have two mutually exclusive options:

$$1 \text{ in A}, \quad 2 \text{ in B}; \quad 1 \text{ in B}, \quad 2 \text{ in A}.$$

The initial condition that provides with certainty the first result is described by ket $|\mathrm{A1}, \mathrm{B2}\rangle$; the initial condition that provides with certainty the second result is described by ket $|\mathrm{A2}, \mathrm{B1}\rangle$. An initial condition which gives the first result with probability $|\alpha|^2$ and the second with probability $|\beta|^2$, under the condition $|\alpha|^2 + |\beta|^2 = 1$, is described by the ket $\alpha|\mathrm{A1}, \mathrm{B2}\rangle + \beta|\mathrm{A2}, \mathrm{B1}\rangle$. It is understood, of course, that previously, in the localization of one of the two particles, the position of both of them is indefinite and not merely unknown.

In the quantum potential associated with $\alpha|\mathrm{A1}, \mathrm{B2}\rangle + \beta|\mathrm{A2}, \mathrm{B1}\rangle$, nonlocal correlation between particles 1, 2 will appear. These correlations, which form the basis for the EPR phenomenon, derive from the constraints imposed on the localization of the two particles.

3. Quantum Jumps

For the stationary states of one-dimensional systems, as the one-dimensional harmonic oscillator [or the hydrogenoid atom, if we limit our consideration to spherically symmetric states], the action variable takes the form $S(x, t) = -Et$, with E constant. The first of Eq. (2) then becomes $V + Q = E$. The condition $Q = 0$ holds in points where the classical motion is inverted. These points approximately mark the boundary of the bound state, and therefore the zone where the system is delocalized. An alternative way of expressing the same concept is through the introduction of the so-called "Bell length" L, defined by the relation [20]:

$$Q = \hbar^2/2mL^2. \tag{18}$$

The inversion points are then those at the edge of the existence domain of L, where $L = \infty$. For example, for the orbital (4) is:

$$L = \frac{1}{\sqrt{\left(\frac{Z}{a_0}\right)\left[\frac{2}{r} - \left(\frac{Z}{a_0}\right)\right]}}, \tag{19}$$

whose existence domain is $0 \leq r \leq 2a_0/Z$. L increases from 0 for $r = 0$ to ∞ for $r = 2a_0/Z$. Substituting this last value of r in V, the eigenvalue $E = -Z^2 R_\infty$ is obtained. Similarly, for the n-th stationary state of the one-dimensional harmonic oscillator $(V = m\omega^2 x^2/2)$

one has:

$$Q = \left(n + \frac{1}{2}\right)\hbar\omega - \frac{1}{2}m\omega^2 x^2. \tag{20}$$

In this case, the existence domain of L is [18]:

$$x^2 \geq \frac{2\hbar}{m\omega}\left(n + \frac{1}{2}\right). \tag{21}$$

and $Q = 0$, $L = \infty$ when the equality holds. Substituting this extreme value of x^2 in V, the eigenvalue $E = (n + \frac{1}{2})\hbar\omega$ is derived.

When the system leaps from one stationary state to another through a QJ, the quantum potential undergoes a discontinuous change $\Delta Q = \Delta E$, that is $Q \to Q' = Q + \Delta E$, where ΔE does not depend on coordinates. This variation does not change the gradient of Q, that is the repulsive force which must compensate the gradient of V (which itself remains unchanged). However, as is easy to verify directly from (18), L changes and its existence domain changes accordingly. In other words, the status of the system delocalization changes.

In particular, the points of inversion of the classical motion change all together at the same instant of time, that is the instant of the jump. This statement implicitly requires a notion of absolute simultaneity, as is natural if a null duration of the QJ is assumed in all frames of reference. Thus, the instantaneous change of size of the allowed region for the system is not a relativistically covariant concept and therefore cannot represent a real physical phenomenon. This is the modern reformulation of the original Rutherford objection to Bohr, reported in the opening of this chapter. The information transfer between the initial and final states during the QJ must be instantaneous, and this request cannot be satisfied in a relativistically covariant description.

4. Transactions: An Overview

With reference to interpretation of quantum mechanics (QM) the term "transaction" firstly appeared in J. G. Cramer's papers [11, 12]. Cramer was the father of the — often misunderstood — "transactional interpretation", usually indicated as TI or TIQM. This interpretation assumes the propagation of the advanced wave function, with negative energy, to be an actual physical reality. The original Cramer formulation was still classical, even though innovative, because modeled on time-symmetric Wheeler–Feynman electrodynamics. The possibility of the QM emergence from an a-temporal background was not contemplated. This possibility, on the other hand, has been discussed in more recent formulations, such as R.E. Kastner's "possibilist" TI (PTI, see [15], in particular Chapters 4 and 8).

In this section, the concept of transaction is essentially presented as an aspect of current QM formalism, and interpretational issues are not discussed in detail.

We assume that the only truly existent "thing" in the physical world are the events of creation and destruction (or, if one prefers, physical manifestation and demanifestation) of certain qualities. In the language of QFT, these events are the "interaction vertices", while the different sets of manifested/demanifested qualities in the same vertex are the "quanta".

As an example, in a certain vertex a photon $(E, \boldsymbol{p}, \boldsymbol{s})$ can be created, where E is the (created) energy of the photon, $\boldsymbol{p}$ is the (created) impulse of the photon and $\boldsymbol{s}$ is the (created) spin of the photon. In a subsequent event, this photon can be absorbed and this corresponds to a packet of properties $(-E, -\boldsymbol{p}, -\boldsymbol{s})$ where $-E$ is the (absorbed) energy of the photon, $-\boldsymbol{p}$ is the (absorbed) impulse of the photon, and $-\boldsymbol{s}$ the (absorbed) spin of the photon; $(-E, -\boldsymbol{p}, -\boldsymbol{s})$ is the absorbed, i.e. destroyed, photon. It is assumed that the first event chronologically precedes the second, and that E is positive. A process is therefore being described in which a positive quantity of energy is created and then destroyed. In an absolutely symmetrical manner, it can be said that we are describing a process which proceeds backward in time and during which a share of negative energy $-E$ is firstly created (in the second event) and then destroyed (in the first event). The two descriptions are absolutely equivalent.

There are two reasons for which, in the process described, one cannot have $E < 0$, i.e. the propagation of a positive energy photon towards the past. A photon with $E > 0$ which retropropagates towards the past, yielding energy to the atoms of the medium through which it travels (such as, say, a photon X which ionizes the matter through which it travels), is seen by an observer proceeding forward in time as a photon with $E < 0$ which *absorbs* energy from the medium. This photon would be spontaneously "created" by subtracting energy from the medium, through a spontaneous coordination of uncorrelated atomic movements which is statistically implausible, and has never been observed experimentally.

From a theoretical point of view, the probability of the occurrence of a creation/destruction event for a quantum Q in a point event x is linked to the probability amplitude $\Psi_Q(x)$, which can be a spinor of any degree. Each component $\Psi_{Q,i}(x)$ of this spinor satisfies the Klein–Gordon quantum relativistic equation $(-\hbar^2\partial^\mu\partial_\mu + m^2c^2)\Psi_{Q,i}(x) = 0$, where m is the mass of the quantum.[a] At the non-relativistic limit, this equation becomes a pair of Schrödinger equations:

$$-\frac{\hbar^2}{2m}\Delta\Psi_{Q,i}(x) = i\hbar\partial_t\Psi_{Q,i}(x), \tag{22}$$

$$-\frac{\hbar^2}{2m}\Delta\Psi^*_{Q,i}(x) = -i\hbar\partial_t\Psi^*_{Q,i}(x). \tag{23}$$

The first equation has only retarded solutions, which classically correspond to a material point with impulse $\boldsymbol{p}$ and kinetic energy $E = \boldsymbol{p}\cdot\boldsymbol{p}/2m > 0$. The second equation has only advanced solutions, which correspond to a material point with kinetic energy $E = -\boldsymbol{p}\cdot\boldsymbol{p}/2m < 0$. Thus, there are no true causal propagations from the future.

The creation of the $E > 0$ energy followed by its subsequent absorption and, conversely, the creation of a $-E < 0$ energy preceded by its destruction are clearly two different descriptions of the same process. This however is true so long as the interaction events are considered, i.e. the true substance of the physical world. From the point of view of the dynamical laws for the probability amplitudes of these events, matters are quite different, however. The creation

[a]It is necessarily positive, because it is a measure of the energy which must be released in order to create the quantum; a physical entity can certainly not be created in the vacuum by subtracting energy from it.

of quality Q is associated with the initial condition for $\Psi_{Q,i}(x)$ in Eq. (22); the destruction of quality Q is associated with the "initial", actually the final, condition for $\Psi^*_{Q,i}(x)$ in Eq. (23). In general, however, the two conditions are different and therefore generate different solutions for the two equations, which are not necessarily mutual complex conjugates.

It is a fundamental fact that the destruction event is not described by Eq. (22) and that the creation event is not described by Eq. (23); this remains true even if the Hamiltonians of interaction with the remaining matter are introduced into the two equations. Thus, at the dynamic laws level, the process of the creation and destruction of Q is completely described solely by the loop:

$$\text{creation of } Q \quad \underset{\text{evolution according to Eq. (23)}}{\overset{\text{evolution according to Eq. (22)}}{\rightleftarrows}} \quad \text{destruction of } Q$$

and not only by the upper or lower half-loop.

More generally speaking, we shall have at $t = t_1$ the event of the creation–destruction of a quality Q ($|Q\rangle\langle Q|$) and at $t = t_2$ the event of the creation–destruction of a quality R ($|R\rangle\langle R|$). These two processes will be linked by a time evolution operator S according to the ring:

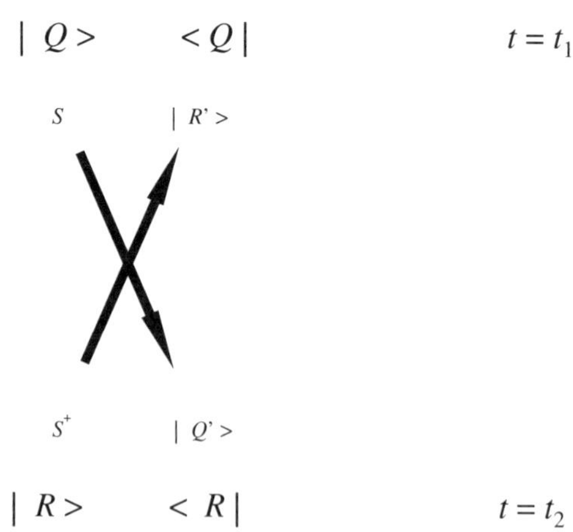

(24)

In other words, $|Q\rangle$ is transported from S into $|Q'\rangle$ and projected onto $\langle R|$, $|R\rangle$ is transported by S^+ into $|R'\rangle$ and projected onto $\langle Q|$. The amplitudes product:

$$\langle R|S|Q\rangle\langle Q|S^+|R\rangle = |\langle R|S|Q\rangle|^2$$

is immediately obtained, which is the probability of the entire process. If quality Q is constituted by a complete set of constants of motion, then $R = Q$ and this is the type of process which can describe the propagation of a photon-type quantum, otherwise it is the generic process of the creation of a quality Q causally linked (by means of S) to the destruction of a quality R. Moving to the representation of the coordinates, by substituting bras and kets with wave functions, we once again obtain as a particular case the result already seen with the non-relativistic expressions (22) and (23). The ring process described above will be called *transaction*. The transaction exists if the propagation S^+ is as "real" as the S; to ascertain this, one must see whether experimental situations exist in which the initial condition $|Q\rangle$ and the final one $\langle R|$ are connected in a nonlocal way. As is well known, the answer is affirmative: the phenomena EPR are highly valid examples of quantum-mechanical predictions which violate locality.

Some remarks.

(1) Nonlocal effects do not follow from the forward propagation S only, yet we see nonlocal effects, which implies that another factor is involved; the existence of advanced propagation S^+ is a logical inference since it is reflected in the Born probability and in the probability amplitudes.
(2) In the same hypothesis, i.e. considering only the forward propagation, one cannot explain the destruction of the "quantum state" as a phenomenon that takes place at a defined instant t_2. Well known paradoxes, such as that of Schrödinger's cat, derive from this.
(3) From an algebraic point of view, the transactional ring is a sort of identity operator, because $SS^+ = S^+S = 1$ and the qualities Q, R are simultaneously created and destroyed. One has the

impression that every quantum process (therefore all matter) and time itself are emitted from an invariant substratum and re-absorbed within it. This substratum or "background" cannot be observed because it is invariant and outside of space and time, which originate from it: a sort of "motionless motor".

The transactional ring may be described by circular inference rules which establish a self-consistency rather than, as in traditional logic, by linear inference rules which establish a deduction. A circular logic, not a linear one, is applied to a ring; it appears therefore that individual quantum processes are self-generated in a non-causal manner through an extra-spacetime mechanism.

Self-generation implies acausality. At the same time, however, the creation/destruction events that occur at $t = t_1$, $t = t_2$ take place because of interactions with other rings. This implies the existence of rules on how the rings connect and therefore the acausality does not turn into complete arbitrariness. This appears to be a natural and entirely convincing explanation of the simultaneous presence of causality and acausality in quantum processes.

(4) The energy is propagated only in one time direction, and the causal effects thus proceed from the past towards the future.

To sum up, the two extreme events of a transaction correspond to two reductions of the two state vectors which describe the evolution of the quantum process in the two directions of time. They constitute the "**R** processes" (**R** stands for *reduction*) of the Penrose terminology [21, 22] and, from our perspective, are the only real physical processes. They are constituted of interaction vertices in which real quantum states (e.g. elementary particles) are created or destroyed; these interactions are not necessarily acts of preparation or detection of a quantum state in a measurement process.

The evolution of probability amplitudes in the two directions of time constitutes, in Penrose terminology [21, 22], a **U** process (where **U** stands for *unitary*). From our ontological viewpoint, **U** processes are not real processes: both the amplitudes and the time evolution operators which act on them are mathematical inventions whose sole purpose is to describe the causal connection between the extreme events of the transaction, i.e. between **R** processes. This connection is possible because the two events derive from the transformation of the same a-spatial and a-temporal "background". As a specific consequence of this assumption, all virtual processes contained in the expansion of the time evolution operator are deprived of physical reality.

According to this approach, therefore, the history of the Universe, considered at the basic level, is given neither by the application of forward causal laws at initial conditions nor by the application of backward causal laws at final conditions. Instead, it is assigned as a whole as a complete network of past, present, and future **R** processes. Causal laws are only rules of coherence which must be verified by the network and are *per se* indifferent to the direction of time; this is the so called *block universe* in Putnam and Rietdijk's conception [23]. From this point of view, the "spatiotemporal handshake", introduced by Cramer and made famous by Kastner [15], resembles the mutual creation of two hands represented in a famous drawing of M. Escher.

A transaction that begins with the creation of quality q and ends with the destruction of quality r can be represented simply by the form $(r \mid q)$. We remark that a symbol as, e.g. $|q)$ represents an operator belonging to a suitable algebraic structure and should be not confused with the associated ket $|q\rangle$ which is instead an amplitude.

As an example of a transactional network, let us consider a process constituted by the decay of a microsystem, prepared in the initial state 1, into two microsystems 2, 3 which are subsequently detected. The preparation consists of the destruction of quality 1 [which we shall indicate with $(1|$] which closes the transaction which precedes it, and of the creation of quality 1 [which we shall indicate with $|1)$] which opens a new transaction. It will be represented by the form $|1)(1|$. The decay consists of the destructions of qualities 2, 3 which close the transaction that began with the preparation, and of the creations of qualities 2, 3 which open a new transaction which will be closed with the detection of microsystems 2, 3. It will be represented by the form $[|2)|3)]\ [(2|(3|]$.

The detection of microsystems 2, 3 will be constituted by the destructions of qualities 2, 3 which close the transaction that began with the decay, and by the creations of qualities 2, 3 which open subsequent transactions. It will be represented by the interaction events $|2)(2|, |3)(3|$.

The double transaction described here corresponds to the process usually associated with the probability amplitude $\langle 2, 3|S|1\rangle$.

Another example is Young's classical double slit experiment. The preparation of the initial state of the particle can be represented by the form $|1)(1|$, following the same reasoning as in the previous case. Instead of the decay, here we have the encounter with slits 2 and 3, i.e. the interaction between a single particle and a double slit screen represented by $[|2)|3)]\ [(2|(3|]$. Instead of the detection of the two particles created in the decay, here we have the sole event of the detection of the particle on the second screen at a certain position 4, i.e. : $|4)(4|$. Two transactions are involved: the first starts with the preparation of the particle and ends with its interaction with the first screen; the second begins with this second interaction and ends with the interaction of the particle with the second screen. The latter interaction then constitutes the beginning of the following transaction. The process is that which corresponds to the probability amplitude $\langle 4|S|1\rangle$.

We note that the forward time evolution of the amplitudes, represented with $S|1\rangle$, contains both the kets $|2\rangle$, $|3\rangle$; nevertheless, processes relating to the passage through the individual slit a (where $a = 2, 3$) do not exist. Such processes would require an intermediate event represented by $|a)(a|$, which effectively does not take place. It is in this sense that processes that can be associated with compound probability amplitudes $\langle 4|S|a\rangle\langle a|S|1\rangle$ are "virtual" and not real. The process of the crossing of one of the two slits becomes real when the other slit is closed.

The relevance of the concept of transaction in relation to the problems highlighted in Section 3 is related to the following fact. If we read the transactional ring (24) only in the forward direction, the event at $t = t_2$ is "read" as a discontinuous jump $|Q'\rangle \to |R\rangle$. This is precisely the experimentally observed QJ. However, the actual physical process that occurs at $t = t_2$ consists of the action of the projector $|R\rangle \langle R|$. This projector can be seen as the emission of the "state" R towards the future, with the concomitant emission of its conjugate towards the past [or, alternatively, the absorption of R coming from the past as a component of Q']. In other words, when the transition between two stationary states occurs, the single experimental run is not looking at the forward amplitude given by the time-dependent superposition of initial and final states, nor the backward conjugate amplitude. The single experimental run *only sees the final state and its conjugate*. Since the wave function conjugation leaves the quantum potential and the Bell length unchanged, the problem identified in Section 3 is completely resolved. The perplexity of Rutherford finds its solution in that the event takes place in an a-temporal domain, where no discontinuity of the dynamics occurs. Such discontinuity is the consequence of the fact that the observer records the facts according to his own particular time in sequential order. In relation to this chronology, the statistics of QJs is instead properly expressed by the time evolution of the forward wave function or that of its conjugate.

The presence of a timeless and spaceless background connected to the spacetime domain through an dynamical causation process may disturb some readers. In order to quiet, at least in part, this anxiety in the next section, we provide a preliminary model of this level of physical reality, trying to show how it reconnects to the level immediately above: that of elementary particles.

5. A Background Model

To illustrate better the concept of background, in this section we present a "toy model" for the annihilation/creation of the wave function of a single elementary particle [10, 16]. In quantum formalism, the quantum leap that brings the state Ψ of a particle in the state Ψ' is represented by the action of the projector $|\Psi'\rangle\langle\Psi'|$ on the ket $|\Psi\rangle$, which leads to a new ket proportional to $|\Psi'\rangle$. The projector $|\Psi'\rangle\langle\Psi'|$ may in turn be considered as the entry (represented by $\langle\Psi'|$) of the component Ψ' of Ψ in an a-temporal and a-spatial background condition, followed by the exit (represented by $|\Psi'\rangle$) of the new state Ψ' from this same condition. In this sense, we have the annihilation ($\langle\Psi'|$) of the state Ψ' followed by the creation ($|\Psi'\rangle$) of a new state Ψ'.

We assume that in the background condition, the information associated with the state Ψ' is encoded in a kind of "internal wave function" inaccessible by direct observation. Firstly, we postulate the existence of an "internal time" variable of the background which will be denoted by τ'. The internal wave function associated with the particle will contain a factor $\Phi(\tau')$, real and harmonic in τ', null at the boundary and outside the interval $[-\theta_0/2, +\theta_0/2]$. The reality condition must be satisfied for the absence of a temporal direction, which implies $\Phi = \Phi^*$. The following equation is a consequence of this postulate:

$$\begin{cases} -\hbar^2 \dfrac{\partial^2}{[\partial(2\pi\tau')]^2}\Phi = (M_{sk}c^2)^2\Phi & \text{for } \tau' \in [-\theta_0/2, +\theta_0/2], \\ \Phi = 0 & \text{otherwise,} \end{cases} \tag{25}$$

where M_{sk} is the original (skeleton) component of the particle mass. From (25), it follows that:

$$M_{sk}c^2 = n'\frac{\hbar}{\theta_0}, \tag{26}$$

where $n' = 0, 1/2, 1, 3/2, \ldots$ is an integer for odd solutions, a half-integer for even solutions.

A second postulate states that each oscillating solution of (25) is globally characterized by a second variable T (which is dimensionally a temperature) such that the not normalized probability of a given value of that variable is expressed by

$$\begin{cases} \exp[-(\hbar/\theta_0)/kT] & \text{for } kT \geq \hbar/\theta_0, \\ 0 & \text{otherwise,} \end{cases} \tag{27}$$

where k is the Boltzmann constant and is $T \geq 0$. Assuming $\tau'' = \hbar/kT$, this probability becomes the square modulus of a factor:

$$\Lambda(\tau'') = \exp[-(\hbar/\theta_0)/2kT] = \exp(-\tau''/2\theta_0). \tag{28}$$

If we assume that while in the background condition the particle is in a state of superposition of different values of τ'', we have:

$$\begin{cases} -i\hbar\dfrac{\partial}{[\partial(i\tau'')]}\Lambda = \dfrac{\hbar}{2\theta_0}\Lambda & \text{for } 0 \leq \tau'' \leq \theta_0, \\ \Lambda = 0 & \text{otherwise.} \end{cases} \tag{29}$$

The creation of the wave function associated with the particle state Ψ' exiting from the background can be considered as the passage from Eq. (27) to a probability identically equal to 1 for any value t of the "external" time accessible to the observer. This probability will be the square modulus of a factor that we can write as:

$$\Lambda = \exp\left(-\frac{i\tilde{\tau}}{2\theta_0}\right). \tag{30}$$

We can imagine this factor to derive from Eq. (28) by the Wick rotation $\tau'' \to i\tilde{\tau}$; simultaneously to this transformation, the factor $\Phi(\tau')$ must disappear.

The third postulate states that the quantity $\tilde{\tau}/2\theta_0$ is merely the external time t measured in units of the oscillation period of the old factor Φ, i.e. θ_0/n'. Therefore:

$$\frac{\tilde{\tau}}{2\theta_0} = \pm\frac{t}{\left(\frac{\theta_0}{n'}\right)} \Rightarrow \tilde{\tau} = \pm 2n't. \tag{31}$$

And thus:

$$\Lambda \to \exp\left(\pm\frac{-in't}{\theta_0}\right) = \exp\left(\mp iM_{sk}c^2\frac{t}{\hbar}\right) \tag{32}$$

i.e. the de Broglie's temporal phase factor. Annihilation is represented by the inverse process.

It is immediately clear that (25) and (29) are two wave equations, respectively, in the real component τ' and the imaginary component $i\tau''$ of a complex internal time $\tau = \tau' + i\tau''$, which thus becomes the "precursor" of time t measured in the laboratory; this last physical quantity is thus emerging. The frequency of the "hidden" oscillation Φ becomes the frequency of the de Broglie's phase factor, experimentally accessible as the particle mass.

Clearly, the entire model dissolves in the limit $\theta_0 \to 0$. The interval θ_0 must be a time scale characterizing the realm of elementary particles, and this leads to the conclusion that it is $c\theta_0 \approx 10^{-13}$ cm, and therefore $\hbar/\theta_0 \approx 50$–$100$ MeV. Consideration should be given to the fact that this time scale is not a minimum time interval, nor does $c\theta_0$ define a minimum spatial interval.

A unitary time evolution process connecting the initial state $\Psi_{\rm in}$ to the final (post-collapse) state $\Psi_{\rm fin}$ through the transition amplitude $\langle\Psi_{\rm in}|S|\Psi_{\rm fin}\rangle$ really connects two quantum leaps (clicks) $|\Psi_{\rm in}\rangle\langle\Psi_{\rm in}|$ and $|\Psi_{\rm fin}\rangle\langle\Psi_{\rm fin}|$. We remind to [10] for details on this connection and the derivation of the Born rule. We also note that the "collapse" $(|\Psi'\rangle\langle\Psi'|)\ |\Psi\rangle = \langle\Psi'|\Psi\rangle|\Psi'\rangle$ is generally an a-spatial process because Ψ and Ψ' are generally non-position eigenstates. In the case where Ψ and Ψ' are represented on a same finite dimensional vector basis, it is also possible to retrieve the Born rule through a geometric collapse model, based on a generalization of the Bloch sphere [1, 2]. This generalization is also incompatible with spatial representation.

6. Conclusions

Since its inception, QM was characterized by the "debate on the fundamentals" that seems now become a separate discipline. In fact, the demand for a different understanding comes from quantum gravity and the continuous development of the field of quantum information. While the Copenhagen interpretation seems increasingly unsatisfactory, the hypothesis of a subquantum "mechanical" medium is completely inadequate to face the problems posed by particle physics [4]. A reconciliation between foundational aspects and current physics research requires novel approaches that currently lack. However, some proposals that express the need for such an integration made recently their appearance, such as that of G. 't Hooft which postulates QM emergence from a primitive quantization on the Planck scale [26].

During a debate on the BBC with M. Pryce in 1962, Bohm offered an interesting reflection:

> *We wondered what actually an electron does. What would it do while it is passing from the source to the slit? That's the point. Well, I could propose, for example, that the electron is not a particle in the sense it is currently meant, but an event. I assume such event happens in a generic medium — a "field" — we can suppose in this field there's an impulse, a wave moving forward and converges in a point so producing a very strong impulse and then diverges and scatters away. Let's imagine these impulses in a series all reaching a line there producing a series of intense pulse The impulses will be so close one to the other that they will look like a particle. In most cases, all that will behave just like a particle and it will behave differently when goes through the two slits, because each impulse will come out according to the way the incident wave passes the slits. The result is that we are looking at something it's neither a wave nor a particle. If you wonder how the electron has actually passed the slit and if it has really passed one slit or the other, I would reply that probably is not that kind of thing which can pass a slit or the other one. Actually, it is something which forms and dissolves continuously and that can be the way it really acts* [7].

This testifies, against the readings of some of his followers, that Bohm was ready to give up radically the classical image of the world and in particular the pitfalls contained in the concept of "particle"! We have done it, proposing QJs emerging from a background that is a system with internal vibratory degrees of freedom, in the spirit of Garrett–Lisi universal action reservoir [13], but also the latest Bohm work on implicate/explicate order [6]. We also sought to clarify the link between QJs, self-conjugate projection operators and transactions, the context within which it is possible to deduce the QM formalism and so bring the quantum potential to the statistics of dialogue with the background [9]. It is clear that the price to pay is the resignation of the classic styling of the Bohm trajectories, whose ontological status is definitely closer to that of Feynman paths [8, 19]. In other words, faced with the three Bell incompatible options (completeness, realism, locality) [5] we have chosen to give up the concept of the semi-classical particle as a persistent object, describing it rather as an aspect of the dialectic between the background and events, which yields the appearance of locality as its particular emerging manifestation. Non-locality is then reconciled with relativity; it is no longer an unexpected visitor but the expression of the fundamental condition of *informed wholeness* within which, as instantiations, the *events* that form the substance of the physical world occur.

References

[1] Aerts, D. and Sassoli de Bianchi, M. (2014). The extended Bloch representation of quantum mechanics and the hidden-measurement solution to the measurement problem. *Ann Phys.*, 351: 975–1025.

[2] Aerts, D. and Sassoli de Bianchi, M. (2015). Do spins have directions? *Soft Computing*, pp. 1–22.

[3] Aharonov, Y., Albert, D. Z. and Vaidman, L. (1988). How the result of a measurement of a component of the spin of a spin-12 particle can turn out to be 100. *Physics Review Letters*, **60**: 1351.

[4] Bacciagaluppi, G. and Valentini, A. (2009). *Quantum theory at the crossroads: Reconsidering the 1927 Solvay Conference*, Cambridge Univ. Press.

[5] Bell, J. S. (1988). *Speakable and Unspeakable in Quantum Mechanics.* Cambridge University Press.

[6] Bohm, D. (1980). *Wholeness and the Implicate Order*, Routledge.

[7] Bohm-Pryce debate (1962). In *Quanta and Reality. A Symposium*, Toulmin, S. (ed.), Hutchinson & Co.

[8] Boscà, M. C. (2013). Some observations upon "Realistic" trajectories in Bohmian quantum mechanics. *Theoria*, **28**(1): 45–60.

[9] Chiatti, L. (2013). The transaction as a quantum concept. *IJRAS*, **16**(4), 28–47. arXiv:1204.6636 [gen-ph].

[10] Chiatti, L. and Licata, I. (2014). Relativity with respect to measurement: collapse and quantum events from fock to cramer. *Systems*, **2**(4): 576–589.

[11] Cramer, J. G. (1980). Generalized absorber theory and the Einstein–Podolsky–Rosen paradox. *Physics Review D*, **22**(2): 362–376.

[12] Cramer, J. G. (1986). The transactional interpretation of quantum mechanics. *Reviews of Modern Physics*, **58**: 647–687.

[13] Garrett-Lisi, A. (2006). Quantum mechanics from a universal action reservoir, arXiv:physics/0605068.

[14] Itano, W. M., Bergquist, J. C. and Wineland, D. J. (2015). Early observations of macroscopic quantum jumps in single atoms. *International Journal of Mass Spectrometry*, **377**: 403–409.

[15] Kastner, R. E. (2013). *The Transactional Interpretation of Quantum Mechanics: The Reality of Possibility*. Cambridge: Cambridge University Press.

[16] Licata, I. and Chiatti, L. (2015). Timeless approach to quantum jumps. *Quanta*, **4**: 10–26.

[17] Licata, I. and Fiscaletti, D. (2014). *Quantum Potential. Physics, Geometry and Algebra*. Springer.

[18] Licata, I. and Fiscaletti, D. (2014). Bell length as mutual information in quantum interference. *Axioms*, **3**: 153–165.

[19] Licata, I. and Fiscaletti, D. (2014). Bohm trajectories and Feynman paths in light of quantum entropy. *Acta Phys. Pol B*, **45**(4): 885–904.

[20] Licata, I. and Fiscaletti, D. (2015). Bell length in the entanglement geometry. *Int. J. Theor. Phys.*, **54**(7): 2362–2381.

[21] Penrose, R. (1989). The Emperor's New Mind, Oxford University Press.

[22] Penrose, R. (2004). The Road to Reality, Oxford.

[23] Petkov, V. (2006). Is there an alternative to the block Universe view? in *The Ontology of Spacetime Vol.* 1 (*Philosophy and Foundations of Physics*) D. Dieks (ed.), Elsevier, pp. 207–228.

[24] Salesi, G. (1996). Spin and Madelung fluid. *Mod. Phys. Lett.*, **A11**: 1815.

[25] Svensson, B. E. Y. (2013). What is a quantum-mechanical "weak value" the value of? *Foundational Physics*, **43**(10): 1193–1205.

[26] 't Hooft, G. (2015). The cellular automaton interpretation of quantum mechanics, arXiv:1405.1548 [quant-ph].

[27] Tsekov, R. (2011). Quantum diffusion. *Phys. Scr.*, **83**: 035004.

Chapter 9

Minkowski Spacetime and QED from Ontology of Time

C. Baumgarten

5244 Birrhard, Switzerland
christian-baumgarten@gmx.net

Classical mechanics, relativity, electrodynamics, and quantum mechanics are often depicted as separate realms of physics, each with its own formalism and notion. This remains unsatisfactory with respect to the unity of nature and to the necessary number of postulates. We uncover an intrinsic connection of these areas of physics and describe them using a common symplectic Hamiltonian formalism. Our approach is based on a proper distinction between variables and constants, i.e. on a basic but rigorous ontology of time. We link these concepts with the obvious conditions for the possibility of measurements. The derived consequences put the measurement problem of quantum mechanics and the Copenhagen interpretation of the quantum mechanical wave function into perspective. According to our (onto-) logic, we find that spacetime cannot be fundamental. We argue that a geometric interpretation of symplectic dynamics emerges from the isomorphism between the corresponding Lie algebra and the representation of a Clifford algebra. Within this conceptional framework, we derive the dimensionality of spacetime, the form of Lorentz transformations and of the Lorentz force and fundamental laws of physics as the Planck–Einstein relation, the Maxwell equations, and finally the Dirac equation.

Keywords: Philosophy of Physics; Lorentz Transformation; Electrodynamics; Dirac Equation; Hamiltonian Mechanics; Coupled Oscillators.

1. Introduction

1.1. Spacetime versus proper time

Schrödinger once wrote that "In Einstein's theory of gravitation matter and its dynamical interaction are based on the notion of an intrinsic geometric structure of the spacetime continuum" [83]. What we will discuss in this chapter suggests to conjecture the reverse statement, i.e. that the intrinsic geometric structure of spacetime is based on the very notion of matter and its dynamical interaction. The idea that spacetime is not fundamental but emergent has been proposed in the past by several authors [2, 45, 53, 55, 66, 75–77, 87, 91, 99]. Some discussed the relation between spacetime and quantum communication [46]. Our conjecture results from a different, almost classical, notion of quantum mechanics, closely connected to the phase space picture of classical statistical mechanics. A significant number of publications support our directions of thought [3, 13–18, 34, 37, 80, 90, 94].

We shall start with the distinction of variables and constants, i.e. from an (onto-)logic of time. Consider the basic quantum-mechanical relationship

$$i\hbar\,\partial_t\psi = E\,\psi\,. \tag{1}$$

The left side is the rate of change of a wave function ψ and the equation expresses that this rate of change is equal to the energy of the system. "Energy" is probably the most fundamental concept in physics. The conservation of energy has no serious exception and physics assigns to the energy the role of substance. Any entity that falls under the notion of "object" is therefore "charged" with a certain amount of energy and is therefore subject to change with a frequency $\omega = E/\hbar$. The rate of change is what quantifies the "passage of time". This is the meaning of saying that time and energy are conjugate quantities. The passage of time is measured by clocks, i.e. by the rate of change of a reference device. And any system that can be described by Eq. (1) is a clock in itself. Metaphorically we say it exists *in time*.

Seen by light, Eq. (1) is nothing but the equation of motion (EQOM) of an harmonic oscillator. If we write the real and imaginary part of the so-called "wave function" separately $\psi = X + i\,Y$, then we obtain with $E/\hbar = \omega$:

$$\begin{aligned} \dot{\psi} &= \dot{X} + i\,\dot{Y} = -i\omega(X + iY), \\ \dot{X} &= \omega\,Y, \\ \dot{Y} &= -\omega\,X, \end{aligned} \tag{2}$$

where the dot indicates the temporal derivative. In matrix form, this reads[a]:

$$\begin{pmatrix} \dot{X} \\ \dot{Y} \end{pmatrix} = \omega \begin{pmatrix} 0 & 1 \\ -1 & 0 \end{pmatrix} \begin{pmatrix} X \\ Y \end{pmatrix}. \tag{3}$$

The interpretation of the use of the unit imaginary in Eq. (1) seems to be the litmus test of our attitude towards quantum mechanics. It is as often presented as a necessary ingredient as its necessity is strictly denied. We believe that the unit imaginary is nothing mysterious or magical that distinguishes quantum from classical mechanics. It is just a compact form of writing Hamilton's EQOM of a classical harmonic oscillator (CHO). However, Eq. (3) is the *normal form* of an algebraically more general equation and it is in this respect an unmotivated limitation of the EQOM — as we are going to show in this chapter.

The difference between the two interpretations is in some correspondence with two conflicting attitudes towards the wave function. Some scientist believe it is a mere mathematical tool while others tend to interpret its components as the true dynamical variables. We will argue that both attitudes miss the point. It is the *pure form* of physical theories — i.e. the EQOM — that requires the definition of some fundamental variables. And it is a *deformation* to believe that we can directly assign physical meaning to these variables. That is the "classical" form of physical theories is — though mathematically sound — conceptually inconsistent or at least incomplete. Let us briefly explain why this is so.

[a]The equality of Eqs. (1) and (3) is known for long (see for instance [80]), but the way of understanding and teaching quantum mechanics has not changed.

Equation (1) tells us that the wave function of a system with energy *changes* at every time. Instead of *postulating* the validity of Eq. (1), we again reverse the argument: We postulate that *change* is immanent to physical reality [78]. The permanent change is the physical mode of existence. Again metaphorically we say that all objects of physics *exist in time*. Any fundamental physical model of reality has to represent this continuous change — because it is the fundamental form of a physical description of reality. This is the essence of Eq. (1): Variables ψ that represent *existing* fundamental quantities have to vary continuously. We show in the following that Eq. (1) can be derived from this postulate. But before doing so, let us briefly describe why classical physics is conceptionally incomplete. The most trivial flaw is the inability of classical physics to give an account of its measurement standards. The most trivial being the length. Classically, we take the existence of a solid rod for granted. Einstein made an enlightening remark: "One should always be aware that the presupposition of the existence in principle of rigid rods is a presupposition suggested by approximate experience but is, in principle, arbitrary"[33]. Why is this so? Because classical physics taken as classical mechanics, electrodynamics, and relativity cannot explain the existence of a finite and fixed atomic radius and hence cannot explain the existence of a measuring rod. Bohr's orbital theory tried to establish fixed radii by an *ad-hoc* postulate — but the idea of definite orbitals conflicts with electrodynamics. We will not derive a finite radius in the following, instead we will make the presumption explicit and incorporate it as a general principle in what follows.

On the fundamental level existence implies continuous change. But though all fundamental quantities continuously change, physics as an experimental science requires the constancy of measurement standards. Without constant reference standards, we could not test physical models. These two apparently trivial facts are the starting point of the ontology of time. Our line of reasoning is in contrast to concepts that postulate spacetime to be fundamental. Equation (1) does not require a concept of spacetime. If, in the following, we speak of time then it is always understood as a rate of change.[b] However, we shall argue in this essay that and how the conjugate concept of spacetime — namely the energy–momentum relationship — and Maxwell equations emerge in an algebraically simple and elegant way from a reinterpretation of Eq. (1), i.e. from a proper distinction between those quantities that change and those that stay constant.

1.2. The form of physics

Usually textbooks on (classical) physics begin with the EQOM of point-masses, the definition of positions, velocities, and accelerations, etc. This means that the fundamental variables of the theory are usually directly charged with physical meaning, a meaning that is supposed to have its origin in our "classical" macroscopic experience. However, this "method" implies that we have to presuppose a considerable number of mechanical concepts. A proper definition of these concepts however relies on and refers to an elaborated theory. Like Baron Munchhausen, who claimed to have pulled himself by the hair out of the swamp, the theory is based upon notions that become meaningful only in the context of the spelled-out physical theory [26]. It is the fate of human existence to be thrown into an already existing world and it seems to be the fate of physics students to be thrown right into a framework of concepts without the chance for a stepwise systematic and logical (*re-*)*construction.* The classical theory implicitly claims that these concepts are in some way *ad-hoc* derivatives from everyday experience — though seen by light Hamiltonian and Lagrangian mechanics can rarely be applied directly to everyday empirical evidence.

We invite the reader to follow a different path and to put on hold with the conceptions of mass, point particles, wave functions, and also spacetime. A renewed analysis of spacetime is required as soon as we understand that the rod of constant length — and with it spacetime — is a mere postulate. But if quantum mechanics

[b]In accelerator physics, this is called *phase advance.*

is required to explain the existence of rods of constant length, then the same is true for the concept of spacetime: It requires an explanation — in the optimal case in form of a derivation.[c]

Physics is an "empirical" science, based on objective measurements. A measurement is the comparison of *variable* properties of objects (measurands) with the corresponding constant properties of reference objects. Before we are able to measure distances, we need to have a ruler, for instance a solid rod. Certainly we can *think* about (the concept of) length without having a ruler, i.e. we can develop geometry. But we cannot perform a measurement nor predict a measurement outcome without the ruler. The material *object* that is used as a ruler (the rod) is the fundament of *objectivity* of measurement. Only with the possibility in principle of the existence of reference objects like the prototype meter, it becomes physically meaningful to speak about the length of an object or the distance between objects, i.e. space. In practice, we say almost unreflectively that the length $\mathbf{L}$ is $\mathbf{x}$ inch, meter, or lightyears. However, the *production* of a ruler is the first thing we have to do before we can measure length. *It is not sufficient to define* it — it has to be physically made. It is a necessary though not a sufficient condition for a measurement. The *handbook of metrology and testing* describes a measurement as follows: "Measurement begins with the definition of the measurand, the quantity intended to be measured. The specification of a measurand requires knowledge of the kind of quantity and a description of the object carrying the quantity. When the measurand is defined, it must be related to a measurement standard, the realization of the definition of the quantity to be

[c]There are more good physical reasons to reconsider the concept of spacetime. The general theory of relativity (GTR) claims that the geometry of spacetime is determined by the distribution of the masses "in it". If this is true, then matter and spacetime are not separable entities. Secondly, it is a well-established part of quantum mechanics that the wave function can either be defined as a function of space and time or equivalently as a function of energy and momentum. Both representations of the wave function are related to each other via a Fourier transformation. If the most fundamental theories suggest that spacetime cannot be consistently understood independent from matter and its dynamics, then we should not ignore this fact.

measured." [23]. Theory should follow practice: before we postulate spacetime, we should explain how a measurement of distances is at all possible.

This chapter is not about metrology. But what is important about the measurement standard is that it must be realized and *may not change with time*. The measurement of a property of a physical entity requires the existence of material entities that stay constant (the rulers) and other entities — *of the same type of properties* — (the measurands) that vary or at least *can* vary. If the measurand changes with time, it is the purpose of a physical model to *predict* the time-dependence of the measurand. When we say that the ruler must not change with time, then we include other invariances as well. Wigner pointed this out by referring to *invariance* as a fundamental principle [100]. Hence, we can say that physics is irresolvably committed to time, i.e. to continuous change *and* to constancy. This is not postulate, it is a *conditio sine qua non.*[d]

The unity of nature is next major premise: To be part of the physical world implies the possibility in principle of (direct or indirect) interaction with all other things that are part of this world and therefore we (have to) presume the existence of a common denominator, a *fundamental level.*[e] Physical models are based on a quantitative description of reality. Whatever a physical theory considers to be fundamental must essentially be representable by quantities. We call these quantities "variables". The mathematical model of these fundamental entities is hence based on a list of variables ψ of (yet) unknown dimension, which all continuously change at all times. But if the components of ψ change at all times and if they are fundamental, then there is no way to define a measurement standard for them. Nevertheless, there might be functions of the fundamental variables that are constant in time: such

[d]Sir Hamilton was not only aware of the intimate connection between physics and time — he even had the vision to develop algebra as the science of pure time [42] and it was his deep belief that "that the intuition of time is more deep-seated in the human mind than the intuition of space" [71].

[e]The basis of this believe is much the same as most theorists believe that it should finally be possible to describe all forces of nature within a grand unified theory.

constants are called *constants of motion* (COMs). These functions may not include linear terms — otherwise it is easy to show that the functions cannot be constants.[f] Hence, the dimension (unit) of these functions is different from the dimension (unit) of the fundamental variables. And therefore they cannot serve as rulers for the considered fundamental variables. The logical consequence is that a direct measurement of the fundamental variables of ψ is not possible. The above considerations explain why the components of the quantum mechanical wave function have to vary at all times, why the quantum mechanical wave function implies an interpretational problem and why QM has to postulate that measurement devices must be "classical" or "macroscopic": Due to the absence of constant rulers, the fundamental variables (the components of the wave function) cannot be directly measured.[g]

The definition of general and abstract quantities like the fundamental variables and their evolutions in time will not suffice to make up a physical theory. At some stage, we need an interpretation that maps quantities of the model to measurable quantities in the world in order to obtain physical meaning. The physical meaning can then only be induced from the mathematical structure of the physical quantity. The interpretation can only refer to already known physical laws in order to identify possible interpretations of the quantities and the relations between them. We can only recognize fundamental physical meaning of certain algebraic relationships because an elaborated theory of elementary particles that has been shown to be experimentally successful in the description of fundamental physical phenomena is already available.

1.3. Outline

The chapter is organized as follows: In Section 2, we describe our understanding of fundamental physical entities. In Section 3, we

---

[f] Dragt *et al.* have shown in [29] that there are no nonzero first order moments in linear Hamiltonian systems.

[g] Or with the words of S. Fortin and O. Lombardi: "The difficulties can be overcome once it is recognized that *classicality is a property of the observables*" [84].

derive Hamilton's EQOM and the basic properties of the symplectic unit matrix. In Section 4, constants of motion are introduced with the help of Lax Pairs. We review important algebraic relations that indicate the construction principles of spacetime: The required congruence of the algebraic structure of (skew-) Hamiltonian matrices with the basic elements (i.e. generators) of Clifford algebras. We describe the basic measurable entities in a world based on time: (Second) moments of fundamental variables. Since second moments can be represented by expectation values of matrix operators, the relations between these matrices are the relations between the observables (i.e. the laws of physics).

In Section 5, we derive the conditions for the emergence of a geometric space from symplectic dynamics. Observables are expectation values of matrix operators and hence spacetime geometry should be representable by a system of basic matrices with certain properties. Such matrix systems indeed exist and are known as (representations of) Clifford algebras.

In Section 6, we describe symplectic transformations as structure preserving transformations. These transformations are isomorphic to Lorentz transformations and are the basis for the apparent geometry of spacetime. We explore the meaning of structure defining transformations.

In Section 7, we give a short overview over the basic properties of Clifford algebras in general and especially of $Cl_{N-1,1}(\mathbb{R})$. We derive conditions that limit the possible dimensionality for an emergent spacetime. Further analysis of the properties of these Clifford algebras allows to restrict spacetime to 3+1 dimensions. Hence, the appropriate algebra is the algebra of the real Dirac matrices.

In Section 8, we describe Lorentz transformations — boosts and rotations — as structure preserving (symplectic) transformations. We present an interpretation of the Dirac matrix system called the "electromechanical equivalence" (EMEQ). Guided by the EMEQ, we derive the Lorentz force and in Section 9, we derive (Quantum-) electrodynamics. We give arguments why momentum and energy should be related to spatial and temporal derivatives and sketch the path towards the Dirac equation, a representation of massive spinors and then describe the significance of CPT-transformations in our approach.

Section 10 finalizes the discussion and a summary is given in Section 11.

2. Obligations of a Fundamental Theory

The basic variables of a fundamental theory are continuously varying quantities[h] — they are dynamical variables. Since quantum theory postulates that the wave function is fundamental, it is nearby to assume that we finally have to identify the components of the wave function with (parts of) our list of "fundamental variable(s)". The Copenhagen interpretation avoids the reference problem by assigning a mere probabilistic meaning to the wave function, i.e. by postulating that (a) the wave function has only a probabilistic meaning and (b) the wave function gives a complete description of reality, i.e. is fundamental. Most discussions of the Copenhagen interpretation focus on the question, if the probability interpretation is correct, or whether the probability is "classical" or "quantum" in nature and what the ontological status of the wave function is. We do not address these questions. Instead our claim is, that *if* the wave function is truly fundamental, then a meaningful *physical* interpretation of the wave function is impossible. This does not imply that it is *less real* than the macroscopic quantities derived from it. However, a meaningful answer to the question *what* varies requires the existence of some constant entity for comparison, i.e. a reference for a measurement. If we say that an entity represented by a variable in our equation is a length, then we need to have, in principle, a unit (for instance the meter) to refer to and we need to have an idea of spacetime. Not only that this reference has to be of the same type (i.e. dimension), but it also has to be macroscopic as "meaning" can only be derived from

[h]The physical constants ($\varepsilon_0, \mu_0, c, \hbar \dots$) as listed in handbooks are not *material* but rather *theoretical* objects. Some of these are merely conversion factors for units. If we refer to "constants", then we address properties of material objects that are potentially useful as measurement standards. For instance, the orbital radii of hydrogen atoms.

everyday (macroscopic) experience. Physical meaning and objectivity depend on a constant quantity of the same dimension that can be used as a reference. We have an idea and a concept of length as we are surrounded by objects with this property. The lack of reference is according to our premises a logical consequence of fundamentality. With respect to the fundamental variables, there is no meaningful answer to the question *what* varies, but only *how* they vary.[i]

This is a logical limit in physics: A fundamental physical theory can describe how entities change but it cannot tell what the entities are that are changing. Hence, the only possible *physically real* rulers, the only possible measurement standards are constants of motion: the prototype meter is an object composed of atoms and molecules which consist of "elementary particles". The motion of these components is stable and the length of the reference object "meter" is a constant of motion of these particles. But at the level of these particles — or at the level of their constituents — at some level we face the pure variation of abstract fundamental variables. We conclude that if it is possible to directly measure (the value of) a variable, it cannot be fundamental.

There is no doubt that we can measure distances and angles *since we have rods and goniometers*. Hence, space (or spacetime, respectively) is not fundamental. Spacetime is a construction for our theoretical needs — it is not itself a "thing" but a formalism to express the relations between things. This is not to say that spacetime does not "exist". If things exist and relations between things exist, then we can speak of the space that exists. However, it is *not fundamental*, which basically means that it can be *derived*. To say this implies the obligation to derive the dimensionality of spacetime, the Lorentz transformations and the role played by electrodynamics, i.e. the speed of light. We shall argue that spacetime is based on the isomorphism of a Lie algebra with (reps of) real Clifford algebras $Cl(N-1,1)$. The "perception" of spacetime is a result of pattern recognition in loose analogy to the perception of

[i]In other words: "The wave function does not describe matter, it describes how matter behaves" [1].

acoustic signals in terms of sound and music.[j] Our ansatz is based on a list of fundamental and abstract variables, the values of which cannot be directly measured. Such a measurement would require constants of motion from "more" fundamental entities. But then the more fundamental variables could not be measured, etc. Hence, it is impossible to directly assign physical meaning to the fundamental dynamical variables. In this respect, fundamental variables are not identical to degrees of freedom DOFs in classical mechanics as DOFs are usually assumed to be measurable at least in principle. But in most other aspects, fundamental variables are similar to DOFs in classical mechanics and at first sight, there is no reason to refuse the possibility of a classical description of the dynamics of these variables.

E. T. Jaynes wrote that "Because of their empirical origins, QM and QED are not physical theories at all. In contrast, Newtonian celestial mechanics, Relativity, and Mendelian genetics are physical theories, because their mathematics was developed by reasoning out the consequences of clearly stated physical principles from which constraint the possibilities". And he continues "To this day we have no constraining principle from which one can deduce the mathematics of QM and QED; [...] In other words, the mathematical system of the present quantum theory is [...] unconstrained by any physical principle" [52]. We agree with this statement insofar as we think that it is important to constraint possibilities and to develop a theory according to well-defined principles. The introduction of the unit imaginary into quantum mechanics is not of this kind. No physical principle has ever been formulated that explains the necessity for the use of the unit imaginary in QM. There are just *ad-hoc* postulates. We believe that the use of the unit imaginary as it is usually done, is a mistake. It introduces a structure without clear and explicit motivation. Hence, we will avoid or at least delay the introduction of

[j]We should be aware that "physically" there are just wavelike density fluctuations in air. It is our mind that *processes* the perceived data and interprets them as sound, language or music. This interpretation is indeed adequate. Nevertheless, it remains an interpretation.

the unit imaginary up to the point, when its use becomes *reasonable* and legitimate. In the meantime, we restrict our considerations to the reals. Or more precisely: we demand that all fundamental variables are of the same type, as we consider it scientifically not legitimate to introduce an asymmetry like the asymmetry between the reals and the imaginary numbers *without reason.* Hence, all fundamental variables are either all real or all imaginary. The latter might be preferable to indicate their unmeasurability, however, as long as there is no need to introduce the distinction between real and imaginary variables, Ockham's razor demands to stay with reals.[k]

It follows from this conception that fundamental "material objects" like for instance electrons have to be described by their structure. Since the location in spacetime and hence a continuous trajectory is according to our principles not fundamental and may hence not serve to define or verify *sameness*, elementary particles have no material identity, but only a structural identity. The only structure at hand is the structure of the variations of the variables — the structure of their dynamics. The fundamental objects (elementary particles) are not directly represented by the fundamental variables, but rather by the dynamical structure of the variations, by the patterns of motion. Physics can be described as a method to analyze patterns of motion, it is to some degree identical to pattern recognition.[l] If "objects" do not change their structure (i.e. their "identity") in some interaction, then the involved dynamical processes must be structure preserving: If a material entity like an elementary particle is defined and understood by its structure, then its continuous existence requires — besides a continuous variation of its constituents (the variables) — that the dynamical structure must be preserved. However, if objects can only be identified by a certain dynamical structure then two objects with the same structure cannot

[k]For discussions about the use of real and/or complex numbers generally in physics and specifically in quantum mechanics, see also [22, 37, 47, 94].

[l]Recall that the birth of modern physics is connected with names like Kepler, who recognized the "true" pattern of planetary motion.

be distinguished. If we could distinguish particles experimentally that are indistinguishable in our physical model, then the model would be incomplete. Thus, if quantum mechanics is indeed fundamental, then it must logically include a concept of identical particles — as particles in a fundamental physical theory have to be represented by structures.

3. Variables, Constants and the Hamiltonian

3.1. Variables

We suggest the following principles for an ontologically proper basis of physics:

1. Existence happens in time. The "time" that is meant here does not have to be identical to the time that an observer would measure using a clock, but we insist that there is a bijective functional relationship.[m]
2. Existence in time is manifest by variation. All quantifiable properties of all fundamental physical entities continuously change and are representable by real quantities that continuously vary, i.e. by "variables".
3. Measurements require constant references (rulers), i.e. a physical model requires constants.

From these axioms, it follows that

1. There are no other (physical) constants than COMs.
2. The fundamental variables have no measurement standard, i.e. cannot be directly measured.

The value of a single variable varies. Hence, a single variable cannot generate COM. Therefore, we start with an arbitrary number $k > 1$ of dynamical variables.[n] According to the axioms, there exist one (or a set of) constant function(s) $\mathcal{H}(\psi_1, \psi_2, \ldots, \psi_k) = \text{const}$

[m]We explicitly include the possibility of regularizing transformations [12, 101].
[n]Ockham's razor commits us to determine the minimum number of variables required to generate objects with spacetime properties.

of the dynamical variable list ψ. Hence, we may write (with the presumption $\mathcal{H}$ = const we imply here $\frac{\partial \mathcal{H}}{\partial t} = 0$):

$$\frac{d\mathcal{H}}{dt} = \frac{\partial \mathcal{H}}{\partial \psi_1}\dot{\psi}_1 + \frac{\partial \mathcal{H}}{\partial \psi_2}\dot{\psi}_2 + \cdots + \frac{\partial \mathcal{H}}{\partial \psi_k}\dot{\psi}_k = 0, \tag{4}$$

or in vector notation:

$$\frac{d\mathcal{H}}{dt} = (\nabla_\psi \mathcal{H})^T \cdot \dot{\psi} = 0. \tag{5}$$

The (simplest) general solution is given by

$$\dot{\psi} = \mathbf{S}\,(\nabla_\psi \mathcal{H}), \tag{6}$$

where $\mathbf{S}$ is a $k \times k$ non-singular skew-symmetric real matrix, i.e. $\mathbf{S}^T = -\mathbf{S}$. The skew-symmetry of $\mathbf{S}$ is sufficient to solve Eq. (6) so that $\mathcal{H}$ is a constant of motion. According to linear algebra, there exists a non-singular matrix $\mathbf{Q}$ such that[o]:

$$\mathbf{Q}^T\,\mathbf{S}\,\mathbf{Q} = \mathrm{diag}(\lambda_0\,\eta_0, \lambda_1\,\eta_0, \lambda_2\,\eta_0, \ldots, 0, 0, 0), \tag{7}$$

where λ_k are real nonzero constants and

$$\eta_0 = \begin{pmatrix} 0 & 1 \\ -1 & 0 \end{pmatrix}, \tag{8}$$

is the basic form of the symplectic unit matrix. Since we have no reason to assume different λ_k, we assume that they are all equal to unity. According to the axioms, the matrix must have full rank $2\,n \le k$, i.e. all variables vary and hence there are no constants in the state vector such, i.e.

$$\mathbf{Q}^T\,\mathbf{S}\,\mathbf{Q} = \mathrm{diag}(\eta_0, \eta_0, \eta_0, \ldots, \eta_0) \equiv \gamma_0. \tag{9}$$

The transformation Eq. (9) is a linear change of variables which is used to find (or recognize or define) the "natural" or *normal*

[o]See for instance Ref. ([86]) and App. (E.4).

variables. We call such transformations *structure defining* (see below and App. (E)) and rewrite Eq. (5) accordingly

$$\begin{aligned}\frac{d\mathcal{H}}{dt} &= (\mathbf{Q}^{-1}\,\nabla\,\mathcal{H})^T\,\mathbf{Q}^T\,\mathbf{S}\,\mathbf{Q}(\mathbf{Q}^{-1}\,\nabla\,\mathcal{H}) = 0 \\ &= (\mathbf{Q}^{-1}\,\nabla\,\mathcal{H})^T\,\gamma_0\,(\mathbf{Q}^{-1}\,\nabla\,\mathcal{H}) = 0. \end{aligned} \tag{10}$$

In the following, we assume that the transformation into the normal variables has been done such that the normal variables are given by ψ and that $\mathbf{S}$ has between transformed into γ_0 so that

$$\dot{\psi} = \gamma_0\,\nabla_\psi\,\mathcal{H}. \tag{11}$$

The matrix γ_0 has the even dimension $2n \times 2n$: in any time-like physical world the dynamical variables come in pairs.[p] It is therefore sensible to refer to a pair of variables when we speak of a degree of freedom. We call $\psi_{2j} = q_j$ the j-th *canonical coordinate* and $\psi_{2j+1} = p_j$ the j-th *canonical momentum*, i.e. $\psi = (q_1, p_1, \dots, q_n, p_n)^T$, but this nomenclature is purely formal as long as the variables ψ are fundamental. It is just the structure of the matrix γ_0 that leads to this distinction. The matrix γ_0 is called the *symplectic unit matrix* and the EQOM have Hamilton's form:

$$\begin{aligned}\dot{q}_k &= \frac{\partial\mathcal{H}}{\partial p_k}, \\ \dot{p}_k &= -\frac{\partial\mathcal{H}}{\partial q_k}. \end{aligned} \tag{12}$$

Whenever we have constants (of motion) of the dynamical variables, we can derive Hamilton's EQOM in some way. We can interpret the abstract and intrinsically unmeasurable basic variable list ψ as "spinors", i.e. as "objects" in a phase space. Just formally, we

[p]The unit imaginary has indeed significance in quantum mechanics as it ensures an even number of fundamental variables and a continuous variation of these variables due to Eq. (1). However, our approach contradicts the frequently expressed opinion that QM inherently requires the use of complex numbers in the sense that it could not be formulated without the unit imaginary.

call the components "coordinates" and "momenta". But in fact every ψ is a point in an abstract $2n$-dimensional phase space.[q] However, the Hamiltonian formalism itself does not require that the variables are fundamental and therefore this formalism can be applied to any system with dynamical constants. The difference to the "classical" Hamiltonian formalism is solely that the classical state vector ψ describes the (average) properties of systems as the position of the center of mass or the average momentum and does not consist of fundamental variables. Thus, the components of a classical state vector ψ can — at least in principle — be measured. But if the state vector represents fundamental quantities, then the components cannot be directly and individually measured. Before we are able to present an interpretation of ψ, we first have to construct COM and the corresponding observables. One COM has been introduced already: The Hamiltonian function $\mathcal{H}$. In classical physics, this function most often represents the energy of a physical system. We will suggest a similar interpretation in what follows.

3.2. The Hamiltonian

We introduced the "Hamiltonian" $\mathcal{H}$ as (an arbitrary) COM. Typically there are several COM and hence the function $\mathcal{H}$ is not yet well defined. In the following, we assume that $\mathcal{H}$ is positive (semi-) definite with respect to the variables ψ. This restriction is neither arbitrary nor weak: If the constructed COM $\mathcal{H}$ is a reference of existing measurable things, then there must be a Hamiltonian function $\mathcal{H}$ that reflects the amount of something, for instance, the amount of substance. In a physical theory, that is free of *ad-hoc* postulates, the constant existence of a substance must be represented by a positive (semi-) definite Hamiltonian *as positivity is immanent to the notion of substance.*

[q]In classical statistical mechanics, this kind of space is called Γ-space [81] but the intimate relation to quantum mechanics is known as for instance described by Kim and Noz [54].

We write the "Hamiltonian" $\mathcal{H}$ as a Taylor series of the $2\,n$ variables ψ_k:

$$\mathcal{H}(\psi) = \mathcal{H}_0 + \varepsilon^T\,\psi + \frac{1}{2!}\,\psi^T\,\mathbf{A}\,\psi + \frac{1}{3!}\,\mathbf{B}_{ijk}\,\psi_i\,\psi_j\,\psi_k + \cdots\,, \tag{13}$$

where ε is a $2\,n$-dimensional vector and $\mathbf{A}$ is a symmetric $2\,n \times 2\,n$ matrix and $\mathbf{B}$ is a tensor that is symmetric in all indices. We assume that $\mathcal{H}$ has a local minimum *somewhere*. Further, we may set $\mathcal{H}_0 = 0$ as it has no influence on the EQOM. A typical method in physics is to stepwise study the solutions of Eq. (13), starting with low amplitudes first. Hence, the first step is to neglect higher order terms and to focus on small amplitude solutions, i.e. to write

$$\mathcal{H}(\psi) \approx \varepsilon^T\,\psi + \frac{1}{2!}\,\psi^T\,\mathbf{A}\,\psi. \tag{14}$$

After the truncation to second order, an offset $\tilde{\psi} = \psi - \psi_0$ of size $\psi_0 = -\mathbf{A}^{-1}\,\varepsilon$ $(\tilde{\psi} = \psi - \psi_0)$ enables to get rid of the linear term, so that:

$$\mathcal{H}(\psi) \approx \frac{1}{2}\,\psi^T\,\mathbf{A}\,\psi. \tag{15}$$

The Hamiltonian EQOM (11) can then be written as a product of a matrix $\mathbf{F}$ and the vector ψ:

$$\dot{\psi} = \gamma_0\,\nabla_\psi\,\mathcal{H} = \mathbf{F}\,\psi, \tag{16}$$

where we defined the matrix

$$\mathbf{F} \equiv \gamma_0\,\mathbf{A}. \tag{17}$$

4. Constants of Motion

We define the dyad $\Sigma_{ij} \equiv \psi_i\,\psi_j$ which can be understood as a matrix of all possible quadratic forms. If we optionally consider some type of averaging, then this matrix becomes a matrix of second moments Σ:

$$\Sigma = \langle \psi\psi^T \rangle, \tag{18}$$

where the angles $\langle\rangle$ indicate some sort of averaging. We define the matrix $\mathbf{S}$ by $\mathbf{S} \equiv \Sigma\,\gamma_0$, so that:

$$\begin{aligned}\dot{\mathbf{S}} &= \dot{\Sigma}\,\gamma_0 = \dot{\psi}\,\psi^T\,\gamma_0 + \psi\,\dot{\psi}^T\,\gamma_0 \\ &= \mathbf{F}\,\psi\,\psi^T\,\gamma_0 + \psi\,\psi^T\,\mathbf{F}^T\,\gamma_0 \\ &= \mathbf{F}\,\mathbf{S} - \mathbf{S}\,\gamma_0\,\mathbf{F}^T\,\gamma_0,\end{aligned} \tag{19}$$

where we use the fact that $\gamma_0^2 = -1$ and $\gamma_0 = -\gamma_0^T$. From Eq. (17), it follows that

$$\mathbf{F}^T = \mathbf{A}^T\,\gamma_0^T = (\gamma_0)^2\,\mathbf{A}\,\gamma_0 = \gamma_0\,\mathbf{F}\,\gamma_0, \tag{20}$$

where we used the fact that $\mathbf{A}$ is symmetric. We then obtain

$$\dot{\mathbf{S}} = \mathbf{F}\,\mathbf{S} - \mathbf{S}\,\mathbf{F}. \tag{21}$$

That is, if $\mathbf{F}$ and $\mathbf{S}$ commute, then the matrix $\mathbf{S}$ is constant, while ψ varies. Operators $\mathbf{S}$ and $\mathbf{F}$ that fulfill Eq. (21) form a so-called "Lax Pair". For such pairs, it can be shown that

$$Tr(\mathbf{S}^k) = \text{const}, \tag{22}$$

holds for any natural number k [19, 61, 93]. Hence, the bilinear form (18) is the basis for a set of constants as given by Eq. (22). Note that the EQOM (Eq. (21)) of the second moments fulfills the requirement to represent observable properties of a physical system. If $\mathbf{S}$ and $\mathbf{F}$ are similar, i.e. if they *share a system of eigenvectors*, then $\mathbf{S}$ is constant. If this is not the case, then we still have the COM given by Eq. (22) as reference quantity.

The average of the quadratic form equals a matrix of second moments, if summed over an ensemble of N spinors:

$$\Sigma = \frac{1}{N}\sum_{k=1}^{N}\psi_k\,\psi_k^T, \tag{23}$$

or — which is a complementary description — if we use a "density" $\rho(\psi)$ to describe the distribution of states in Γ-space. In this case, the matrix of second moments is given by

$$\Sigma = \frac{\int \rho(\psi)\,\psi\,\psi^T\,d^{2n}\psi}{\int \rho(\psi)\,d^{2n}\psi}. \tag{24}$$

Equation (21) holds also for a single Γ-space trajectory $\psi(t)$, but for a single trajectory one finds for all k that $Tr(\mathbf{S}^k) = 0$, so that the constancy is trivial: The eigenvalues of a matrix of second moments Σ depend on whether they are computed for a single vector ψ (which results in a matrix Σ with vanishing determinant and vanishing eigenvalues), for two linear independent vectors (ψ_1, ψ_2) (Σ has vanishing determinant, but two nonzero eigenvalues) or $\geq 2\,n$ linear independent vectors ψ_k as in Eq. (23). Only in the latter case, Σ is non-singular.[r] This implies that our ansatz unfolds to full generality only if we consider ensembles. Or more precise, the observables are averaged over ensembles.

Since $\mathbf{S}$ (and $\mathbf{F}$) are by definition the product of a symmetric matrix Σ and the skew-symmetric matrix γ_0, it follows that

$$Tr(\mathbf{S}) = Tr(\mathbf{F}) = 0, \tag{25}$$

so that the simplest meaningful COM are given as

$$Tr(\mathbf{S}^2) = \text{const.} \tag{26}$$

Equation (15) describes a n-dimensional harmonic oscillator. The truncation of the Hamiltonian to second order is not arbitrary — it guarantees stability. It is well known that nonlinearities yield (in many or even most cases) unstable or chaotic behavior. Therefore, in order to establish a system of stable references linear systems are preferable. This restriction is *per se* not problematic, since the equations that we will derive are also linear.[s] We

[r]In accelerator physics, the eigenvalues of $\mathbf{S} = \Sigma\,\gamma_0$ are the "emittances" (times the unit imaginary). It is evident that the position of a single particle at time t defines a point but not an area in phase space [6, 8].

[s]We refer to the well-known regularization methods that allows to map the Kepler problem and a large variety of other central potential problems to the harmonic oscillator [10].

might also interpret this restriction as a low-energy approximation as in [53].

The COM that can be used as references are quadratic forms (22). If a reference is based on a quadratic form, the variable compared with the reference (the measured variable) also must be a quadratic form. Therefore, the "dynamical variables" q_i and p_i cannot be measured directly, but only (functions of) quadratic forms based on these variables. We call these (functions of) quadratic forms observables.

Matrices that fulfill Eq. (20) are called "Hamiltonian" or sometimes "infinitesimally symplectic". In our opinion, these names are misleading, the former mainly because the matrix $\mathbf{F}$ does not appear in the Hamiltonian and the latter since $\mathbf{F}$ is neither symplectic nor infinitesimal. Therefore, we use the term *symplex* (plural *symplices*) [6, 8].

A *cosymplex* (or "antisymplex" or "skew-Hamiltonian" matrix) is a matrix $\mathbf{C}$ that holds[t]:

$$\mathbf{C}^T = -\gamma_0\,\mathbf{C}\,\gamma_0. \tag{27}$$

A cosymplex $\mathbf{C}$ can always be written as a product of γ_0 and a skew-symmetric matrix. The sums of (co-) symplices are (co-) symplices, i.e. the superposition principle holds. Hence, (co-) symplices form a linear vector space and any (co-) symplex can be written as a linear combination of "basic" (co-) symplices. The algebra of (co-) symplices is the Lie algebra $sp(2\,n)$.

Denoting symplices by $\mathbf{S}$ and cosymplices by $\mathbf{C}$ (optionally with subscript), it is easily shown that the anticommutator of two

[t]The nomenclature of (co-) symplices combines the terms "symplectic", "(co-) sine", and "matrix". The connection to (co-) sine will become obvious in Section 6.1. Furthermore, in geometry a "1-simplex" is a line, a "2-simplex" is a triangle, and a "3-simplex" is a volume (tetrahedron). Equivalently, the elements of a Clifford algebra (for instance the Dirac algebra) include vectors (vertices), bi-vectors (edges), and tensors (faces). Pascal's triangle can be used for both — the number of k-simplices of an n-simplex ($k <= n$) and to the number of traceless k-(co-) symplices of an N-dimensional Clifford algebra. Hence, the Clifford algebra $Cl_{3,1}$ represented by the real Dirac-matrices can be compared to a regular 3-simplex (tetrahedron). See Fig. 3 in the Appendix.

symplices is a cosymplex:

$$\begin{aligned}(\mathbf{S}_1\,\mathbf{S}_2+\mathbf{S}_2\,\mathbf{S}_1)^T &= \mathbf{S}_2^T\,\mathbf{S}_1^T+\mathbf{S}_1^T\,\mathbf{S}_2^T \\ &= \gamma_0\,\mathbf{S}_2\,\gamma_0\,\gamma_0\,\mathbf{S}_1\,\gamma_0+\gamma_0\,\mathbf{S}_1\,\gamma_0\,\gamma_0\,\mathbf{S}_2\,\gamma_0 \\ &= -\gamma_0\,(\mathbf{S}_2\,\mathbf{S}_1+\mathbf{S}_1\,\mathbf{S}_2)\,\gamma_0. \end{aligned} \tag{28}$$

The following general rules can be derived:

$$\left.\begin{array}{c}\mathbf{S}_1\,\mathbf{S}_2-\mathbf{S}_2\,\mathbf{S}_1\\ \mathbf{C}_1\,\mathbf{C}_2-\mathbf{C}_2\,\mathbf{C}_1\\ \mathbf{C}\,\mathbf{S}+\mathbf{S}\,\mathbf{C}\\ \mathbf{S}^{2\,n+1}\end{array}\right\}\Rightarrow\ \text{symplex,}$$

$$\left.\begin{array}{c}\mathbf{S}_1\,\mathbf{S}_2+\mathbf{S}_2\,\mathbf{S}_1\\ \mathbf{C}_1\,\mathbf{C}_2+\mathbf{C}_2\,\mathbf{C}_1\\ \mathbf{C}\,\mathbf{S}-\mathbf{S}\,\mathbf{C}\\ \mathbf{S}^{2\,n}\\ \mathbf{C}^{n}\end{array}\right\}\Rightarrow \text{cosymplex.} \tag{29}$$

The relations Eq. (29) define the structure of a specific algebra, called the Lie algebra $sp(2\,n)$ of the symplectic group $Sp(2\,n)$. As the matrix of generators $\mathbf{F}$, also the matrix of observables $\mathbf{S}$ of Eq. (22) is a symplex. According to Eq. (29), odd powers of $\mathbf{S}$ are symplices with zero trace. Hence, the nonzero (i.e. useful) constants of motion given by Eq. (22) are of even order.

Since the matrix $\mathbf{A}$ is symmetric (Eq. (15)), the maximal number ν of free parameters of a symplex for n degrees of freedom (DOF) is

$$\nu_s=\frac{2\,n\,(2\,n+1)}{2}. \tag{30}$$

If the matrix $\mathbf{F}$ can be expressed as a sum of basic matrices, the basis ν_s elements. A basis for all real $2n\times 2n$ matrices is completed by ν_c cosymplices:

$$\nu_c=\frac{2\,n\,(2\,n-1)}{2}. \tag{31}$$

4.1. Measurable quantities: Observables

Usually the second moments are assumed to be average values of N identical systems as in Eq. (23), but the equations hold equivalently for a single system. However, it should be noted that for a vector of $2\,n$ fundamental variables, $2\,n$ linear independent "samples" are required, if the matrix $\mathbf{S} = \Sigma\,\gamma_0$ is supposed to have a non-vanishing determinant. Later, we will come back to this point.

We can define the Σ-matrix as well using a (normalized) probability density ρ:

$$\Sigma = \int \rho(\psi)\,\psi\,\psi^T\,d^{2n}\psi \equiv \langle\psi\,\psi^T\rangle, \tag{32}$$

with the normalization according to:

$$1 = \int \rho(\psi)\,d^{2n}\psi, \tag{33}$$

i.e. the density $\rho(\psi)$ is defined as a function of the *fundamental variables* or *phase space* variables.

We introduced the ("S-matrix") $\mathbf{S} = \Sigma\gamma_0$ and derived Eq. (21). Let the *adjunct* vector $\bar{\psi}$ be defined by

$$\bar{\psi} \equiv \psi^T\,\gamma_0, \tag{34}$$

so that $\mathbf{S} = |\psi\rangle\langle\bar{\psi}|$ and the expectation value of an operator $\mathbf{O}$:

$$\langle\mathbf{O}\rangle \equiv \langle\bar{\psi}\,\mathbf{O}\,\psi\rangle. \tag{35}$$

Now consider that $\mathbf{O}$ is a cosymplex, i.e. can be written as $\gamma_0\,\mathbf{B}$ with some skew-symmetric matrix $\mathbf{B}$, then the expectation value of $\mathbf{O} = \gamma_0\,\mathbf{B}$ vanishes:

$$\begin{aligned}\langle\mathbf{O}\rangle &= \langle\psi^T\,\gamma_0^2\mathbf{B}\,\psi\rangle\\ &= -\langle\psi^T\,\mathbf{B}\,\psi\rangle = 0.\end{aligned} \tag{36}$$

Hence, only symplices represent measurable properties of (closed) systems. Cosymplices $\mathbf{C}$ only yield a non-vanishing result when they appear between different (i.e. linearly independent) states:

$$\bar{\phi}\,\mathbf{C}\,\psi \neq 0 \Rightarrow \phi \neq \lambda\,\psi, \tag{37}$$

with some arbitrary (nonzero) factor λ. Vice versa — if two (normalized) states ϕ and ψ may not "generate" non-vanishing expectation values of the cosymplex type, then it follows that:

$$\phi = \pm\psi. \tag{38}$$

Since we assume that physical reality can be described by the totality of all fundamental variables, the coefficients of the matrix $\mathbf{F}$ are functions of (internal or external) measurable quantities, i.e. of some symplex $\mathbf{S}$ generated from fundamental variables. And since cosymplices have a vanishing expectation value, they cannot contribute. We will use this important result in Section 9 in the derivation of MWEQ. Furthermore, if a structure that represents a "particle" is preserved in a physical process, then the involved driving terms must be symplices. Otherwise, the evolution in time would not be symplectic, i.e. *structure preserving*, and vice versa. If the particle structure (particle type) is transformed in a process, then there must be a contribution from cosymplices as for instance *axial vector currents* as in the V-A theory of weak interaction.

This means that *if we focus on a closed system*, then (the expectation values of) all cosymplices have to vanish.

The properties of "open" and "closed" have a relative meaning which depends on the context. If we consider two separate "particles" described by state vectors ψ and ϕ and let them get in contact, then we would for instance combine the state vectors into a larger state vector $\Psi = (\psi, \phi)^T$. In App. (C.2), we give the explicit form of general $2\,n \times 2\,n$-symplices. It is shown that non-diagonal sub-blocks may *per se* have arbitrary terms. Hence, the question, if a (sub-) matrix is a symplex or a cosymplex depends on the context. However, *all* quantities — including the matrix $\mathbf{F}$ — must by some means result from "fundamental variables", i.e. from state vectors. Also a static electromagnetic field has its origin in charges and currents of "particles" that are described by state vectors. If reality is described by the totality of the fundamental variables, then also the symplex $\mathbf{F}$

must somehow be generated from ψ. Since the matrix of observables $\mathbf{S} = \Sigma\,\gamma_0$ is a symplex as well, the functional relationship might be simple: consider (for instance) a Hamiltonian of the form

$$\mathcal{H} \propto \psi^T\,\Sigma^{-1}\,\psi, \tag{39}$$

which implies $\mathbf{F} \propto \gamma_0\,\Sigma^{-1}$. In this special case, a Boltzmann distribution $\rho(\psi) \propto \exp\left(-\frac{\mathcal{H}}{k\,T}\right)$ is identical to a multivariate Gaussian density distribution [7]. In general, the functional relationship between $\mathbf{F}$ and $\mathbf{S}$ might be more involved, but obviously there is a kind of *duality* between the matrix of observables $\mathbf{S} = \Sigma\,\gamma_0$ and the matrix of generators $\mathbf{F} = \gamma_0\,\mathbf{A}$ that can be illustrated by Eq. (39). This duality is well-known in classical mechanics. The Hamiltonian is an observable and the generator of translations in time (evolution in time), the angular momentum is an observable and the generator of rotations, the momentum is the generator of spatial translations [40].

The time derivative of an expectation value is (assuming the observable $\mathbf{O}$ does not explicitly depend on time):

$$\begin{aligned}
\langle\dot{\mathbf{O}}\rangle &= \langle\dot{\bar{\psi}}\,\mathbf{O}\,\psi\rangle + \langle\bar{\psi}\,\mathbf{O}\,\dot{\psi}\rangle \\
&= \langle\dot{\psi}^T\,\gamma_0\,\mathbf{O}\,\psi\rangle + \langle\bar{\psi}\,\mathbf{O}\,\mathbf{F}\,\psi\rangle \\
&= \langle\psi^T\,\mathbf{F}^T\,\gamma_0\,\mathbf{O}\,\psi\rangle + \langle\bar{\psi}\,\mathbf{O}\,\mathbf{F}\,\psi\rangle \\
&= \langle\psi^T\,\gamma_0\,\mathbf{F}\,\gamma_0^2\,\mathbf{O}\,\psi\rangle + \langle\bar{\psi}\,\mathbf{O}\,\mathbf{F}\,\psi\rangle \\
&= \langle\bar{\psi}\,(\mathbf{O}\,\mathbf{F} - \mathbf{F}\,\mathbf{O})\,\psi\rangle.
\end{aligned} \tag{40}$$

This general law connects the commutator of an operator $\mathbf{O}$ and $\mathbf{F}$ with its time derivative. This is the linear form of the Poisson bracket and it is purely classical. One conclusion from Eq. (40) in combination with Eq. (29) is that the time evolution is such that (co-) symplices stay (co-) symplices. They belong to different domains.

5. From Lie Algebra to Spacetime Geometry

From the (anti-) commutation properties described by Eq. (29), it follows that any fundamental system of (co-) symplices requires that the basic (co-) symplices either commute or anticommute: Only in this case, the product is a pure symplex or cosymplex, i.e. either observable or not. If we consider pure (co-) symplices $\mathbf{A}$ and $\mathbf{B}$, then

$$(\mathbf{A}\,\mathbf{B})^T = \mathbf{B}^T\,\mathbf{A}^T = \pm\,\gamma_0\,\mathbf{B}\,\mathbf{A}\,\gamma_0. \tag{41}$$

It follows that the product of pure (co-) symplices is again a pure (co-) symplex, iff $\mathbf{A}\,\mathbf{B} = \pm\mathbf{B}\,\mathbf{A}$. This implies that all (anti-)commutators of pure (co-)symplices are again pure (co-)symplices. Hence, if there are systems of (anti-)commuting matrices, they are expected to play a fundamental role.

In the following, we will show how spacetime structures emerge from the algebra of (co-)symplices (Eq. (29)), i.e. we will explain why and how the spinors ψ — which are composed of fundamental variables — are able to produce the relations between observables that represent the spacetime structures. In order to do this, we have to identify algebraic objects that represent (unit) vectors for the space and time directions (or — equivalently — the energy and momentum directions). For this purpose, we need a definition of "length" (a norm) and of orthogonality. So far we only introduced time and a single symplex: γ_0. Apparently, γ_0 is the natural choice of the "unit vector" in the "time direction".

Usually vector norms in Euclidean spaces represent the length of a vector, i.e. a property that is invariant under rotations and translations. Hence, we prefer an invariant quantity for the desired definition of a norm. Besides the Hamiltonian, the only invariants we introduced so far are given by Eq. (22). The simplest nontrivial k-norm is given for $k = 2$, i.e.:

$$\|\mathbf{A}\|^2 \equiv \frac{1}{2\,n}\,\mathrm{Tr}(\mathbf{A}^2). \tag{42}$$

where the division by $2\,n$ is introduced to let the unit matrix have unit norm. The matrix γ_0 hence has a norm of -1, i.e. the norm (42) is not positive definite.

Two elements of an algebra are said to be orthogonal, if their inner product vanishes. Since we have no inner product defined (yet), we use an alternative *geometric* definition which we consider to be equivalent in *flat spacetime* — the *Pythagorean theorem*. Two vectors

$\vec{v}$ and $\vec{w}$ are orthogonal, iff [96]

$$\|\vec{v}+\vec{w}\|^2 = \|\vec{v}\|^2 + \|\vec{w}\|^2, \tag{43}$$

which implies (in conventional vector notation) that the inner product vanishes:

$$(\vec{v}+\vec{w})^2 = \vec{v}^2 + \vec{w}^2 + 2\,\vec{v}\cdot\vec{w}. \tag{44}$$

For two elements $\mathbf{A}$ and $\mathbf{B}$ of the Lie algebra $sp(2\,n)$, the use of Eq. (42) then yields

$$\begin{aligned}\|\mathbf{A}+\mathbf{B}\|^2 &= \frac{1}{2\,n}\,\mathrm{Tr}\left[(\mathbf{A}+\mathbf{B})^2\right]\\ &= \frac{1}{2\,n}\,\mathrm{Tr}(\mathbf{A}^2+\mathbf{B}^2+\mathbf{AB}+\mathbf{BA})\\ &= \|\mathbf{A}\|^2+\|\mathbf{B}\|^2+\frac{1}{2\,n}\,\mathrm{Tr}(\mathbf{AB}+\mathbf{BA}).\end{aligned} \tag{45}$$

As we argued above, all elements of the desired algebra either pairwise commute or anticommute. Two anticommuting elements $\mathbf{A}$ and $\mathbf{B}$ with

$$\mathrm{Tr}\,(\mathbf{A}\,\mathbf{B}+\mathbf{B}\,\mathbf{A}) = 0. \tag{46}$$

are then orthogonal in the sense that they obey the Pythagorean theorem:

$$\|\mathbf{A}+\mathbf{B}\|^2 = \|\mathbf{A}\|^2 + \|\mathbf{B}\|^2. \tag{47}$$

Hence, the (trace of the) anticommutator is structurally equal to the inner product[u]:

$$\mathbf{A}\cdot\mathbf{B} \equiv \frac{\mathbf{A}\,\mathbf{B}+\mathbf{B}\,\mathbf{A}}{2}. \tag{49}$$

[u]Since the anticommutator does not necessarily result a scalar, one may distinguish between inner product and scalar product. The scalar product can then be defined by the trace of the inner product:

$$(\mathbf{A}\cdot\mathbf{B})_S \equiv \frac{1}{2\,n}\mathrm{Tr}\left(\frac{\mathbf{A}\,\mathbf{B}+\mathbf{B}\,\mathbf{A}}{2}\right), \tag{48}$$

where the index "S" indicates the scalar part.

The above arguments and the fact that γ_0 seems to have the form of a unit vector suggests to search for other unit vectors γ_k for $k \in [1 \dots N-1]$, especially other unit vectors that anticommute with γ_0. For such unit vectors, we find by the use of $\gamma_0^2 = -1$:

$$\begin{aligned} \gamma_k\,\gamma_0 + \gamma_0\,\gamma_k &= 0, \\ \gamma_k\,\gamma_0^2 + \gamma_0\,\gamma_k\,\gamma_0 &= 0, \\ \gamma_k &= \gamma_0\,\gamma_k\,\gamma_0, \end{aligned} \tag{50}$$

so that in accordance with Eqs. (20) and (27) γ_k $(k \neq 0)$ is a symplex, iff it is symmetric. It is a cosymplex, iff it is skew-symmetric. This fact is remarkable. Since up to now we just defined time, it is nearby to interpret γ_0 as the unit vector "in the direction of time". Then from Eq. (50), we can take that all other unit vectors which are perpendicular to γ_0 and that correspond to driving terms of the Hamiltonian, are symmetric and hence formally different from "time". In other words, there is only one observable direction of time. Using the algebraic identity

$$\mathrm{Tr}(\mathbf{A}\,\mathbf{A}^T) = \sum_{i,j} a_{ij}^2,$$

where a_{ij} are the elements of the matrix $\mathbf{A}$ — it follows, that (with an appropriate normalization) symmetric unit "vectors" $\mathbf{A}$ square to $+1$ and skew-symmetric $\mathbf{B}$ square to -1:

$$\begin{aligned} \mathrm{Tr}(\mathbf{A}^2) &= \mathrm{Tr}(\mathbf{A}\,\mathbf{A}^T) = \sum_{i,j} a_{ij}^2 \geq 0, \\ \mathrm{Tr}(\mathbf{B}^2) &= -\mathrm{Tr}(\mathbf{B}\,\mathbf{B}^T) = -\sum_{i,j} b_{ij}^2 \leq 0. \end{aligned} \tag{51}$$

Therefore, there are two types of measurable unit vectors: γ_0 that represents time and squares to -1 and a number of symmetric symplices γ_k (anticommuting with γ_0 and with each other) that square to $+1$. We interpret the latter (if existent) as the unit vectors in space-like directions. Skew-symmetric elements that are orthogonal to γ_0 are cosymplices and have a vanishing expectation value.

In summary: if a (flat) spacetime emerges from dynamics, all symplices that represent orthogonal directions have to *anticommute*. Therefore, the obvious choice of the *inner product* or *scalar product* of two elements of a dynamically generated geometric algebra is the anticommutator. The unit vector in the time-direction squares to -1 and all vector-type symplices γ_x which are orthogonal to γ_0 are symmetric and square to $+1$.

Given that we can find a (sub-) set of N pairwise orthogonal symplices γ_k with $k \in [0, 1, \dots, N-1]$ in $sp(2\,n)$, then the product of two of them holds:

$$\begin{aligned} \gamma_i\,\gamma_j &= \frac{\gamma_i\,\gamma_j + \gamma_j\,\gamma_i}{2} + \frac{\gamma_i\,\gamma_j - \gamma_j\,\gamma_i}{2} \\ &= \frac{\gamma_i\,\gamma_j - \gamma_j\,\gamma_i}{2}, \end{aligned} \tag{52}$$

for $i \neq j$. It is easy to prove that the product $\gamma_i\,\gamma_j$ with $i \neq j$ is orthogonal to both — γ_i and γ_j. Hence, we identify the *exterior product* with the commutator of $\mathbf{A}$ and $\mathbf{B}$:

$$\mathbf{A} \wedge \mathbf{B} = \frac{\mathbf{A}\,\mathbf{B} - \mathbf{B}\,\mathbf{A}}{2}. \tag{53}$$

A system of N pairwise anticommuting real matrices (that square to ± 1) represents a Clifford algebra $Cl_{p,q}$, $N = p + q$. p is the number of unit vectors that square to $+1$ and q the number of unit vectors that square to -1. We found that in our case $q = 1$. If (as we presumed) spacetime emerges from the dynamics of fundamental variables, then the apparent geometry of spacetime is described by the (representation of) a Clifford algebra $Cl_{N-1,1}$. The unit vectors are called *generators* of the Clifford algebra.

There are $\binom{N}{k}$ possible combinations (without repetition) of k elements from a set of N generators. Then there are $\binom{N}{2}$ products of 2 basic matrices (named *bi-vectors*), $\binom{N}{3}$ products of 3 basic matrices (named *trivectors*), and so on. The product of all N basic matrices is called *pseudoscalar*. The N anticommuting generators are the (unit) *vectors*. The total number of all k-vectors then is[v]:

$$\sum_{k=0}^{N} \binom{N}{k} = 2^N. \tag{54}$$

As we have no reason to allow more elements in the algebra than 2^N, the vector space $sp(2\,n)$ and $Cl_{N-1,1}$ have the same dimension:

$$2^N = (2n)^2, \tag{55}$$

which — if fulfilled — is nothing but the condition for *completeness*.

If such a basis of anticommuting unit matrices exists, we know that

$$\begin{aligned}
\gamma_k^2 &= \pm 1 && \text{for all } k \in [0, \dots, 4n^2 - 1],\\
\gamma_k &= \pm \gamma_k^T && \text{for all } k \in [0, \dots, 4n^2 - 1],\\
\gamma_k\,\gamma_j &= \pm \gamma_j\,\gamma_k && \text{for all } j \neq k, \in [0, \dots, 4n^2 - 2].
\end{aligned} \tag{56}$$

From Eq. (56) above, we readily conclude:

$$\begin{aligned}
\gamma_0\,\gamma_k\,\gamma_0 &= \pm\,\gamma_k^T,\\
\gamma_k\,\gamma_0\,\gamma_k^T &= \pm\,\gamma_0,
\end{aligned} \tag{57}$$

so that all γ_k are either symplices or cosymplices and they are either symplectic ($\gamma_k\,\gamma_0\,\gamma_k^T = \gamma_0$) or "cosymplectic" ($\gamma_k\,\gamma_0\,\gamma_k^T = -\gamma_0$). Since $\mathrm{Det}(\mathbf{AB}) = \mathrm{Det}(\mathbf{A})\mathrm{Det}(\mathbf{B})$ and since $\gamma_k^2 = \pm\mathbf{1}$ it follows that all basic matrices have a determinant of ± 1.

Given such a matrix system exists for some n and N, then it has remarkable symmetry properties. Note that we derived these properties without fixing the size of the matrices. If (as we demand) Eq. (55) holds, then the system is a complete *basis* of the vector space of real $2n \times 2n$-matrices and therefore these matrices $\mathbf{M}$ can be written as a linear combination of the γ_k according to:

$$\mathbf{M} = \sum_{k=0}^{(2n)^2-1} m_k\,\gamma_k, \tag{58}$$

[v]The case $k = 0$ can be identified with the unit matrix.

so that with the signature s_j defined by $s_j \equiv \mathrm{Tr}(\gamma_j^2)/(2n)$ we have

$$\begin{aligned}\frac{1}{4n}\,\mathrm{Tr}(\mathbf{M}\,\gamma_j + \gamma_j\,\mathbf{M}) &= \tfrac{1}{4n}\,\mathrm{Tr}\left(\sum_{k=0}^{2n-1} m_k\,(\gamma_j\,\gamma_k + \gamma_k\,\gamma_j)\right)\\ &= s_j\,m_j.\end{aligned} \tag{59}$$

Given that γ_k with $k \in [0,\ldots,N-1]$ are the generating elements of the algebra, then the *vector* $\mathbf{A} = \sum_{k=0}^{N-1} a_k\,\gamma_k$ squares to a scalar:

$$\mathbf{A}^2 = \sum_{k=0}^{N-1} s_k\,a_k^2 = \sum_{k=1}^{N-1} a_k^2 - a_0^2. \tag{60}$$

Hence, the algebra of a dynamical system generates a flat Minkowski spacetime with proper observables if the corresponding real matrix algebra is a representation of a Clifford algebra in which all generating elements are symplices. The idea to relate Clifford algebras and symplectic spaces is not new (see, for instance, [4, 11]) and references therein), however (to the knowledge of the author) it has not been shown yet that the concept of a geometric (Clifford) algebra can be derived from the concept of a dynamically emergent spacetime.

6. Structure Preservation

6.1. Transfer matrices and symplectic transformations

The solution of the Hamiltonian EQOM (Eq. (16)) for the linearized Hamiltonian (with constant matrix $\mathbf{A}$) are given by the matrix exponential:

$$\psi(t) = \exp\left(\mathbf{F}\,(t - t_0)\right)\psi(t_0) = \mathbf{M}(t,t_0)\,\psi(t_0). \tag{61}$$

The matrix $\mathbf{M}(t,t_0) = \exp\left(\mathbf{F}\,(t-t_0)\right)$ is usually called *transfer matrix* and it is well-known to be symplectic if $\mathbf{F}$ is a symplex ("Hamiltonian"), i.e. $\mathbf{M}$ fulfills the following (equivalent) equations [68]:

$$\begin{aligned}\mathbf{M}\,\gamma_0\,\mathbf{M}^T &= \gamma_0,\\ \mathbf{M}^{-1} &= -\gamma_0\,\mathbf{M}^T\,\gamma_0,\end{aligned}$$

$$\begin{aligned} (\mathbf{M}^{-1})^T &= -\gamma_0\,\mathbf{M}\,\gamma_0, \\ \mathbf{M}^T &= -\gamma_0\,\mathbf{M}^{-1}\,\gamma_0. \end{aligned} \tag{62}$$

Since the matrix exponential of $\mathbf{F}$ is quite complicated in general, it is more convenient to consider transformations generated by basic symplices γ_j, since all γ_j square to ± 1:

$$\begin{aligned} \mathbf{R}_j &= \exp\left(\gamma_j\,\varepsilon\right) \\ &= \begin{cases} 1\,\cos\left(\varepsilon\right) + \gamma_j\,\sin\left(\varepsilon\right) & \text{for } \gamma_j^2 = -1, \\ 1\,\cosh\left(\varepsilon\right) + \gamma_j\,\sinh\left(\varepsilon\right) & \text{for } \gamma_j^2 = 1. \end{cases} \\ \mathbf{R}_j^{-1} &= \exp\left(-\gamma_j\,\varepsilon\right) \\ &= \begin{cases} 1\,\cos\left(\varepsilon\right) - \gamma_j\,\sin\left(\varepsilon\right) & \text{for } \gamma_j^2 = -1, \\ 1\,\cosh\left(\varepsilon\right) - \gamma_j\,\sinh\left(\varepsilon\right) & \text{for } \gamma_j^2 = 1. \end{cases} \end{aligned} \tag{63}$$

The solutions are the sum of the unit matrix and γ_j with (hyperbolic) (co-) sine coefficients. From Eq. (63), we find that skew-symmetric symplices (those that square to -1) are generators of orthogonal transformations, since

$$\begin{aligned} \mathbf{R}_j &= 1\,\cos\left(\varepsilon\right) + \gamma_j\,\sin\left(\varepsilon\right), \\ \mathbf{R}_j^T &= 1\,\cos\left(\varepsilon\right) - \gamma_j\,\sin\left(\varepsilon\right), \\ &\Rightarrow = \mathbf{R}_j\,\mathbf{R}_j^T = 1. \end{aligned} \tag{64}$$

Hence, skew-symmetric symplices are generators of (symplectic) *rotations*. The symplectic transformations generated by symmetric symplices are essentially *boosts*. As we will show in more detail in Section 8.1, it is legitimate to call the structure preserving property of symplectic transformations the *principle of (special) relativity* since Lorentz transformations of "spinors" ψ in Minkowski spacetime are a subset of the symplectic transformations for two DOFs.

The structure of the matrix exponential of a symplex $\mathbf{F}$ and hence the structure of a symplectic matrix is given by:

$$\mathbf{M}(t) = \exp\left(\mathbf{F}\,t\right) = \mathbf{C} + \mathbf{S}, \tag{65}$$

where the (co-) symplex $\mathbf{S}$ ($\mathbf{C}$) is given by:

$$\begin{aligned}\mathbf{S} &= \sinh(\mathbf{F}\,t),\\ \mathbf{C} &= \cosh(\mathbf{F}\,t),\end{aligned} \tag{66}$$

since (the linear combination of) all odd powers of a symplex are symplices and all even powers are cosymplices. The inverse transfer matrix $\mathbf{M}^{-1}(t)$ is then given by:

$$\mathbf{M}^{-1}(t) = \mathbf{M}(-t) = \mathbf{C} - \mathbf{S}. \tag{67}$$

For the matrix of second moments Σ and $\mathbf{S} = \Sigma\,\gamma_0$ Eqs. (61) and (62) yield:

$$\begin{aligned}\Sigma(t) &= \psi(t)\,\psi^T(t) = \mathbf{M}(t,t_0)\,\psi(t_0)\,\psi(t_0)^T\,\mathbf{M}^T(t,t_0),\\ \Sigma(t)\,\gamma_0 &= -\mathbf{M}(t,t_0)\,\Sigma(t_0)\,\gamma_0^2\,\mathbf{M}^T(t,t_0)\,\gamma_0,\\ \mathbf{S}(t) &= \mathbf{M}(t,t_0)\,\mathbf{S}(t_0)\,\mathbf{M}^{-1}(t,t_0),\end{aligned} \tag{68}$$

so that the evolution of $\mathbf{S}$ in time is a *symplectic similarity transformation*. Hence, all *eigenvalues* of $\mathbf{S}$ are constants.

From Eq. (61), it is obvious that the column vectors of the transfer matrix are the solutions for initial conditions $\psi(0)$ that are equal to unit vectors. The first column $\psi_1(t)$ of $\mathbf{M}(t,t_0)$ is given by:

$$\psi_1(t) = \mathbf{M}(t,t_0)\,(1,0,0,0)^T. \tag{69}$$

Furthermore, we note that if a symplex $\mathbf{F}$ is symplectically similar to γ_0, then it follows from Eq. (62)

$$\mathbf{F} \propto \mathbf{M}\,\gamma_0\,\mathbf{M}^{-1} = \mathbf{M}\,\mathbf{M}^T\,\gamma_0, \tag{70}$$

so that $\mathbf{F}$ equals the matrix $\mathbf{S} = \Sigma\,\gamma_0$ if the second moments are the average over the orthogonal unit vectors as exemplified by Eq. (69):

$$\Sigma = \mathbf{M}\,\mathbf{M}^T = \sum_k \psi_k(t)\psi_k^T(t). \tag{71}$$

This is another example for the mentioned duality between observables and generators (see Eq. (39) and Ref. [40]).

6.2. Time-dependent Hamiltonian

We started out from a Hamiltonian that has no explicit time dependence. Let us see, if we can still use the same concepts if the driving term in Eq. (15) is time-dependent:

$$\begin{aligned} \frac{d\mathcal{H}}{dt} &= \frac{\partial\mathcal{H}}{\partial t} + \nabla_\psi\mathcal{H}\,\dot{\psi} = 0, \\ 0 &= \frac{1}{2}\psi^T\,\dot{\mathbf{A}}\psi + \psi^T\,\mathbf{A}\,\dot{\psi} = 0. \end{aligned} \tag{72}$$

We use the Ansatz:

$$\dot{\psi} = (\gamma_0\,\mathbf{A} + \mathbf{G})\,\psi, \tag{73}$$

with a general symplex $\mathbf{G}$ and plug it into Eq. (72)

$$\begin{aligned} 0 &= \frac{1}{2}\psi^T\,\dot{\mathbf{A}}\psi + \psi^T\,\mathbf{A}\,(\gamma_0\,\mathbf{A} + \mathbf{G})\,\psi, \\ 0 &= \psi^T\left(\frac{1}{2}\dot{\mathbf{A}} + \mathbf{A}\,\gamma_0\,\mathbf{A} + \mathbf{A}\,\mathbf{G}\right)\psi. \end{aligned} \tag{74}$$

As we know that $\psi^T\mathbf{A}\,\gamma_0\,\mathbf{A}\psi = 0$ due to the skew-symmetry of $\mathbf{A}\,\gamma_0\,\mathbf{A}$, the remaining matrix has to vanish separately, i.e. has to be skew-symmetric. Furthermore, we have to assume that $\dot{\mathbf{A}} = \dot{\mathbf{A}}^T$ as the symmetry is logically required in the Hamiltonian for a single system:

$$\begin{aligned} 0 &= \left(\frac{1}{2}\dot{\mathbf{A}} + \mathbf{A}\,\mathbf{G}\right)^T + \frac{1}{2}\dot{\mathbf{A}} + \mathbf{A}\,\mathbf{G}, \\ 0 &= \dot{\mathbf{A}} + \mathbf{G}^T\,\mathbf{A} + \mathbf{A}\,\mathbf{G}, \\ 0 &= \dot{\mathbf{A}} + \gamma_0\,\mathbf{G}\,\gamma_0\,\mathbf{A} + \mathbf{A}\,\mathbf{G}, \\ 0 &= \gamma_0\,\dot{\mathbf{A}} - \mathbf{G}\,\gamma_0\,\mathbf{A} + \gamma_0\,\mathbf{A}\,\mathbf{G}, \\ \dot{\mathbf{F}} &= \mathbf{G}\,\mathbf{F} - \mathbf{F}\,\mathbf{G}. \end{aligned} \tag{75}$$

This shows us that as long as $\dot{\mathbf{A}} = \dot{\mathbf{A}}^T$ holds, then $\mathcal{H}$ remains constant if the time-dependence of $\mathbf{F}$ is the result of symplectic motion. Since

$\mathcal{H}$ is positive definite, we may write $\mathbf{F}$ in the general form

$$\begin{aligned}
\mathbf{F} &\equiv \mathbf{B}\,\mathbf{B}^T\,\gamma_0,\\
\dot{\mathbf{F}} &= \dot{\mathbf{B}}\,\mathbf{B}^T\,\gamma_0 + \mathbf{B}\,\dot{\mathbf{B}}^T\,\gamma_0,\\
\dot{\mathbf{F}} &= \dot{\mathbf{B}}\,\mathbf{B}^{-1}\,\mathbf{B}\,\mathbf{B}^T\,\gamma_0 + \mathbf{B}\,\mathbf{B}^T\,(\mathbf{B}^T)^{-1}\dot{\mathbf{B}}^T\,\gamma_0,\\
\dot{\mathbf{F}} &= (\dot{\mathbf{B}}\,\mathbf{B}^{-1})\,\mathbf{F} - \mathbf{F}\,(\gamma_0\,(\mathbf{B}^T)^{-1}\dot{\mathbf{B}}^T\,\gamma_0).
\end{aligned} \tag{76}$$

The last lines of Eqs. (75) and (76) are equivalent, if

$$\begin{aligned}
\mathbf{G} &= \gamma_0\,(\mathbf{B}^T)^{-1}\dot{\mathbf{B}}^T\,\gamma_0,\\
\gamma_0\,\mathbf{G}\,\gamma_0 &= (\mathbf{B}^T)^{-1}\dot{\mathbf{B}}^T,\\
\mathbf{G}^T &= (\mathbf{B}^T)^{-1}\dot{\mathbf{B}}^T,\\
\mathbf{G} &= \dot{\mathbf{B}}\,\mathbf{B}^{-1},\\
\dot{\mathbf{B}} &= \mathbf{G}\,\mathbf{B}.
\end{aligned} \tag{77}$$

Hence we again find a spinor-type of linear equation with a symplex $\mathbf{G}$ as driving term:

$$\dot{\mathbf{B}} = \mathbf{G}\,\mathbf{B}, \tag{78}$$

which tells us that symplectic motion can be regarded as a kind of "closed system", if all spinors change according to Eq. (78) and all symplices change according to Eq. (76). According to classical statistical mechanics, the phase space density $\rho(\psi)$ is a constant, if it is a function of the Hamiltonian (or — more generally — of the COM) only. In our case, we can expect a system in thermodynamical equilibrium, if the density is a function of $\psi^T\,\mathbf{A}\,\psi$, i.e. a function of the bilinear forms that we identified as measurable quantities [81]. In Section 8.1, we will show that the COM can be interpreted by the *mass* of the system.

6.3. Cosymplices as generators of transformations

The definition of what we consider to be a symplex or a cosymplex depends on the choice of the symplectic unit matrix γ_0. Equation (63) holds also for (skew-) symmetric cosymplices as it holds for symplices. Consider a skew-symmetric cosymplex γ_j (which we know to anticommute with γ_0), then the matrix exponential is an orthogonal similarity transformation, so that the transformed matrix γ_0 is skew-symmetric and squares to the negative unit matrix:

$$\begin{aligned}\tilde{\gamma}_0^T &= (\mathbf{R}_j\,\gamma_0\,\mathbf{R}_j^T)^T = \mathbf{R}_j\,\gamma_0^T\,\mathbf{R}_j^T \\ &= -\mathbf{R}_j\,\gamma_0\,\mathbf{R}_j^T = -\tilde{\gamma}_0, \\ \tilde{\gamma}_0^2 &= (\mathbf{R}_j\,\gamma_0\,\mathbf{R}_j^T)^2 = \mathbf{R}_j\,\gamma_0^2\,\mathbf{R}_j^T \\ &= -\mathbf{R}_j\,\mathbf{R}_j^T = -\mathbf{1}.\end{aligned} \tag{79}$$

The transformed matrix *differs* from γ_0, but has the same properties.

If we consider on the other hand a symmetric cosymplex γ_j (which we know to commute with γ_0), then the matrix exponential is also symmetric and commutes with γ_0 so that

$$\begin{aligned}(\mathbf{R}_j\,\gamma_0\,\mathbf{R}_j^T)^T &= \mathbf{R}_j\,\gamma_0^T\,\mathbf{R}_j^T = -\mathbf{R}_j^2\,\gamma_0, \\ (\mathbf{R}_j\,\gamma_0\,\mathbf{R}_j^T)^2 &= \mathbf{R}_j^4\,\gamma_0^2 = -\mathbf{R}_j^4\,\mathbf{1}.\end{aligned} \tag{80}$$

The transformed matrix is still antisymmetric but (in general) it does not square to the (negative) unit matrix.

One may conclude that skew-symmetric cosymplices are generators of time-like rotations: they change the "orientation" of γ_0, the system is rotated towards a new "time direction" $\tilde{\gamma}_0 = \mathbf{R}\,\gamma_0\,\mathbf{R}^T$, but remains structurally similar (though not the same).

The *physical meaning* of such non-symplectic transformations can be related to a variety of phenomena. In [9] for instance, non-symplectic transformations have been shown to be related to the appearance of vector potentials and to the transformation into curvilinear coordinate systems. In App. C.3, we give an example which is directly related to the structure of a system. Furthermore, we explicitly give a matrix that transforms between the six possible choices for γ_0 of the Dirac algebra in App. E.4.

7. The Dimensionality of Spacetime

7.1. Clifford algebras

Sets of anticommuting non-singular matrices are well-known as generators of *Clifford algebras*. The inner products of the basic elements are the elements of the *metric tensor*. The vector space constructed in this way is Euclidean only for diagonal metric tensors with signatures $s_1 = s_2 = \cdots = s_N = 1$, i.e. Euclidean unit vectors square to $+1$. Therefore, generators that square to $+1$ are *spatial* or *space-like* and generators that square to -1 *time-like*.[w] A Clifford algebra with p space-like and q time-like generator matrices is indicated by $Cl_{p,q}$.

From Eq. (55), it follows that such matrix systems can only be complete for even $N = 2\,M$.[x] Then:

$$n = 2^{M-1}. \tag{81}$$

Any real matrix system that fulfills these requirements is based on a number of DOFs which is a power of 2. The simplest system hence is the one for which $M = 1$ ($N = 2$, $n = 1$) holds, i.e. it consists of 2×2-matrices.

As mentioned in Section 4, only those elements γ_k that are symplices have a non-vanishing expectation value $\bar{\psi}\,\gamma_k\,\psi$ and are therefore *observables*. Hence, we restrict ourselves to Clifford algebras with generators that can be represented by symplices. This includes γ_0 as the direction of time. Since any generator that anticommutes with γ_0 is symmetric and therefore squares to $+1$, we have $q = 1$, i.e. the Clifford algebra that represents spacetime is of the type $Cl_{p,1}(\mathbb{R})$.

If $\gamma_0, \gamma_1, \ldots, \gamma_{N-1}$ are the generators of a Clifford algebra $Cl_{N-1,1}$ with metric

$$\gamma_\mu\,\gamma_\nu + \gamma_\nu\,\gamma_\mu = 2\,\mathrm{Diag}(-1, 1, \ldots, 1), \tag{82}$$

then the bi-vector elements $\gamma_0\,\gamma_k$ with $k = 1, 2, \ldots\, N-1$ are symmetric and therefore generators of boosts ("boosters") while the bi-vector elements $\gamma_j\,\gamma_k$ with $j \neq k$ and $j, k \neq 0$ are the skewsymmetric generators of spatial rotations ("rotators" or "gyrators"). We come back to this in the next subsection.

[w] This is only meaningful for *real* representations, since the case of complex representations allows to switch the sign of matrices by the multiplication with the unit imaginary.

[x] A Clifford algebra also can be defined for odd N, but these algebras are not isomorphic to a representation in form of $2n \times 2n$ real matrices.

7.2. The number of space dimensions

We have shown that it is possible to construct spacetime geometry on the basis of (Hamiltonian) dynamics, i.e. on the geometric *interpretation* of the matrix operators that represent the observables. The possible numbers of spacetime dimensions $N = p + q$ in this approach is given by the representation theory of real matrices [72]:

$$p - q = 0 \text{ or } 2 \bmod 8. \tag{83}$$

As we have derived above, there is only one single skew-symmetric (time-like) symplex-generator of the Clifford algebra (i.e. γ_0), i.e. we have to set $q = 1$ and therefore there are $p = N - 1$ space dimensions so that:

$$N - 2 = 0 \text{ or } 2 \bmod 8. \tag{84}$$

Hence, the dimensionality of dynamically emergent spacetime geometries are algebraically limited to $1 + 1$, $3 + 1$, $9 + 1$, $11 + 1$, $17 + 1$, $19 + 1$, $25 + 1$, $27 + 1$, $\dots$ dimensions, corresponding to symplices of size 2×2, 4×4, 32×32, 64×64, $\dots$ [37].

Symplices of size 2×2 have *either* real *or* pure imaginary eigenvalues and are therefore not able to represent the general (i.e. complex) case. But there is another argument, why one degree of freedom is not sufficient. A single degree of freedom does not enable to describe interaction. We cannot expect to derive physics based on non-interaction. Especially spacetime can only be explained based on interaction. Furthermore, one oscillating degree of freedom implies turning points, where the momentum (or kinetic energy) is zero. At these points, there is a moment without change. An hypothetical observer could not experimentally decide, whether the direction of time has changed at these points or not. Only in case of two (or more) dimensions, there is an additional COM — the angular momentum — that (if it is nonzero) ensures that there *is change at all times*. In case of two dimensions and spin, time reversal is always connected to a nonzero *action*. Therefore, the 2×2 case cannot be generalized and is rejected as too simple.

Certainly there are several different lines of argumentation, why spacetime has $3+1$ dimensions. However, the algebraic framework described here has the advantage that it is under certain constraints maximal abstract and does not require to postulate and establish an elaborated physical framework based on which the arguments achieve explanatory power. Precisely speaking, our considerations are less concerned with "the world" but instead with the most simple general interpretation of clusters of variables as objects of "a world".

1. The eigenvalues of $2n \times 2n$ matrices over the reals are (for $n > 1$) pairs of complex conjugate eigenvalues. Since N is even, all representations for $N > 1$ have matrix dimensions that are multiples of 4 and can therefore always be block-diagonalized to 4×4-blocks. Therefore, higher dimensional cases based on $4^M \times 4^M$-matrices can be split into M objects with representation 4×4, i.e. into M objects in $Cl_{3,1}$.
2. The number of variables of the spinor should be equal to the dimension of the geometry constructed from it. The Pauli and the Dirac algebra are the only cases where the state vector ψ has the same dimension as the spacetime constructed from it, i.e. the only case where $2\,n = N$ so that
$$2^N = N^2. \tag{85}$$
This criterium is extremely strong and restricts N to $N = 2$ or $N = 4$. We will discuss this point in Section 9.3.
3. Since N is even in all possible dynamically generated spacetime dimensions, the dimension of the spinor $2\,n$ is always a multiple of 4. Therefore, all higher-dimensional spinors (and hence spacetime geometries) can be re-interpreted as a number of interacting 4-spinors.
4. In 3 space dimensions, the number of rotation axis equals the number of spatial directions since one can form 3 rotator pairs $\gamma_j\,\gamma_k$ (both space-like vectors) without repetition from 3 spatial directions. Both — rotators and boosters — mutually anticommute with each other and hence generate an orthogonal space of the same dimension. In more than 3 space dimensions, this

changes: The number of boosts stays proportional to the number of spatial dimensions, but the number of (symplectic, i.e. binary) rotators grows quadratically with the number of spatial directions $N-1$ since one can select $\binom{N-1}{2} = (N-1)(N-2)/2$ rotators (see Table 2). For $N-1=4$, space dimensions this gives six rotators and not all of them are orthogonal to each other. In more than 3 space dimensions, some rotators commute, for instance in a 4-dimensional space the rotators $\gamma_1\gamma_2$ and $\gamma_3\gamma_4$ commute [70]. This means that (dynamically generated) spacetimes with more than 3 spatial dimensions are topologically not homogeneous. More precisely: $D=3$ is the only space dimension in which all rotational axis have the same relation to all others.

Hence, the simplest possible generally useful dimensionality of "objects" is $3+1$.[y]

In Section 4, we raised the question of measurability. We emphasized that all unit vectors have to be symplices in order to be measurable. Since all basic elements (unit vectors) anticommute, also the bi-vectors — i.e. the products of two unit vectors — are symplices. Geometrically bi-vectors are "oriented faces". However, products of three orthogonal symplices - which would correspond to a volume — are cosymplices and therefore have vanishing expectation values (see also App. C.1)). It follows that in a dynamically emerging spacetime "objects" can have positions (or directions), but they have no measurable volume, i.e. fundamental (measurable) objects have — interpreted as objects in spacetime — no volume, they are point-like particles.

7.3. The real Dirac matrices

In the previous section, we have shown that the case of 4×4-matrices that represent a geometric (Clifford) algebra is unique and fundamental. It is the simplest and most general algebra that allows for construction of spacetime. And it is the only algebra

[y]Other authors also considered it worthwhile to analyze aspects of the dimensionality of spacetime from a Lie-algebraic point of view [21].

that generates a spacetime *as we experience it.* With respect to the basic variables ψ, it represents systems with two DOFs. It can also be regarded as "fundamental" as it describes the smallest possible oscillatory system with (internal) coupling or "interaction".[z]

An appropriate choice of the basic matrices of the Clifford algebra $Cl_{3,1}(R)$ are the well-known Dirac matrices. Ettore Majorana was the first who described a system of Dirac matrices that contains exclusively imaginary (or exclusively real) values [65]. We call these matrices the "real Dirac matrices" (RDMs) and since we are — on the fundamental level — committed to the reals, we will use the RDMs instead of the conventional system.[aa]

Often the term "Dirac matrices" is used in more restrictive way and designates only four matrices, i.e. the generators of the Clifford algebra.[bb] The RDMs as we define them here include the complete system of 16 real matrices and include all elements of the Clifford algebra generated by γ_μ:

$$\begin{aligned}
\gamma_{14} &= \gamma_0\,\gamma_1\,\gamma_2\,\gamma_3; & \gamma_{15} &= 1 \\
\gamma_4 &= \gamma_0\,\gamma_1; & \gamma_7 &= \gamma_{14}\,\gamma_0\,\gamma_1 = \gamma_2\,\gamma_3 \\
\gamma_5 &= \gamma_0\,\gamma_2; & \gamma_8 &= \gamma_{14}\,\gamma_0\,\gamma_2 = \gamma_3\,\gamma_1 \\
\gamma_6 &= \gamma_0\,\gamma_3; & \gamma_9 &= \gamma_{14}\,\gamma_0\,\gamma_3 = \gamma_1\,\gamma_2 \\
\gamma_{10} &= \gamma_{14}\,\gamma_0 = \gamma_1\,\gamma_2\,\gamma_3 \\
\gamma_{11} &= \gamma_{14}\,\gamma_1 = \gamma_0\,\gamma_2\,\gamma_3 \\
\gamma_{12} &= \gamma_{14}\,\gamma_2 = \gamma_0\,\gamma_3\,\gamma_1 \\
\gamma_{13} &= \gamma_{14}\,\gamma_3 = \gamma_0\,\gamma_1\,\gamma_2.
\end{aligned} \tag{86}$$

[z]The role of the Dirac matrices as generators of the symplectic group $Sp(4, R)$ has been described before, see for instance [63]. It has been noted already in [28].

[aa]According to the fundamental theorem of the Dirac matrices by Pauli, we are free to choose any system of Dirac matrices, since all systems are algebraically similar, i.e. can be mapped onto each other via similarity transformations [73]. The transformation to the conventional system is given in App. E.5. The isomorphism of this matrix group to the 3 + 2-dimensional de Sitter group has been pointed out by Dirac in [28].

[bb]Note that we define $\gamma_{14} = \gamma_0\,\gamma_1\,\gamma_2\,\gamma_3$, which is labeled γ_5 in QED.

There are $2\,n! = 24$ possible permutations of the variables in the state vector ψ, but since a swap of the two DOFs does not change the form of γ_0, there are 12 possible basic systems of RDMs [6]. Since γ_0 is antisymmetric, there are only six possible choices for γ_0. Each of them allows to choose between two different sets of the "spatial" matrices.[cc] The explicit form of the RDMs is given in [6, 8]. The four basic RDMs are anti-commuting

$$\gamma_\mu\,\gamma_\nu + \gamma_\nu\,\gamma_\mu = 2\,g_{\mu\nu}\,\mathbf{1}, \tag{87}$$

and the "metric tensor" $g_{\mu\nu}$ has the form[dd]

$$g_{\mu\nu} = \mathrm{Diag}(-1, 1, 1, 1). \tag{88}$$

Any real 4×4 matrix $\mathbf{A}$ can be written as a linear combination of RDMs according to

$$\mathbf{A} = \sum_{k=0}^{15} a_k\,\gamma_k, \tag{89}$$

where a quarter of the trace $\mathrm{Tr}(\mathbf{A})$ equals the *scalar component* a_{15}, since all other RDMs have zero trace. The RDM-coefficients a_k are given by the *scalar product*

$$a_k = \mathbf{A}\,\cdot\,\gamma_k = \frac{\mathrm{Tr}(\gamma_k^2)}{4}\,\mathrm{Tr}\left(\frac{\mathbf{A}\,\gamma_k + \gamma_k\,\mathbf{A}}{8}\right). \tag{90}$$

Any real 4×4 Hamiltonian matrix (symplex) $\mathbf{F}$ can be written as a linear combination in this basis

$$\mathbf{F} = \sum_{k=0}^{9} f_k\,\gamma_k. \tag{91}$$

If Minkowski spacetime is interpreted as a physical reality without intrinsic connection to quantum mechanics, then it appears to be a jolly contingency that the (real) Dirac matrices have the geometric

[cc] Given one system is known as γ_μ, then the alternative system is γ_0, $\gamma_0\gamma_1$, $\gamma_0\gamma_2$, $\gamma_0\gamma_3$.

[dd] The convention of QED ($g_{\mu\nu} = \mathrm{Diag}(1, -1, -1, -1)$) cannot be represented with real Dirac matrices [72, 82].

properties that exactly fit to the needs of spacetime geometry [20, 43]. In [20] the authors write: "Historically, there have been many attempts to produce the appropriate mathematical formalism for modeling the nature of physical space, such as Euclid's geometry, Descartes' system of Cartesian coordinates, the Argand plane, Hamilton's quaternions, Gibbs' vector system using the dot and cross products. We illustrate however, that Clifford's geometric algebra (GA) provides the most elegant description of physical space." If spacetime emerges from dynamics, then there is deeper reason for this elegance. Geometric algebra (GA) — if introduced by postulating abstract mathematical anticommuting objects — leaves a mark of "artificial construction" that has few explanatory power. But GA is way more than an elegant way to write down equations: if *spacetime emerges from dynamics* in the form of the Dirac algebra, then we have the *missing link*, the *physical reason* for the congruence of GA with 3 + 1-dimensional Minkowski spacetime. Then the relativistic covariance of the Dirac algebra is the only logical explanation for the form of the Lorentz transformations. This implies a reversal of common understanding. Neither the experimentally found or postulated properties of spacetime nor the postulated constancy of the speed of — yet undefined — light waves are the proper reason for the form of the Lorentz transformations, but the algebraic structure of Hamiltonian dynamics of the fundamental variables. It is a consequence of the physical restriction that COM are required as references for measurements. Measurable physical properties of physical objects are those properties that appear or can be related to (functions of) constants of motion — in some (other) physical system. Non-measurable (Boolean) properties are typically due to symmetry properties — again determined by the properties of the Dirac algebra. We will show this in detail in the next section.

8. Lorentz Transformations are Symplectic Transformations

Structure preservation refers to Eq. (7). That is: the transformation of the skew-symmetric matrix $\mathbf{S}$ to γ_0 is the exact contrary of structure preserving — it is *structure defining*. Only transformations that preserve the form of γ_0 are called structure preserving. And since γ_0 represents the form of Hamilton's equations, structure preserving transformations are *canonical*. *Linear* structure preserving transformations are called *symplectic*.

However, the form of γ_0 — as a result of Eq. (7) — is not unique in the Dirac algebra. *Any* skew-symmetric γ-matrix squares to -1 and is equally useable as symplectic unit matrix and can equally well represent the time direction. The use of the other skew-symmetric matrices $(\gamma_{10}, \gamma_{14}, \gamma_7, \gamma_8, \gamma_9)$ implies a permutation of the variables in ψ and a permutation of the indices of the other matrices. There are six different equally legitimate systems [6]. If we consider interactions between spinors, we have to take the possibility into account that not all (sub-) systems are described by the same structure defining transformation according to Eq. (7) — though we will not discuss this in detail here.

Using Eq. (62), it is easy to show that symplectic transformations are *structure preserving*, i.e. a transformation of the dynamical variables ψ that does not change the form of the equations, neither it changes the form of γ_0. For the matrix of observables $\mathbf{S} = \Sigma\gamma_0$, we expect a similarity transformation. The linear transformation $\mathbf{R}$ is given as

$$\tilde{\psi} = \mathbf{R}\,\psi. \tag{92}$$

Hence,

$$\begin{aligned} \mathbf{S} &= \langle \psi\psi^T \rangle\,\gamma_0, \\ \tilde{\mathbf{S}} &= \langle \tilde{\psi}\tilde{\psi}^T \rangle\,\gamma_0 \\ &= \langle \mathbf{R}\,\psi\psi^T\,\mathbf{R}^T \rangle\,\gamma_0 \\ &= -\mathbf{R}\,\langle \psi\psi^T \rangle\,\gamma_0\,\gamma_0\,\mathbf{R}^T\gamma_0 \\ &= -\mathbf{R}\,\mathbf{S}\,\gamma_0\,\mathbf{R}^T\gamma_0, \end{aligned} \tag{93}$$

so that the conditions are fulfilled if $\mathbf{R}$ is symplectic (fulfills Eq. (62)):

$$\begin{aligned} -\gamma_0\,\mathbf{R}^T\gamma_0 &= \mathbf{R}^{-1}, \\ \mathbf{R}\,\gamma_0\,\mathbf{R}^T &= \gamma_0. \end{aligned} \tag{94}$$

Then, it is easy to show that symplectic transformations are *structure preserving*, i.e. a symplex $\mathbf{F}$ transformed by a symplectic similarity transformation $\mathbf{R}$ remains a symplex:

$$\begin{aligned} \tilde{\mathbf{F}} &= \mathbf{R}\,\mathbf{F}\,\mathbf{R}^{-1}, \\ \Rightarrow \tilde{\mathbf{F}}^T &= (\mathbf{R}^{-1})^T\,\mathbf{F}^T\,\mathbf{R}^T \\ &= (-\gamma_0\,\mathbf{R}\,\gamma_0)\,(\gamma_0\,\mathbf{F}\,\gamma_0)\,(-\gamma_0\,\mathbf{R}^{-1}\,\gamma_0) \\ &= \gamma_0\,\mathbf{R}\,\mathbf{F}\,\mathbf{R}^{-1}\,\gamma_0 \\ &= \gamma_0\,\tilde{\mathbf{F}}\gamma_0. \end{aligned} \tag{95}$$

In combination with Eq. (68), one finds that a transformation is symplectic if it is possible to find a (constant) Hamiltonian from which the transformation can be derived: a transformation is symplectic, if it can be expressed as the result of a *possible evolution in time*. This implies that not all similarity transformations are possible evolutions in time.

As we show in the following, it is legitimate to call the structure preserving property of symplectic transformations the *principle of (special) relativity* since Lorentz transformations (LTs) of spinors in Minkowski spacetime are a subset of the possible symplectic transformations for two DOFs. It is well-known that the matrix exponential of a "Hamiltonian matrix" (i.e. a symplex) is a symplectic transformation. It remains to be shown which symplices generate LTs. Consider we transform a matrix $\mathbf{X} = \mathcal{E}\,\gamma_0 + P_x\,\gamma_1 + P_y\,\gamma_2 + P_z\,\gamma_3$ using $\mathbf{R}_4 = \exp(\gamma_4\,\varepsilon/2)$, then we obtain after decomposition of the transformed matrix $\mathbf{X}'$ into the RDM-coefficients:

$$\begin{aligned} \mathbf{X}' &= \mathbf{R}_4\,\mathbf{X}\,\mathbf{R}_4^{-1}, \\ \mathbf{X}' &= \mathcal{E}'\,\gamma_0 + P_x'\,\gamma_1 + P_y'\,\gamma_2 + P_z'\,\gamma_3, \\ \mathcal{E}' &= \mathcal{E}\,\cosh(\varepsilon) + P_x\,\sinh(\varepsilon), \\ P_x' &= P_x\,\cosh(\varepsilon) + \mathcal{E}\,\sinh(\varepsilon), \\ P_y' &= P_y, \\ P_z' &= P_z. \end{aligned} \tag{96}$$

With the usual parametrization where ε is the "rapidity" ($\beta = \tanh(\varepsilon)$, $\gamma = \cosh(\varepsilon)$, $\beta\gamma = \sinh(\varepsilon)$), one finds

$$\begin{aligned}\mathcal{E}' &= \gamma\,\mathcal{E} + \beta\,\gamma\,P_x,\\ P_x' &= \gamma\,P_x + \beta\,\gamma\,\mathcal{E},\end{aligned} \tag{97}$$

which is a Lorentz boost along x. Same transformation, but $\mathbf{X} = E_x\,\gamma_4 + E_y\,\gamma_5 + E_z\,\gamma_6 + B_x\,\gamma_7 + B_y\,\gamma_8 + B_z\,\gamma_9$:

$$\begin{aligned}\mathbf{X}' &= \mathbf{R}\,\mathbf{X}\,\mathbf{R}^{-1},\\ E_x' &= E_x,\\ B_x' &= B_x,\\ E_y' &= E_y\,\cosh(\varepsilon) + B_z\,\sinh(\varepsilon),\\ E_z' &= E_z\,\cosh(\varepsilon) - B_y\,\sinh(\varepsilon),\\ B_y' &= B_y\,\cosh(\varepsilon) - E_z\,\sinh(\varepsilon),\\ B_z' &= B_z\,\cosh(\varepsilon) + E_y\,\sinh(\varepsilon),\end{aligned} \tag{98}$$

which is again a Lorentz boost along x, but now it shows the transformation behavior of electromagnetic fields. Obviously, the algebraic structure of the Dirac matrices in combination with symplectic dynamics results directly in Lorentz transformations. Indeed the transformations above are identical to the Lorentz transformations of the Dirac equation.[ee]

Our approach does not refer to the speed of light, to time dilation or with alike. N.D. Mermin once wrote: "Relativity is not a branch of electromagnetism" [67]. Our approach goes beyond this statement. We did not directly introduce spacetime. But we introduced Lorentz transformations in direct combination with the electromagnetic fields. Lorentz transformations, electrodynamics, spinors, and the mass–energy relation all together are facets of the same phenomenon. Spacetime is generated by transformation operators, which also represent the electromagnetic interaction. The conventional approach

[ee] Most textbooks on relativistic QM contain inadequately short and/or incomprehensible descriptions of the covariance of the Dirac equation. The best depiction found by the author is given in [85].

derives LTs from the "constancy of the speed of light". This is a phenomenological Ansatz which certainly has a legitimization in the history of physics, but is *logically misleading*: the "constancy" of the speed of light is neither the cause nor the reason for, but just one aspect of (special) relativity. *Relativity* as such is nothing but structure preservation, i.e. it corresponds to the linear canonical transformations of classical mechanics. Singh continued the work of Mermin in the (algebraic) direction [89]. At about the same time, Penrose and Rindler wrote in [74]: "The formalism most commonly used for the mathematical treatment of manifolds and their metrics is, of course, the tensor calculus (or such essentially equivalent alternatives as Cartan's calculus of moving frames). But in the specific case of four dimensions and Lorentzian metric, there happens to exist — by accident or providence — another formalism which is in many ways more appropriate, and that is the formalism of 2-spinors." What Rindler formulates in terms of complex 2-spinors, can (more "classically") be expressed by real 4-spinors as we present it here. In this way, it becomes more evident that the emergence of spacetime is no accident, but the simplest method of pattern recognition of abstract and general symplectic dynamics. Furthermore, by the use of real 4-spinors, we discover that the electromagnetic fields are intrinsically connected to the emergent spacetime geometry. This is the reason why the speed of light is related to the geometry of spacetime. This is explained in more detail in Section 8.1.

The possibility to derive electrodynamics from relativity alone has been questioned [35]. But in the context of symplectic dynamics as suggested here, there are additional restrictions and MWEQs can indeed be derived. It is an experimental fact that electromagnetic interactions are structure preserving. No electromagnetic process known to the author is able to transform a fermion into another. There is pair production and particle–antiparticle annihilation, but it is conventional wisdom to identify the antiparticle with the particle going "backwards in time" [85]. The fact that Lorentz transformations of the Dirac spinor are indeed linear canonical transformations, is long known [28, 51]. But it is rarely mentioned in discussions on relativity — likely because relativity is almost exclusively depicted as a theory of space and time. However — as Edward J. Gillis remarked, "relativity is not about spacetime" [38].

8.1. The electromechanical equivalence (EMEQ)

We started our considerations with abstract entities. It is therefore unavoidable to present an *interpretation* of the terms (i.e. the observables) derived from these entities. The interpretations themselves can neither be "derived" nor be proven to be "correct". We can only show that they are consistent and meaningful and that they have explanatory power. Interpretations make sense or they don't — but they do not exclude other interpretations.

The transformation properties (96) and (98) suggest the introduction of the electromechanical equivalence (EMEQ), as presented in Refs. [6, 8]. In those papers, the EMEQ was merely used as a *formal tool* that allows to obtain a descriptive interpretation of symplectic transformations. Here, we argue substantially and present a derivation of MWEQs.

In Section 6.1, we have shown that the fundamental solution of the EQOM is given by a symplectic transfer matrix. In order to produce COM, the transfer matrix should represent a strongly stable system, i.e. all of its eigenvalues lie on the unit circle in the complex plane. In this case, the symplex $\mathbf{F}$ has exclusively purely imaginary eigenvalues. We analyzed the structure of $\mathbf{F}$ by introducing the RDMs which are a representation of the Clifford algebra $Cl_{3,1}$. Therefore, the list of unit symplices includes the four basic elements $\gamma_0 \dots \gamma_3$ and the six bi-vectors. The other six members of the group of the RDMs are cosymplices. The analysis of the transformation properties indicates a possible physical interpretation of the coefficients of the algebra: we associate the f_k (Eq. (91)) as follows:

$$\begin{aligned} f_0 &\equiv \mathcal{E} && \text{Energy} \\ (f_1, f_2, f_3)^T &\equiv \vec{P} && \text{Momentum} \\ (f_4, f_5, f_6)^T &\equiv \vec{E} && \text{Electric field} \\ (f_7, f_8, f_9)^T &\equiv \vec{B} && \text{Magnetic field} \end{aligned} \tag{99}$$

This association is not arbitrary as the transformation behavior under rotations and Lorentz boosts of the f_k has to fit to the corresponding physical quantities. However, it is not unique. Besides the electromagnetic fields, any second rank tensor would fulfill the formal requirements. If we would not be interested in electrodynamics but for instance in hydrodynamics or the motion of rigid bodies, we would have to use a different scheme of interpretation. This ambiguity is unavoidable since our starting point was the abstract structure of Hamiltonian symplectic dynamics.[ff]

However, our choice is legitimate if the EMEQ is consistent, meaningful and if it helps in structural analysis as we will show in what follows. Dimensional problems should not appear as long as we can find physical constants that allow to translate energy, momentum, electric and magnetic fields into frequencies (the dimension of **F** is a frequency). For the energy, the required factor is $\hbar$:

$$\begin{aligned} \mathcal{E}/\hbar &= \text{frequency}, \\ |\vec{P}|\, c/\hbar &= \text{frequency}, \\ \frac{e}{m_e\, c}\, |\vec{E}| &= \text{frequency}, \\ \frac{e}{m_e}\, |\vec{B}| &= \text{frequency}. \end{aligned} \tag{100}$$

We assume that in our unit system, all these units have the numerical value of unity, i.e. $m_e = e = \hbar = c = 1$.

With this nomenclature, the eigenvalues $\pm\, i\, \omega_1$, $\pm\, i\, \omega_2$ of **F** are given as:

$$\begin{aligned} \omega_1 &= \sqrt{K_1 + 2\,\sqrt{K_2}}, \\ \omega_2 &= \sqrt{K_1 - 2\,\sqrt{K_2}}, \end{aligned}$$

[ff] And it tells us that major parts of the laws of physics are the result of pattern recognition in the same sense as the interpretation of acoustic signals in terms of sound and music is. Because, that is what we do here: We analyze the patterns of fundamental variables in Hamiltonian motion.

$$
\begin{aligned}
K_1 &= -\mathrm{Tr}(\mathbf{F}^2)/4 = \mathcal{E}^2 + \vec{B}^2 - \vec{E}^2 - \vec{P}^2, \\
K_2 &= \mathrm{Tr}(\mathbf{F}^4)/16 - K_1^2/4, \\
&= (\mathcal{E}\,\vec{B} - \vec{P} \times \vec{E})^2 - (\vec{E} \cdot \vec{B})^2 - (\vec{P} \cdot \vec{B})^2, \\
\mathrm{Det}(\mathbf{F}) &= \omega_1^2\,\omega_2^2 = K_1^2 - 4\,K_2.
\end{aligned} \tag{101}
$$

Bi-vectors form a so-called *even subalgebra* of the Dirac algebra.[gg] We have shown that any Clifford algebra $\mathrm{Cl}_{p,q}$ which is able to represent spacetime has an even number of generators N. This holds also for the Dirac algebra. Even dimensional algebras have *even subalgebras*, which means that even elements generate exclusively even elements, whereas odd elements can be used to generate the full algebra: Fermions can generate electromagnetic fields, but photons cannot generate fermions. Consequently, there should be (differential-) equations that allow to derive the bi-vector fields $\vec{E}$ and $\vec{B}$ from vectors, but not vice versa. This means that the description of a free particle without external fields should not refer to bi-vectors. Hence, it is natural to distinguish the vector and bi-vector components and for vanishing bi-vectors ($\vec{E} = \vec{B} = 0$) one finds:

$$
\begin{aligned}
K_1 &= \mathcal{E}^2 - \vec{P}^2, \\
K_2 &= 0, \\
\omega_1 &= \omega_2 = \sqrt{K_1} = \sqrt{\mathcal{E}^2 - \vec{P}^2},
\end{aligned} \tag{102}
$$

which are the relativistic invariants of matter-fields. At the end of Section 4, we argued that a constant measurement reference can be constructed as a quadratic form. Here, we have an example, since $\sqrt{K_1} = m = \sqrt{\mathcal{E}^2 - \vec{P}^2}$: *The mass is a COM and has the physical meaning and unit of frequency.* This is an important result and a significant step towards quantum mechanics: It is the combination of the Planck–Einstein relation $E = \hbar\,\omega$ and Einstein's $E = m\,c^2$. It follows that the "time" parameter as we introduced it in Eq. (12),

[gg] We define elements to be *even*, if they do not change sign when all *basic elements* $\gamma_0, \ldots, \gamma_3$ change sign. They are odd otherwise, i.e. bi-vectors are even, vectors and tri-vectors are odd.

corresponds to the *eigentime* or *proper time*. The eigenfrequency that defines the scale of the eigentime is the mass and hence a massive particle is described by an inertial frame of reference in combination with an oscillator of constant frequency (i.e. a clock). This is the deeper meaning of Einstein's clock attached to all inertial frames.

For vanishing vector components ($\mathcal{E} = 0 = \vec{P}$), we have

$$\begin{aligned} K_1 &= \vec{B}^2 - \vec{E}^2, \\ K_2 &= -(\vec{E} \cdot \vec{B})^2, \end{aligned} \tag{103}$$

which are the well-known relativistic invariants of the electromagnetic field. It follows that the frequencies are only real-valued for $\vec{E} \cdot \vec{B} \geq 0$ and $\vec{B}^2 - \vec{E}^2 \geq 0$. Furthermore, the eigenfrequencies of a pure electromagnetic wave vanish since $K_1 = K_2 = 0$. This implies that we cannot transform by whatever means into a system of a pure e.m.-wave: The *ideal* electromagnetic wave has no inertial system, no eigenfrequency and no intrinsic phase advance. An analysis of the transformations generated by the bi-vectors yields that the elements γ_4, γ_5, and γ_6 are responsible for Lorentz boosts, while γ_7, γ_8, and γ_9 are generators of (spatial) rotations. This corresponds to our physical intuition since (charged) particles are *accelerated* by electric fields but their "trajectories" are bended (i.e. rotated) in magnetic fields. It is therefore no surprise that — using the EMEQ — the derivation of the Lorentz force is straightforward (see Section 8.2).

Furthermore, we note that there are vector-elements associated with $\mathcal{E}$ and $\vec{P}$, but no elements associated directly with the spacetime coordinates. For a representation of a single elementary particle, this makes sense insofar as the dynamics of a single particle may not refer to absolute positions (unless via the bi-vector fields). Spacetime coordinates do not refer to particle properties, but to the relative position (relation) of particles. Hence, as long as we refer to a single object based on fundamental variables, we observe the momentum and energy *of* a particle as it behaves in an *external electromagnetic field*.

As we already mentioned in Section (4), the symplex $\mathbf{S} = \Sigma\,\gamma_0$ is singular, if we use only a single real spinor ψ to define

$$\Sigma = \langle\, \psi\, \psi^T\, \gamma_0 \,\rangle, \tag{104}$$

where the angles imply some (unspecified) sort of average. Since the eigenvalues of the symplex **S** yield the "frequency" or "mass" $m = \sqrt{\mathcal{E}^2 - \vec{P}^2}$ of the structure which we identify with a particle, we need at least $N \geq 4$ linear independent spinors to obtain a massive fermion. In the classical Dirac theory, the electron spinor is a linear combination of two complex spinors, i.e. also requires four real spinors. From this point of view, *inertial mass* might be interpreted statistically — as a property that can only be derived using averaging — either over samples or in time. The idea that mass might be a statistical phenomenon is neither new nor extraordinarily exotic, see for instance [60, 98] and references therein. But here we do not refer to entropy but rather to dynamical or algebraic properties of real spinors.

8.2. The Lorentz force

The four-momentum of a particle (fermion) is defined by vectors. The vector components of **S** are associated with the four-momentum:

$$\mathbf{P} = \mathcal{E}\,\gamma_0 + P_x\,\gamma_1 + P_y\,\gamma_2 + P_z\,\gamma_3 = \mathcal{E}\,\gamma_0 + \vec{P}\,\vec{\gamma}, \tag{105}$$

where $\mathcal{E}$ is the energy and $\vec{P}$ the momentum. Accordingly, the fields are associated with the bi-vectors of the "force" matrix **F**. Then, Eq. (19) is (in appropriate units) the Lorentz force equations

$$\frac{d\mathbf{P}}{d\tau} = \dot{\mathbf{P}} = \frac{q}{2\,m}(\mathbf{F}\,\mathbf{P} - \mathbf{P}\,\mathbf{F}), \tag{106}$$

where τ is the proper time and $\frac{q}{2m}$ is a relative scaling factor. In the lab frame time $dt = \gamma\,d\tau$, Eq. (106) yields (setting $c = 1$):

$$\begin{aligned}
\frac{d\mathcal{E}}{d\tau} &= \frac{q}{m}\,\vec{P}\,\vec{E}, \\
\frac{d\vec{P}}{d\tau} &= \frac{q}{m}\,(\mathcal{E}\,\vec{E} + \vec{P}\times\vec{B}), \\
\gamma\,\frac{d\mathcal{E}}{dt} &= q\,\gamma\,\vec{v}\,\vec{E},
\end{aligned}$$

$$\begin{aligned}
\gamma\,\frac{d\vec{P}}{dt} &= \frac{q}{m}\left(m\,\gamma\,\vec{E} + m\,\gamma\,\vec{v}\times\vec{B}\right),\\
\frac{dE}{dt} &= q\,\vec{v}\,\vec{E},\\
\frac{d\vec{P}}{dt} &= q\left(\vec{E}+\vec{v}\times\vec{B}\right).
\end{aligned} \tag{107}$$

(Note that Eq. (106) allows to add arbitrary multiples of $\mathbf{P}$ to $\mathbf{F}$ without an effect on $\dot{\mathbf{P}}$.)

9. Electrodynamics from Symplectic Spacetime

The electromagnetic terms of the symplex $\mathbf{F}$ are bi-vectors. It has been pointed out in Section 8.1 that bi-vectors form an *even subalgebra* which implies that they are generated by the product of two vectors. Up to now, only a single type of four-vector has been introduced, namely the energy–momentum four-vector of a single system ("particle") of interest. Using a pair of four-vectors, one can construct a bi-vector by the use of the commutator of two vectors. However, in order to have a nonzero bi-vector, these two vectors must be different. This could either be the energy–momentum four-vector of a second system or a four-vector representing something different than energy and momentum. Firstly, we can argue that a second particle did — in contrast to the bi-vectors — not (yet) appear in our considerations and therefore it is unavoidable to introduce a new four-vector. And secondly, the EMEQ introduces momenta without the corresponding coordinates. Hence, we write this four-vector as

$$\mathbf{X} = \gamma_0\,t + \vec{\gamma}\cdot\vec{x}, \tag{108}$$

and express the fields $\vec{E}$ and $\vec{B}$ — which were originally understood as functions of the eigentime τ — by $\vec{x}$ and t:

$$\begin{aligned}
\vec{E}(\tau) &\to \vec{E}(\vec{x},t),\\
\vec{B}(\tau) &\to \vec{B}(\vec{x},t).
\end{aligned} \tag{109}$$

This means that we extend the dynamics of (single) observables to the dynamics of "fields", depending on the new vector type parameters t and $\vec{x}$. For now, we keep open the question, how these vectors are related to $(\mathcal{E}, \vec{P})$. But we cannot resist to cite Einstein here: "Spacetime does not claim existence on its own, but only as a structural quality of the field" [32].

There have been several attempts to "derive" MWEQs in the past [27, 31, 36, 56–58]. None of these succeeded in finding general acceptance. Most textbooks treat just the Lorentz covariance of MWEQs. A derivation of MWEQs from the covariance condition alone is considered impossible [35]. But our ansatz is based on an algebraic framework with additional (i.e. symplectic) constraints. Symplectic constraints are well-known in different context and led (for instance) to the non-squeezing theorem, i.e. to the metapher of the "symplectic egg" [25, 41]. We have shown that Lorentz transformations as well as the action of the Lorentz force are symplectic. The appearance of cosymplices as generators is excluded for the structure-preserving electromagnetic theory.

In order to keep track of the transformation properties with respect to symplectic transformations, the bi-vectors fields can only be expressed (for instance in form of a Taylor series) by algebraic expressions that respect the appropriate transformation properties. According to Eq. (99), the field matrix is given by

$$\mathbf{F} = E_x\,\gamma_4 + E_y\,\gamma_5 + E_z\,\gamma_6 + B_x\,\gamma_7 + B_y\,\gamma_8 + B_z\,\gamma_9. \tag{110}$$

If we write the fields as a Taylor series, the first term *must* have the following form in order to be linear in $\mathbf{X}$ *and* to yield a bi-vector:

$$\mathbf{F} = \mathbf{F}_0 + (\mathcal{D}\mathbf{F})\,\mathbf{X} - \mathbf{X}(\mathcal{D}\mathbf{F}) + \cdots\,, \tag{111}$$

where $\mathcal{D}\mathbf{F}$ is an appropriate derivative taken at $\mathbf{X} = 0$ and *must* be a vector. Alternative forms of the linear term either include an "axial" vector (cosymplex) $\mathbf{V}$ or a pseudoscalar, both having a vanishing expectation value:

$$\begin{aligned} \mathbf{V} &= V_t\,\gamma_{10} + V_x\,\gamma_{11} + V_y\,\gamma_{12} + V_z\,\gamma_{13}, \\ \mathbf{F}_1 &= (\mathbf{V}\,\mathbf{X} + \mathbf{X}\mathbf{V})/2. \end{aligned} \tag{112}$$

Since we can exclude the appearance of cosymplices in structure preserving interactions, the first order term has the form of Eq. (111).

In the following, we analyze first and second order expressions that can be constructed with respect to the transformation properties. We skip constant terms for the moment and analyze the relations of the partial derivatives. Linear terms can exclusively be expressed by the commutator of two vector quantities, i.e. $\mathbf{X}$ and another (constant) four-vector $\mathbf{J} = \rho_0\,\gamma_0 + \vec{j}_0 \cdot \vec{\gamma}$:

$$\mathbf{F}_1 = (\gamma_0\,\vec{E} + \gamma_{14}\,\gamma_0\vec{B}) \cdot \vec{\gamma} = \frac{4\pi}{3}\frac{1}{2}\,(\mathbf{X}\,\mathbf{J} - \mathbf{J}\,\mathbf{X}), \tag{113}$$

so that we obtain

$$\begin{aligned} \vec{E} &= \frac{4\pi}{3}\,(\vec{j}_0\,t - \rho_0\,\vec{x}), \\ \vec{B} &= \frac{4\pi}{3}\,(\vec{x} \times \vec{j}_0), \end{aligned} \tag{114}$$

and find that these linear terms fulfill MWEQs:

$$\begin{aligned} \vec{\nabla} \cdot \vec{E} &= 4\pi\,\rho_0, \\ \vec{\nabla} \cdot \vec{B} &= 0, \\ \vec{\nabla} \times \vec{E} + \partial_t\vec{B} &= 0, \\ \vec{\nabla} \times \vec{B} - \partial_t\vec{E} &= 4\pi\vec{j}_0. \end{aligned} \tag{115}$$

We continue with a second order term[hh]:

$$\mathbf{F}_2 = \frac{1}{4}\,(\mathbf{X}\,\mathbf{J}\,\mathbf{X}\,\mathbf{J} - \mathbf{J}\,\mathbf{X}\,\mathbf{J}\,\mathbf{X}), \tag{116}$$

and it also fulfills MWEQs:

$$\begin{aligned} \vec{\nabla} \cdot \vec{E} &= 4\,\pi\,\rho_1, \\ \vec{\nabla} \cdot \vec{B} &= 0, \end{aligned}$$

[hh]In first order, the two partials of the induction law vanish separately, i.e. $\vec{\nabla}\times\vec{E} = 0 = \partial_t\vec{B}$.

$$\vec{\nabla} \times \vec{E} + \partial_t \vec{B} = 0,$$
$$\vec{\nabla} \times \vec{B} - \partial_t \vec{E} = 4\,\pi\,\vec{j}_1, \tag{117}$$

where

$$\begin{aligned} \rho_1 &= -\vec{j}_0^{\,2}\,t - 3\,t\,\rho_0^2 + 4\,(\vec{j}_0 \cdot \vec{x})\,\rho_0, \\ \vec{j}_1 &= 4\,(\vec{j}_0 \cdot \vec{x} - \rho_0\,t)\vec{j}_0 + (\rho_0^2 - \vec{j}_0^{\,2})\,\vec{x}. \end{aligned} \tag{118}$$

Both orders fulfill the continuity equation:

$$\partial_t\,\rho + \vec{\nabla} \cdot \vec{j} = 0. \tag{119}$$

So far this is not a derivation of MWEQs, but it shows that the transformation properties of bi-vectors and the construction of spacetime imply MWEQs. There is a method to express these ideas mathematically by the use of differential operators. Since the aim is the construction of spacetime, the appropriate generalized derivative must be of vector-type, i.e. we define the covariant derivative by

$$\partial \equiv -\partial_t\,\gamma_0 + \partial_x\,\gamma_1 + \partial_y\,\gamma_2 + \partial_z\,\gamma_3. \tag{120}$$

The non-abelian nature of matrix multiplication requires to distinguish differential operators acting to the right and to the left, i.e. we have ∂ as defined in Eq. (120), $\overrightarrow{\partial}$ and $\overleftarrow{\partial}$ which is written to the right of the operand (thus indicating the order of the matrix multiplication) so that

$$\begin{aligned} \mathbf{F}\overleftarrow{\partial} &\equiv -\partial_t\,\mathbf{F}\,\gamma_0 + \partial_x\,\mathbf{F}\,\gamma_1 + \partial_y\,\mathbf{F}\,\gamma_2 + \partial_z\,\mathbf{F}\,\gamma_3, \\ \overrightarrow{\partial}\mathbf{F} &\equiv -\gamma_0\,\partial_t\,\mathbf{F} + \gamma_1\,\partial_x\,\mathbf{F} + \gamma_2\,\partial_y\,\mathbf{F} + \gamma_3\,\partial_z\,\mathbf{F}. \end{aligned} \tag{121}$$

In order to keep the transformation properties of derivative expressions transparent, we distinguish between the *commutative*

$$\partial \wedge \mathbf{A} \equiv \frac{1}{2}\,(\overrightarrow{\partial}\mathbf{A} - \mathbf{A}\overleftarrow{\partial}) \tag{122}$$

and the *anti-commutative*

$$\partial \cdot \mathbf{A} \equiv \frac{1}{2}\left(\overrightarrow{\partial}\mathbf{A} + \mathbf{A}\overleftarrow{\partial}\right) \tag{123}$$

derivative. Then we find:

$$\begin{aligned}
\frac{1}{2}\left(\overrightarrow{\partial}\ \text{vector}\ - \ \text{vector}\ \overleftarrow{\partial}\right) &\Rightarrow \ \text{bi-vector},\\
\frac{1}{2}\left(\overrightarrow{\partial}\ \text{vector}\ + \ \text{vector}\ \overleftarrow{\partial}\right) &\Rightarrow \ \text{scalar} = 0,\\
\frac{1}{2}\left(\overrightarrow{\partial}\ \text{bi-vector}\ - \ \text{bi-vector}\ \overleftarrow{\partial}\right) &\Rightarrow \ \text{vector},\\
\frac{1}{2}\left(\overrightarrow{\partial}\ \text{bi-vector}\ + \ \text{bi-vector}\ \overleftarrow{\partial}\right) &\Rightarrow \ \text{axial vector} = 0.
\end{aligned} \tag{124}$$

Since (in case of the Dirac algebra) any symplex is the sum of a vector and a bi-vector, we find (for this special case):

$$\frac{1}{2}\left(\overrightarrow{\partial}\ \text{symplex}\ + \ \text{symplex}\ \overleftarrow{\partial}\right) = \partial \cdot \ \text{symplex}\ = 0. \tag{125}$$

Now we return to Eq. (111) and find that $\mathcal{D}\mathbf{F}$ must be a vector and therefore has (in "first order") the form

$$\mathcal{D}\mathbf{F} = \frac{1}{2}\left(\overrightarrow{\partial}\mathbf{F} - \mathbf{F}\overleftarrow{\partial}\right) = 4\,\pi\,\mathbf{J}, \tag{126}$$

which is nothing but a *definition* of the *vector current*

$$\mathbf{J} = \rho\,\gamma_0 + j_x\,\gamma_1 + j_y\,\gamma_2 + j_z\,\gamma_3. \tag{127}$$

Written explicitly in components, Eq. (126) is given by

$$\begin{aligned}
\vec{\nabla} \cdot \vec{E} &= 4\,\pi\,\rho,\\
\vec{\nabla} \times \vec{B} - \partial_t \vec{E} &= 4\,\pi\,\vec{j}.
\end{aligned} \tag{128}$$

We define a vector-potential $\mathbf{A} = \phi\,\gamma_0 + A_x\,\gamma_1 + A_y\,\gamma_2 + A_z\,\gamma_3$ which has (to first order) the form[ii]:

$$\mathbf{A}_1 = \frac{1}{4}\,(\mathbf{X}\,\mathbf{F}_0 - \mathbf{F}_0\,\mathbf{X}), \tag{129}$$

where $\mathbf{F}_0$ is the constant term of Eq. (111). It is easily verified that $\mathbf{F}_0 = \frac{1}{2}\,(\overrightarrow{\partial}\mathbf{A}_0 - \mathbf{A}_0\overleftarrow{\partial})$ and hence the electromagnetic (bi-vector-) fields of the general symplex $\mathbf{F}$ is given by

$$\mathbf{F} = \frac{1}{2}\,(\overrightarrow{\partial}\mathbf{A} - \mathbf{A}\overleftarrow{\partial}), \tag{130}$$

or explicitly in components:

$$\begin{aligned}\vec{E} &= -\vec{\nabla}\phi - \partial_t\vec{A},\\ \vec{B} &= \vec{\nabla}\times\vec{A}.\end{aligned} \tag{131}$$

It is well-known that the homogeneous MWEQs are a direct consequence of Eq. (130):

$$\begin{aligned}\overrightarrow{\partial}\mathbf{F} + \mathbf{F}\overleftarrow{\partial} &= \frac{1}{2}\,(\overrightarrow{\partial}^2\,\mathbf{A} - \mathbf{A}\overleftarrow{\partial}^2) = 0,\\ \vec{\nabla}\cdot\vec{B} &= 0,\\ \vec{\nabla}\times\vec{E} + \partial_t\vec{B} &= 0,\end{aligned} \tag{132}$$

since the squared operators are scalars and commute with $\mathbf{A}$. Accordingly, the continuity equation is a direct consequence of Eq. (126):

$$\begin{aligned}\overrightarrow{\partial}\mathbf{J} + \mathbf{J}\overleftarrow{\partial} &= \frac{1}{8\pi}\,(\overrightarrow{\partial}^2\,\mathbf{F} - \overrightarrow{\partial}\mathbf{F}\overleftarrow{\partial} + \overrightarrow{\partial}\mathbf{F}\overleftarrow{\partial} - \mathbf{F}\overleftarrow{\partial}^2)\\ &= \partial_t\rho + \vec{\nabla}\vec{j} = 0.\end{aligned} \tag{133}$$

[ii]The commutative derivative of the trivial first order form $\mathbf{A} \propto \mathbf{X}$ vanishes and is therefore useless in this context.

The validity of the Lorentz gauge follows from Eq. (124)

$$\frac{1}{2}\left(\overrightarrow{\partial}\mathbf{A} + \mathbf{A}\overleftarrow{\partial}\right) = \partial_t\,\phi + \vec{\nabla}\vec{A} = 0, \tag{134}$$

and one obtains the wave equation of the vector potential $\mathbf{A}$:

$$4\,\pi\,\mathbf{J} = -\overrightarrow{\partial}^2\mathbf{A} = (\partial_t^2 - \vec{\nabla}^2)\,\mathbf{A}. \tag{135}$$

In a "current free region", the fields have to fulfill the homogeneous wave equation:

$$\overrightarrow{\partial}^2\mathbf{F} = (\vec{\nabla}^2 - \partial_t^2)\mathbf{F} = 0. \tag{136}$$

It follows that plane electromagnetic fields "in vacuum" hold $\omega^2 = \vec{k}^2$ so that the fields have no "eigenfrequency" and no dispersion.

Note that Eqs. (132) and (133) are of the same form. The continuity equation (133) describes charge as something with the properties of a substance in spacetime.

In summary: we have shown that Lorentz transformations can indeed be understood as symplectic transformations. The generators of these transformations are electromagnetic fields. Since electrodynamics is a structure preserving interaction, the description of electromagnetic fields requires that exclusively structure preserving terms appear. With these "additional conditions" we were able to derive MWEQs if we interpret the bi-vectors of the matrix $\mathbf{F}$ as functions of vector-type spacetime coordinates of $Cl_{3,1}(\mathbb{R})$. The bi-vector fields in $\mathbf{F}$ fulfill a wave equation that describes the propagation of electromagnetic waves, i.e. of light. One finds that electromagnetic waves "in vacuum" fulfill the relation $\omega^2 = \vec{k}^2$. From the constant group velocity $\frac{\partial\omega}{\partial k} = 1$, we obtain the constancy of the speed of light. Therefore, we did not derive MWEQs from special relativity, but both theories are the result of the construction of spacetime from symplectic dynamics.

Above equations are not restricted to 3 space dimensions. However, the connections between commutators and anti-commutators are way more complex in 10 or 12 dimensions. We had to consider additional "channels": In 3-dimensional space, the symplex $\mathbf{F}$ is composed exclusively of components of the vector and bi-vector type. In higher-dimensional space (10- or 12-dimensional spacetime), the general symplex $\mathbf{F}$ consists (besides vectors and bi-vectors) of n-vectors with $n \in [5, 6, 9, 10, \dots]$ (see App. C.1). Accordingly, we had to include higher-order terms into the Hamiltonian and likely we would have trouble with the stability of the described objects.

9.1. The density

From the combination of Eqs. (109) and (135), it follows that also the (current-) density has also to be constructed as a function of space and time:

$$\begin{aligned} \rho &= \rho(\vec{x}, t), \\ \vec{j} &= \vec{j}(\vec{x}, t), \end{aligned} \tag{137}$$

and it is clear that we have to normalize the density by

$$\int \rho(\vec{x}, t)\, d^3x = const = Q, \tag{138}$$

where Q is the "charge". Especially, the relation between the density $\rho(\vec{x}, t)$ and the "phase space density" $\tilde{\rho}(\psi)$ according to Eq. (32) requires some attention. It is clear that the naive assumption

$$\rho(\vec{x}, t) = \tilde{\rho}(\psi(\vec{x}, t)) \tag{139}$$

cannot be applied directly. According to Eq. (32), we have

$$1 = \int \tilde{\rho}(\psi)\, d^4\psi = \int \tilde{\rho}(\psi(\vec{x}, t))\, \sqrt{g}\, d^3x, \tag{140}$$

where g is the appropriate Gramian determinant. We assume in the following that the normalization has been adjusted accordingly.

If spacetime would be fundamental, we could be sure that the wave function $\psi(\vec{x}, t)$ would be well-defined and single valued for any coordinate of Minkowski spacetime. However, if spacetime emerges from dynamics, then spacetime coordinates might be functions of the phase space position $t(\psi), \vec{x}(\psi)$. This includes the possibility that (a) different phase space positions are mapped to the same spacetime-position and (b) that space and time coordinates are not unique, i.e. particles appear "instantaneously" at different "locations" of spacetime. Finally, this might indeed be the reason why ψ has to be interpreted as a "probability density".

9.2. The wave equations

According to Eq. (135) the vector potential fulfills a wave equation — and also the (electromagnetic) fields can be described by waves. Solutions of wave equations are usually analyzed with the help of the Fourier transformation, i.e. the solutions can be written as superpositions of plane waves

$$\begin{aligned}\phi(\vec{x},t) &= \int \tilde{\phi}(\vec{k},\omega)\, e^{i(\vec{k}\cdot\vec{x}-\omega t)}\, d^3k,\\ \tilde{\phi}(\vec{k},\omega) &= \int \phi(\vec{x},t)\, e^{-i(\vec{k}\cdot\vec{x}-\omega t)}\, d^3x,\end{aligned} \tag{141}$$

where we skipped normalization constants for simplicity. One important feature of the Fourier transform is the replacement of nabla operator $\vec{\nabla}$ with the wave vector $i\vec{k}$ and of the time derivative with the frequency $\partial_t \to -i\,\omega$. The Fourier transformed MWEQs are algebraic conditions for wave functions (we skip the tilde as it is usually clear from the context, if we refer to the fields or their Fourier transform):

$$\begin{aligned}i\,\vec{k}\cdot\vec{B} &= 0,\\ i\,\vec{k}\times\vec{E} - i\,\omega\vec{B} &= 0,\\ i\,\vec{k}\cdot\vec{E} &= 4\pi\,\rho,\\ i\,\vec{k}\times\vec{B} + i\,\omega\vec{E} &= 4\pi\vec{j}.\end{aligned} \tag{142}$$

The second of Eq. (142) implies that all (single) solutions also fulfill $\vec{E}\cdot\vec{B} = 0$. The homogeneous parts of Eq. (142) are structurally an exact copy of the expressions that we obtain from Eq. (101) with the condition $K_2 = 0$:

$$\begin{aligned} \mathcal{E}\,\vec{B} - \vec{P} \times \vec{E} &= 0, \\ \vec{P} \cdot \vec{B} &= 0, \\ \vec{E} \cdot \vec{B} &= 0. \end{aligned} \tag{143}$$

In [8] we made use of the condition $\vec{P} \cdot \vec{B} = 0$ and $\vec{E} \cdot \vec{B} = 0$ which have to be fulfilled in order to block-diagonalize the matrix **F**. But block-diagonalization ("decoupling") did not require the third term $\mathcal{E}\,\vec{B} - \vec{P} \times \vec{E}$ to vanish. However, if the corresponding MWEQ would not yield zero, then we had to deal with a nonzero but divergence-free magnetic current density $4\pi\vec{j}_m = \partial_t\,\vec{B} + \vec{\nabla} \times \vec{E}$, — which has not been found to date. There is another more abstract reason, why we may restrict the solutions to the case of $K_2 = 0$: only with this condition being fulfilled, the symplex **F** is a symplectic similarity transformation of the pure time direction as described by Eq. (70).[jj] The third scalar product $\vec{P} \cdot \vec{E}$ (and the corresponding field equation $\vec{\nabla} \cdot \vec{E}$) is in the general case nonzero and we emphasize that this is another correspondence to the symplectic decoupling formalism ([8]). It is also remarkable, that the so-called duality rotations[kk] of the electromagnetic field do not fit into this approach [6]: Electric and magnetic fields cannot be exchanged or mixed. Though both are bi-vector fields (forming a "second rank" tensor), the special role of the time coordinate breaks the suspected symmetry between $\vec{E}$ and $\vec{B}$. Furthermore, the only generator that might be used for such a rotation is the pseudo-scalar, i.e. a cosymplex. Insofar, the symplectic foundation of electrodynamics has a higher explanatory power than the "conventional" formalism.

If we summarize (and extend) the stability conditions $K_2 \geq 0$ and $K_1 \geq 2\,\sqrt{K_2}$ [8] by the condition $K_2 = 0$, then it is appropriate (or even mathematically inevitable) to use this striking structural

[jj] For $K_2 > 0$, there are two different frequencies while for $K_2 = 0$ the eigenfrequencies are identical as in case of γ_0.

[kk] Concerning the electromagnetic duality rotation see for instance the short theoretical review article of J. A. Mignaco [69] and the latest experimental results [62].

similarity and to *postulate* the equivalence of $\vec{k}$ and $\vec{P}$ (ω and $\mathcal{E}$). That is, we claim that $\vec{P} \propto \vec{k}$ ($\mathcal{E} \propto \omega$) and hence that the momentum and energy equal via the Fourier transform the spatial and time derivatives, respectively. There is still an optional proportionality factor. But for the same reason that we did not refer to the speed of light, we also skip the proportionality factor $\hbar$, since it has no physical significance, but depends on the choice of units [30, 49, 79]:

$$\begin{aligned}\mathcal{E} &= \omega = i\,\partial_t,\\ \vec{P} &= \vec{k} = -i\,\vec{\nabla}.\end{aligned} \tag{144}$$

This means that the constant K_2 in Eq. (101) vanishes for every plane wave solution since all single terms forming K_2 vanish separately:

$$\begin{aligned}K_2 &= (\mathcal{E}\,\vec{B} - \vec{P}\times\vec{E})^2 - (\vec{E}\cdot\vec{B})^2 - (\vec{P}\cdot\vec{B})^2,\\ \mathcal{E}\,\vec{B} - \vec{P}\times\vec{E} &\to \omega\,\vec{B} - \vec{k}\times\vec{E} = 0,\\ \vec{P}\cdot\vec{B} &\to \vec{k}\cdot\vec{B} = 0,\\ \vec{E}\cdot\vec{B} &= 0.\end{aligned} \tag{145}$$

It is easy to show that "in vacuum", we also have $\vec{E}^2 - \vec{B}^2 = 0$ so that also $K_1 = 0$ and there are no eigenfrequencies of electromagnetic waves. Furthermore, they do not generate inertial frames of reference.

As we projected the bi-vector elements (fields) of the general symplex $\mathbf{F}$ into spacetime, we need to do the same with the phase space variables ψ and thus obtain the "wave function" or spinor $\psi(\vec{x},t)$. A normalization might be written as

$$\int \psi(\vec{x},t)^T\,\psi(\vec{x},t)\,d^3x = 1. \tag{146}$$

The phase space density function ρ (and optionally the Gramian, see Eq. (140)) might be "included" into the spinor function, such that

$$\tilde{\psi} = \sqrt{\rho}\,\psi. \tag{147}$$

Since the electric current must be a vector, the simplest Ansatz for the current density of a "particle" described by ψ is[ll]

$$\begin{aligned} \rho &= -\bar{\psi}\,\gamma_0\,\psi, \\ j_x &= \bar{\psi}\,\gamma_1\,\psi, \\ j_y &= \bar{\psi}\,\gamma_2\,\psi, \\ j_z &= \bar{\psi}\,\gamma_3\,\psi. \end{aligned} \tag{148}$$

The EQOM of $\psi(\vec{x},t)$ must result in a current density (Eq. (148)) that fulfills the continuity equation. The simplest possible solution is known to be the Dirac equation. At first sight, Eq. (16) does not appear to be similar to the Dirac equation. But with the help of the Fourier transform, the equivalence becomes obvious. The eigenvalues of $\mathbf{F}$ for a "particle" in a field-free region are given by $\pm\,i\,\sqrt{\mathcal{E}^2 - \vec{P}^2} = \pm\,i\,m$. Together with Eq. (144), we can now flip the "account" of Eq. (16) by the replacement of operators and eigenvalues (and vice versa):

$$\begin{aligned} (\mathcal{E}\,\gamma_0 + P_x\,\gamma_1 + P_y\,\gamma_2 + P_z\,\gamma_3)\,\psi &= \dot{\psi} = \frac{d\psi}{d\tau}, \\ (i\,\partial_t\,\gamma_0 - i\,\partial_x\,\gamma_1 - i\,\partial_y\,\gamma_2 - i\,\partial_z\,\gamma_3)\,\psi &= \pm\,i\,m\,\psi, \\ (\partial_t\,\gamma_0 - \partial_x\,\gamma_1 - \partial_y\,\gamma_2 - \partial_z\,\gamma_3)\,\psi &= \pm\,m\,\psi. \end{aligned} \tag{149}$$

Apart from "missing" unit imaginary, which is "hidden" in the definition of the Dirac matrices γ_μ, Eq. (149) is the Dirac equation. A unitary matrix which transforms between RDMs and the conventional Form of the Dirac matrices ($\tilde{\gamma}_\mu$) is explicitly given in App. E.5. However, unitary transformations do not change the signs of the metric tensor. If we desire to have $\gamma_0^2 = \mathbf{1}$, we still need to multiply the transformed matrices by the unit imaginary, so that:

$$(-i\,\partial_t\,\tilde{\gamma}_0 + i\,\partial_x\,\tilde{\gamma}_1 + i\,\partial_y\,\tilde{\gamma}_2 + i\,\partial_z\,\tilde{\gamma}_3)\,\psi = \pm\,m\,\psi. \tag{150}$$

[ll]This definition looks quite similar to the definition of the momentum by the EMEQ: Mass and charge density are proportional in case of "point particles".

9.3. Lightlike spinors and massive multispinors

So far we described spinors as a single list of fundamental variables of the form $\psi = (q_1, p_1, q_2, p_2)^T$ i.e. as a 4-dim. column vector and the matrix of second moments Σ as $\Sigma = \psi\,\psi^T$ or explicitly $\Sigma_{ij} = \psi_i\,\psi_j$. In this case, the matrix Σ as well as the matrix $\mathbf{S} = \Sigma\,\gamma_0$ have a vanishing determinant and the eigenfrequencies (101) are equally zero. The matrix $\mathbf{S}$ is a symplex with the RDM-coefficients being second order monomials of the Γ space coordinate ψ. In order to apply the EMEQ *and* to distinguish these productions from the "external" forces $\mathbf{F}$, we use different names (i.e. lower case letters and $\mathcal{U}$ instead of $\mathcal{E}$ [28]:

$$\begin{aligned}
\mathcal{U} &\propto \frac{1}{2}\,(q_1^2 + p_1^2 + q_2^2 + p_2^2),\\
p_x &\propto \frac{1}{2}\,(-q_1^2 + p_1^2 + q_2^2 - p_2^2),\\
p_y &\propto (q_1\,q_2 - p_1\,p_2),\\
p_z &\propto (q_1\,p_1 + q_2\,p_2),\\
e_x &\propto (q_1\,q_2 - p_1\,p_2),\\
e_y &\propto (-q_1\,p_2 - p_1\,q_2),\\
e_z &\propto \frac{1}{2}\,(q_1^2 - p_1^2 + q_2^2 - p_2^2),\\
b_x &\propto (q_1\,q_2 + p_1\,p_2),\\
b_y &\propto \frac{1}{2}\,(q_1^2 + p_1^2 - q_2^2 - p_2^2),\\
b_z &\propto (p_2\,q_1 - p_1\,q_2).
\end{aligned} \tag{151}$$

For dimensional reasons, Eq. (151) represent *actions*, we stay with proportionality. Evidently Eq. (151) defines a bilinear mapping $\mathbb{R}^4 \to \mathbb{R}^{10}$, i.e. for every single "point" in 4-D Γ space there is a 3+1-dimensional structure including electromagnetic fields. This mapping has the following general structural properties with respect to every single phase space point:

$$\begin{aligned}
\vec{p}^2 &= \vec{e}^2 = \vec{b}^2 = \mathcal{U}^2, \\
0 &= \vec{e}^2 - \vec{b}^2, \\
\mathcal{U}^2 &= \frac{1}{2}\,(\vec{e}^2 + \vec{b}^2), \\
\mathcal{U}\,\vec{p} &= \vec{e} \times \vec{b}, \\
\mathcal{U}^3 &= \vec{p} \cdot (\vec{e} \times \vec{b}), \\
m^2 &\propto \mathcal{U}^2 - \vec{p}^2 = 0, \\
\vec{p} \cdot \vec{e} &= \vec{e} \cdot \vec{b} = \vec{p} \cdot \vec{b} = 0.
\end{aligned} \tag{152}$$

Since we made no other assumptions about the spinor, it follows that single spinors are lightlike, i.e. have mass zero and the vectors $\vec{e}$, $\vec{b}$, and $\vec{p}$ form a trihedron.

Multiple spinors are required to form the representation of a massive particle, more precisely: A massive object is composed of $\nu \geq 4$ linearly independent spinors. This raises questions about the four-dimensional Γ-space — the "phase space" or "spinor space", for instance: How do we construct and represent these spinors? In a "classical" phase space, we have two general methods to represent an ensemble: The first is given by a classical density distribution $\rho(\psi)$ which describes the number of phase space points per unit volume and the second is sampling, i.e. the representation of the phase space by ν "samples" that could for instance be represented by the columns of a $4 \times \nu$-matrix Ψ. The single spinors are the columns of this matrix and the Σ-matrix is given by[mm]

$$\Sigma = \frac{1}{\nu}\,\Psi\,\Psi^T. \tag{153}$$

Now consider that the density distribution $\rho(\psi)$ has an internal symmetry. In the simplest case, it might have a point-symmetric

[mm]The root of the normalization factor $\frac{1}{\nu}$ could equally well be included into the definition of the spinor.

of the form:

$$\rho(\psi) = \rho(-\psi). \tag{154}$$

In this case, we might replace ρ by

$$\rho \to \frac{1}{2}(\rho(\psi) + \rho(-\psi)). \tag{155}$$

More generally, if the symmetry properties of the four-dimensional phase space can be expressed by a matrix η, then:

$$\rho \to \frac{1}{2}(\rho(\psi) + \rho(\eta\,\psi)), \tag{156}$$

or if there are N symmetries:

$$\rho \to \frac{1}{N}\sum_k (\rho(\psi) + \rho(\eta_k\,\psi)). \tag{157}$$

Instead of plugging the symmetry into the density, it could also be expressed by the spinor, i.e. by replacing the single spinor ψ by a sampling matrix of spinors:

$$\psi \to (\psi, \eta_1\psi, \dots, \eta_N\,\psi). \tag{158}$$

In this form, it is still possible to multiply the matrix of second moments by the density without any restriction to ψ. This is a mixed form that allows to represent the phase space symmetry by the spinor and still write spinor and density as a product. In this case, we compute the matrix of second moments according to

$$\Sigma = \int d\psi^{2n}\,\rho(\psi)\,\Psi\,\Psi^T. \tag{159}$$

The expectation value of an operator $\mathbf{O}$ is then expressed by

$$\langle \mathbf{O} \rangle = \int d\psi^{2n}\,\rho(\psi)\,\mathrm{Tr}(\Psi^T\,\Psi). \tag{160}$$

The highest degree of symmetry is given, if all single spinors (column vectors) that form the multispinor Ψ are pairwise orthogonal.

In this case, the Σ-matrix is proportional to a unit matrix

$$\Psi\,\Psi^T = (q_1^2 + p_1^2 + q_2^2 + p_2^2)\,\mathbf{1}, \tag{161}$$

and describes — according to the EMEQ — a massive particle in its rest frame. If the single spinors of Ψ (i.e. the columns of Ψ) are ψ_k, then such a system can be constructed from a single spinor ψ by the use of linear orthogonal operators $\mathbf{R}_k$ according to

$$\begin{aligned} \psi_k &= \mathbf{R}_k\,\psi, \\ \psi_j \cdot \psi_k &= \psi_j^T\,\psi_k \\ &= \psi^T\,\mathbf{R}_j^T\,\mathbf{R}_k\,\psi = 0. \end{aligned} \tag{162}$$

The orthogonality then requires that the product $\mathbf{R}_j^T\,\mathbf{R}_k$ must be skew-symmetric:

$$\begin{aligned} (\mathbf{R}_j^T\,\mathbf{R}_k)^T &= \mathbf{R}_k^T\,\mathbf{R}_j = -\mathbf{R}_j^T\,\mathbf{R}_k \\ &\Rightarrow \mathbf{R}_k^T\,\mathbf{R}_j + \mathbf{R}_j^T\,\mathbf{R}_k = 0. \end{aligned} \tag{163}$$

Now it is known that the γ-matrices are all orthogonal. If we replace $\mathbf{R}_k$ by γ_k, then we have the following condition:

$$\gamma_k^T\,\gamma_j = -\gamma_j^T\,\gamma_k. \tag{164}$$

If both operators γ_j and γ_k are symmetric or both skew-symmetric, then the condition is equivalent to the requirement that they anticommute:

$$\gamma_k\,\gamma_j + \gamma_j\,\gamma_k = 0. \tag{165}$$

If one is symmetric and the other one is skew-symmetric, then they have to commute:

$$\gamma_k\,\gamma_j - \gamma_j\,\gamma_k = 0. \tag{166}$$

What is remarkable here is the fact that orthogonal operators *which are independent of* ψ exist only in even-dimensional vector spaces. In odd dimensional spaces, such operators are impossible: Though it is always possible to define a linear orthogonal operator $\mathbf{R}$ such that $(\mathbf{R}\,\mathbf{x}) \cdot \mathbf{x} = 0$, only in even dimensional spaces the rotational operators $\mathbf{R}$ can be defined in such a way that they are independent of $\mathbf{x}$, i.e. which rotate *any* spinor in such a way that $\psi^T \mathbf{R}\, \psi = 0$. One such example is γ_0.

Without loss of generality, we can restrict the first operator to equal the identity matrix $\mathbf{R}_0 = \mathbf{1}$. Then all $\mathbf{R}_k$, $k \in [1, 2, \dots, 2\,n-1]$ have to be skew-symmetric to make the product $\psi\, \psi_k = \psi\, \mathbf{R}_k\, \psi$ ($k = 1, 2, 3$) equal to zero. In this case, the product $\psi_j^T\, \psi_k = \psi^T\, \mathbf{R}_j^T\, \mathbf{R}_k\, \psi$ for $j \neq k$ must vanish and therefore the product of the two matrices must also be skew-symmetric. In case of a $2\,n = 4$-dimensional spinor phase space, we can select some of the 15 traceless RDMs for the $\mathbf{R}_k$. In this case, one needs to select three out of six skew-symmetric matrices. The three selected skew-symmetric matrices must pairwise anticommute. There are (up to an orthogonal transformation and up to a signed permutation) only two sets that fulfill this conditions:

$$
\begin{aligned}
\Psi_s &= \frac{1}{2}\,(\psi, \gamma_7\, \psi, \gamma_8\, \psi, \gamma_9\, \psi),\\
\Psi_c &= \frac{1}{2}\,(\psi, \gamma_0\, \psi, \gamma_{10}\, \psi, \gamma_{14}\, \psi).
\end{aligned}
\tag{167}
$$

The two multispinors are then formally orthogonal matrices, i.e.

$$
\begin{aligned}
\Sigma &= \Psi_{s,c}\, \Psi_{s,c}^T = (q_1^2 + p_1^2 + q_2^2 + p_2^2)\, 1,\\
\mathbf{S} &= (q_1^2 + p_1^2 + q_2^2 + p_2^2)\, \gamma_0.
\end{aligned}
\tag{168}
$$

They can also be interpreted as geometric objects, i.e. as representations of a "phase space unit cell" and/or as a certain symmetry pattern of a phase space distribution.[nn] However, the possibility for such a ψ-independent definition of an orthogonal multispinor does not work in arbitrary even dimensional spaces: The next considered Clifford-algebra $Cl_{9,1}$ has a spinor of size $2\,n = 32$. Hence, we would need 31 orthogonal skew-symmetric and pairwise anticommuting matrices. If such a system of matrices would exist, then it would be the basis of a 31-dimensional Clifford algebra, which is obviously impossible. Therefore, in $9+1$ and higher dimensions, we cannot construct an orthogonal massive multispinor in the described

[nn]Note that the four matrices $1, \gamma_7, \gamma_8, y_9$ as well as $1, \gamma_0, \gamma_{10}, \gamma_{14}$ are representations of the quaternions $1, i, j, k$.

way simply by phase space symmetries. Neither a Hurwitz- nor a Kustaanheimo–Stiefel transformation (see App. 11) exists for 32 or more variables. And therefore the wave function cannot be written in a general way as a product of density and 32-component spinor: Only in 2 and 4 dimensions, the spinor space is separable from the phase space in the way described above. Hence the condition expressed by Eq. (85) excludes all but the 4-dimensional phase space to be the simplest possible basis of "objects" of a physical world. If our arguments are cogent, then the corresponding *space of observables of any physical world* must have $3+1$ dimensions.

We argued that the multispinor approach forces the phase space density $\rho(\psi)$ to have a specific symmetry. Then we should also have a method to suppress a symmetry. This is technically difficult with a positive definite density ρ. If the (multi-) spinors are multiplied with $\sqrt{\rho}$, then the option to have a negative $\sqrt{\rho}$ is mathematically not easily expressible. It is preferable to write the density as the square of a function ϕ:

$$\rho = \phi^2(\psi), \tag{169}$$

such that it is possible to express asymmetry without the use of negative densities. If we then write

$$\begin{aligned} \tilde{\phi} &= \frac{1}{2}\,(\phi(\psi) + \phi(\mathbf{X}\,\psi)), \\ \rho = \tilde{\phi}^2 &= \frac{1}{4}\,(\phi^2(\psi) + \phi^2(\mathbf{X}\,\psi) + 2\,\phi(\psi)\phi(\mathbf{X}\,\psi)). \end{aligned} \tag{170}$$

then the density vanishes, where $\phi(\psi) = -\phi(\mathbf{X}\,\psi)$, i.e. where ϕ is purely skew-symmetric with respect to the transformation $\mathbf{X}$.

9.4. Deformation of the phase space unit cell

Now consider a deformed phase space ellipsoid. We give weights to the phase space lattice as defined by the multispinors in Eq. (167)

$$\begin{aligned} \Psi_s &= (a\,\psi, b\,\gamma_7\,\psi, c\,\gamma_8\,\psi, d\,\gamma_9\,\psi), \\ \Psi_c &= (a\,\psi, b\,\gamma_0\,\psi, c\,\gamma_{10}\,\psi, d\,\gamma_{14}\,\psi), \end{aligned} \tag{171}$$

Then the corresponding simplex $\Psi_c \Psi_c^T \gamma_0$ has the following form:

$$\begin{aligned}\Psi_{c,s} \Psi_{c,s}^T \gamma_0 &= \psi^T \, \psi \, \mathrm{Diag}(a^2, b^2, c^2, d^2) \, \gamma_0 \\ &= \psi^T \, \psi \, \big((a^2 + b^2 + c^2 + d^2) \, \mathcal{E} \\ &\quad + (-a^2 + b^2 + c^2 - d^2) \, P_x \\ &\quad + (a^2 - b^2 + c^2 - d^2) \, E_z \\ &\quad + (a^2 + b^2 - c^2 - d^2) \, B_y) \big) /4. \end{aligned} \tag{172}$$

This corresponds to a decoupled oscillator matrix as described in [8]. Any symplex with purely imaginary eigenvalues is symplectically similar to this type of symplex, which represents a deformed phase space lattice. Note that the energy is represented by the squared diagonal of the phase space lattice while the mass is represented by a function of the volume. In the next section, we show how the phase space lattice is related to CPT-transformations.

9.5. CPT-transformations

The use of multispinors imprints a certain phase space symmetry. But this phase space symmetry is correlated with (or can be mapped to) specific "real world" symmetries.

If we consider the difference between the multispinors Ψ_s and Ψ_c, the first thing we may note is that all single spinors (columns) of Ψ_s are constructed by symplectic transformations. In a sufficiently ergodic system, they could, at least in principle, be interpreted as a single phase space trajectory at different times, while Ψ_c is composed of (partially) disjunct components. It is one of the remarkable features of the 4-dim. phase space that there are these disjunct areas. A system with the phase space position ψ at $t = 0$ can by no means be moved symplectically to position $\gamma_{14} \, \psi$ and vice versa. The 4-dim. phase space is split into two symplectically disjunct regions.

In this context, we have to take the effect of certain RDMs into account, which are used to model charge conjugation (γ_{14}), parity conjugation (γ_0), and time reversal γ_{10}:

$$\begin{aligned} \mathbf{F} &= \mathcal{E} \, \gamma_0 + \vec{P} \cdot \vec{\gamma} + \gamma_0 \, \vec{E} \cdot \vec{\gamma} + \gamma_{14} \, \gamma_0 \, \vec{B} \cdot \vec{\gamma}, \\ -\gamma_0 \, \mathbf{F} \, \gamma_0 &= \mathcal{E} \, \gamma_0 - \vec{P} \cdot \vec{\gamma} - \gamma_0 \, \vec{E} \cdot \vec{\gamma} + \gamma_{14} \, \gamma_0 \, \vec{B} \cdot \vec{\gamma}, \end{aligned}$$

$$\begin{aligned}\gamma_{10}\,\mathbf{F}\,\gamma_{10} &= \mathcal{E}\,\gamma_0 - \vec{P}\cdot\vec{\gamma} + \gamma_0\,\vec{E}\cdot\vec{\gamma} - \gamma_{14}\,\gamma_0\,\vec{B}\cdot\vec{\gamma},\\ \gamma_{14}\,\mathbf{F}\,\gamma_{14} &= \mathcal{E}\,\gamma_0 + \vec{P}\cdot\vec{\gamma} - \gamma_0\,\vec{E}\cdot\vec{\gamma} - \gamma_{14}\,\gamma_0\,\vec{B}\cdot\vec{\gamma}.\end{aligned} \tag{173}$$

The expectation values of an operator $\mathbf{F}$ using the multispinor Ψ_c can also be written as

$$\begin{aligned}\langle\mathbf{F}\rangle &= \Psi_c^T\,\gamma_0\,\mathbf{F}\,\Psi_c = \frac{1}{4}\sum_{k=0}^{3}\psi_k^T\,\gamma_0\,\mathbf{F}\,\psi_k\\ &= \frac{1}{4}\,(\psi^T\,\gamma_0\,\mathbf{F}\,\psi + \psi^T\,\gamma_0^T\,\gamma_0\,\mathbf{F}\,\gamma_0\,\psi\\ &\quad + \psi^T\,\gamma_{10}^T\,\gamma_0\,\mathbf{F}\,\gamma_{10}\,\psi + \psi^T\,\gamma_{14}^T\,\gamma_0\,\mathbf{F}\,\gamma_{14}\,\psi)\\ &= \frac{1}{4}\,(\psi^T\,\gamma_0\,\mathbf{F}\,\psi - \psi^T\,\gamma_0\,\gamma_0\,\mathbf{F}\,\gamma_0\,\psi\\ &\quad + \psi^T\,\gamma_0\,\gamma_{10}\,\mathbf{F}\,\gamma_{10}\,\psi + \psi^T\,\gamma_0\,\gamma_{14}\,\mathbf{F}\,\gamma_{14}\,\psi)\\ &= \frac{1}{4}\langle\mathbf{F} - \gamma_0\,\mathbf{F}\,\gamma_0 + \gamma_{10}\,\mathbf{F}\,\gamma_{10} + \gamma_{14}\,\mathbf{F}\,\gamma_{14}\rangle.\end{aligned} \tag{174}$$

If we compare this to Eq. (173), then the result is clearly equivalent to a projection resulting in $\langle\mathbf{F}\rangle = \mathcal{E}$.

9.6. Duality between operators and observables

It is well known that observables have a dual role in both classical as well as in quantum mechanics. Equation (19) describes the change of the second moments of an ensemble of spinors due to the force matrix $\mathbf{F}$. In classical point mechanics, we would have written "external" forces, since self-interaction is almost always divergent in theories that assume spacetime to be fundamental. In our approach, the situation is different and in order to account for the dynamics of fundamental variables, self-interaction should be included. This means that the force matrix $\mathbf{F}$ is the sum of self-forces $\mathbf{F}_s$ and "external" forces $\mathbf{F}_x$. If no external forces are present, the spinor ensemble nevertheless "oscillates" with the frequency determined by the eigenvalues of $\mathbf{F}_s$. Since both, the self-force and the $\mathbf{S}$-matrix of the system are symplices, it is natural to write the functional dependence of the self-force as a matrix-function of $\mathbf{S}$:

$$\mathbf{F}_s = f(\mathbf{S}). \tag{175}$$

Since only odd powers of a symplex are again a symplex, the Taylor series of f may contain odd powers only. It is obvious that the self-force and the $\mathbf{S}$-matrix will commute so that the $\mathbf{S}$-matrix either depends explicitly on time (which implies external influences) or it can assumed to be static:

$$\dot{\mathbf{S}} = \mathbf{F}_s\,\mathbf{S} - \mathbf{S}\,\mathbf{F}_s = 0. \tag{176}$$

Without "external" influences, the matrix of second moments remains static though the spinor itself oscillates, in accordance with the onto-logic of time. Insofar, we have to take care when interpreting Eq. (106). "Classically" the force matrix $\mathbf{F}$ contains only the external forces. But here we use the sum of external and the self-forces. In the absence of external interactions, the oscillation is exclusively due to self-interaction.

10. Anything Else?

10.1. Once more: Cosymplices

Much of what has been derived above is a consequence of the distinction between symplices and cosymplices (or "Hamiltonian" and "Skew-Hamiltonian") matrices and their expectation values. One central argument was that the generators of the considered Clifford algebras have to be symplices *since only simplices have non-vanishing expectation values.* Only symplices are generators of symplectic transformations and hence only symplices represent forces with measurable effects. One might argue that some entity might exist though its expectation value and its obvious consequences remain zero. Maybe it has hidden or indirect effects? For instance, the algebra of $2\,n \times 2\,n = 8 \times 8$ (co-) symplices represents a Clifford algebra $\mathrm{Cl}_{3,3}$ that has — compared to the Dirac algebra — two additional cosymplices as time-like generators. Let the generators ζ_μ with $\mu \in [0, \dots, 5]$ (with the real Pauli matrices η_ν defined in

App. (C.2) be defined by:

$$\begin{aligned} \zeta_0 &= \eta_3 \otimes \eta_3 \otimes \eta_0, \\ \zeta_1 &= \eta_3 \otimes \eta_3 \otimes \eta_1, \\ \zeta_2 &= \eta_3 \otimes \eta_1 \otimes \eta_2, \\ \zeta_3 &= \eta_3 \otimes \eta_2 \otimes \eta_2, \\ \zeta_4 &= \eta_1 \otimes \eta_0 \otimes \eta_2, \\ \zeta_5 &= \eta_2 \otimes \eta_0 \otimes \eta_2. \end{aligned} \tag{177}$$

The first four elements correspond to (a variant of) the Dirac algebra, and the last two represent "hidden" dimensions corresponding to additional energy contributions ε_1 and ε_2, but are cosymplices, if ζ_0 is the symplectic unit matrix. All ζ_μ are pairwise anticommuting and the metric tensor is given by $g_{\mu\nu} = \mathrm{diag}(-1, 1, 1, 1, -1, -1)$. The commutator table of the first four generators then is (as in case of the Dirac matrices) the electromagnetic field tensor [6]. But since (according to Eq. (29)) the commutator of a symplex and a cosymplex is a cosymplex, the mixed terms between symplices and cosymplices have vanishing expectation values:

$$\langle [\zeta_\mu, \zeta_\nu] \rangle = \begin{pmatrix} 0 & E_x & E_y & E_z & 0 & 0 \\ -E_x & 0 & B_x & -B_y & 0 & 0 \\ -E_y & -B_x & 0 & B_z & 0 & 0 \\ -E_z & B_x & -B_y & 0 & 0 & 0 \\ 0 & 0 & 0 & 0 & 0 & \lambda \\ 0 & 0 & 0 & 0 & -\lambda & 0 \end{pmatrix}. \tag{178}$$

Hence, the two additional elements are (in average) "decoupled" from the first four (and all other symplices). The commutator of the two extra cosymplices is a symplex and hence yields a nonzero field value λ, which acts only between the two additional time-like dimensions. Considerations like these are the background for the claim that time might indeed be "multidimensional" — but it remains to be shown that we would be able to notice.

10.2. Higher-dimensional spaces

If spacetime is an emergent phenomenon, then the question arises whether other (high-dimensional) spacetimes may not emerge "in parallel". We have given a number of algebraic arguments why a 3 + 1-dimensional spacetime is a very special case which cannot be replaced easily by other dimensions. Specifically, we believe to have shown that it is not so much a question of the properties of a "real" external world (which is "independent" of the observer) which we experience and investigate. We should always keep in mind that our argumentation is founded simply on the possibility in principle of observation. If our argumentation is sound, then the possibility in principle for the existence of completely different universes with different dimensionality and different laws of nature is much smaller than considered elsewhere [95].

But let us ignore the above given arguments for 3 + 1-dimensions for a moment: Indeed the strongest evidence for a 3 + 1-dimensional spacetime that we have at hand, is based on light and electrodynamics. Chemistry is almost exclusively based on electromagnetic interactions of electrons: we experience the world through electronic interaction. Our world is the world of electrons. What we see, is light emitted, reflected (or absorbed) by electrons. If we touch a solid object, then electrons are "touching" electrons. We can experimentally investigate weak and strong forces, gravitation and so on, but whatever we "see" with our own eyes, is to almost any degree of approximation based of the electromagnetic interaction of electrons.

Assume that a 9 + 1-dimensional systems exist in parallel *by the properties of other (high-dimensional) fields* — then it still remains questionable, if and how observers would interpret these 9 + 1-dimensional objects. Certainly, we cannot expect that 9 + 1-dimensional entities are just "3 times of the same", since the algebraic features (for instance of rotations) are different: an essential feature of three dimensions is that the spatial rotations do not commute. In more than three dimensions, we can form commuting rotators, say $\gamma_1\,\gamma_2$ and $\gamma_3\,\gamma_4$. Hence, in nine dimensions, we sort out three sets of three non-commuting rotators. Consider, we use

$$\begin{aligned} &(1)\ \gamma_2\,\gamma_3,\ \gamma_3\,\gamma_1,\ \gamma_1\,\gamma_2,\\ &(2)\ \gamma_5\,\gamma_6,\ \gamma_6\,\gamma_4,\ \gamma_4\,\gamma_5,\\ &(3)\ \gamma_8\,\gamma_9,\ \gamma_9\,\gamma_7,\ \gamma_7\,\gamma_8. \end{aligned} \tag{179}$$

We then find that any two rotators from different groups commute.[oo] On the level of observables we derive from Eq. (21) that *if two observables commute they are decoupled*. Presumably observers that are socialized in their perception to adapt to a 3+1-dimensional world would not interpret the 9 + 1-dimensional entity correctly. Instead of "seeing" a 9 + 1-dimensional object "parallel" to our "electronic" 3 + 1-dimensional spacetime, an observer might see three objects located in 3 + 1-dimensions. The ontologically motivated idea that our spacetime dimensionality must be a fundamental and a unique property of the world (and may not depend on the type of the interaction), might have guided physics into the wrong direction after all.[pp]

Table 2 summarizes the theoretical number of (co-) symplices that are to be expected in dynamically generated spacetimes based on Clifford algebras. Obviously, all algebraically possible spacetimes beyond 3 + 1-dimensions include high-order symplices, i.e. penta-, hexa-, and deca-vectors. However, we would like to add a remark here: *If* one would argue that only vectors and bi-vectors have non-negligible effects, then we find that the N-dimensional symplex-algebra $Cl_{N-1,1}$ is composed of N vectors, $N-1$ boosts and $(N-1)(N-2)/2$ rotations. With $N = 2n$, this sums up to an

---

[oo]Dynamically emerging spacetimes have their own properties, which are determined by the structure of the corresponding symplectic Clifford algebra. One cannot simply add another dimension as in arbitrary dimensional Euclidean spaces. Insofar *physical* spaces are much more restricted than mathematical spaces.

[pp]There are however suggestions how to derive particle physics and even dark matter from a *triplet algebra* of dimension $2^{12} = 4096$ based on a 64-dimensional phase space [92].

effective number of symplices ν_s^{eff} given by

$$\nu_s^{\text{eff}} = N + N - 1 + \frac{(N-1)(N-2)}{2} = \frac{N\,(N+1)}{2} = \frac{2\,n\,(2\,n+1)}{2}, \tag{180}$$

just as if the phase space dimension would be $N = 2\,n$ and not 2^N. Since only the generators are observables, one might find arguments for an effective dimensional reduction even of higher-dimensional dynamical systems. The *effective* phase space dimension $n_{\text{eff}} = 2\,n$ is then given by [6]:

$$n_{\text{eff}} = \sqrt{2\,\nu_s^{\text{eff}} + \frac{1}{4}} - \frac{1}{2}. \tag{181}$$

11. Summary and Conclusion

Based on the identification of time with change, we introduced fundamental variables, which are defined exclusively by the property of variation in time. Measurability requires reference to invariant constant rulers. Since pure constants are by definition not available on a fundamental level, they have to emerge from dynamics in the form of constants of motion. We have shown that this is the simplest mathematical model that allows us to do so is given by the Hamiltonian formalism. We found that the ostensible necessity of the use of complex numbers in quantum mechanics can be translated into the "classical" finding that dynamical variables come as pairs. The Hamiltonian ansatz resulted in an algebra of symplices and cosymplies. Reasonable assumptions as for instance a proper distinction of measurable and unmeasurable quantities guided us to Lie algebras that are isomorphic to Clifford algebras $\text{Cl}(N-1,1)$. The most fundamental of these algebras is the Dirac algebra represented by real matrices.

Looking back, we can identify pure variables with the components of the Dirac spinor, i.e. with the components of the quantum mechanical wave function. If we include a density function, then the spinors can be interpreted as "probability amplitudes". We argued that the separation into a multispinor and a pure phase space density function only works, if $2^N = (2N)^2$ holds, i.e. if the dimension of Clifford algebra and spinor algebra are the same. This was the last missing argument to show that Hamiltonian dynamics has special properties in four-dimensional phase spaces.

One goal of our Gedanken experiment was to demonstrate the emergence of a Minkowski geometry (i.e. spacetime geometry) through symplectic dynamics. Spacetime can indeed be a mere interpretation of dynamics if the Lie algebraic construction of geometry based on symplices is isomorph to the representation of a Clifford algebra. Most (if not all) arguments that we used in the derivation are related to specific (algebraic) symmetries. Apparently, these symmetries are less obvious in the conventional form of the Dirac matrices. Furthermore, the use of complex numbers in the conventional form of the Dirac theory wrongly suggests that we are in a non-classical domain. We have shown that some of the apparent differences between quantum and classical mechanics are of ontological nature and loose much of their significance in the light of the ontology of existence in time. Non-classicality is in this scheme neither connected to $\hbar$ or the unit imaginary but to the fact that all quantities of the fundamental level vary at all times and to the consequences thereof.

The idea that the elements of Clifford algebras are related to Minkowski spacetime is well-known and has been described by D. Hestenes and others in various publications on *spacetime algebra* [43, 44]: "The Dirac matrices are no more and no less than matrix representations of an orthonormal frame of spacetime vectors and thereby they characterize spacetime geometry". However (to our knowledge), Hestenes never discussed *why* Minkowski spacetime should have these properties. According to our interpretation, Clifford algebras are the optimal mathematical representation of spacetime *because* spacetime emerges by a pattern of special symmetry properties which can be expressed by the isomorphism of the Hamiltonian dynamical structure to Clifford algebras. The methods of GA are fascinating, but the formal elegance of GA remains unexplained unless we understand that both spacetime and electromagnetism emerge from the structure of the Dirac algebra. It is this connection that can explain why GA is the (only) appropriate mathematical tool to represent spacetime [5]. But this connection unfolds its full explanatory power only in combination with fact that the Dirac algebra is also the algebra of symplectic coupling. This can be shown in the simplest and most obvious way by the use of the real Dirac algebra.

We also claimed that these insights are arguments for the (apparent) dimensionality of spacetime. This might appear unacceptable to physicists who are committed to a realistic ontology. However, already the question *why* spacetime *should* be 3 + 1-dimensional, implies the *possibility in principle* of a mechanism which could explain the dimensionality of spacetime. If this question is accepted as legitimate, then it appears to the author that the only logically possible answer must be related to the structure of interaction as suggested by Einstein. We have given a mathematically simple and sound quasi-classical explanation. Our derivation of relativity does not require the principles of relativity, neither the constancy of the speed of light nor the principle that the laws of physics must be the same in all inertial frames. The presented approach allows to *derive* both principles from the conventional classical concept of canonical transformations. The latter is a result of the fact that Lorentz transformations are structure preserving (i.e. symplectic) transformations of the spinor part of the wave function. The former is a result of our derivation of Maxwell's equations. Three central arguments were given in preparation of the derivation of the Lorentz force and electrodynamics in form of MWEQs: The first argument is based on the difference between even and odd elements of the algebra, the second on the fact that the expectation values of cosymplices must vanish in symplectic dynamics for symmetry reasons and the last one on the transformation properties. The wave equations follow from MWEQs and the comparison of the Fourier transformed MWEQs with the structure of decoupling then lead us to the identification of momentum (energy) with space (time) derivatives and hence to the Dirac equation.

It is often claimed that the way in which probabilities appear in quantum mechanics is special and unusual in some unspecified way. This alienation is often related to the "complex probability amplitudes". But if we reinterpret the formalism of quantum mechanics by its close relation to the matrix of second moments, then its form and postulates are quite familiar.

We do not claim that our "derivation" is rigorous in a strict mathematical sense, nor that it is complete. And it cannot be: our abstract ansatz may only include *abstract objects* and becomes a physical theory only with an appropriate interpretation. Interpretations cannot be proven, they can only be adequate and consistent or not.

Our central "interpretation" is the EMEQ which is induced by the isomorphism of certain classical quantities (energy, momentum, electric, and magnetic field components) with the abstract observables of the $3+1$-dimensional Clifford algebra. The fact that we can derive the Lorentz force equations from it, is neither a physical nor a mathematical "proof" of this interpretation, but it is a proof that our interpretation is compatible with this fundamental dynamical law. It appears to be remarkable to the author that a single three-dimensional orbit is mathematically isomorphic to the average envelope of a two-dimensional ensemble. In this sense, we join the view of Mark Van Raamsdonk: "Everything around us, the whole three-dimensional world, is an illusion born from information encoded elsewhere, on a two-dimensional chip" [97]. Most authors that analyzed the structure of the symplectic group $Sp(4, R)$ interpreted this space to be $2+1$-dimensional. However, as we argued above, the space of *observables* that corresponds to $Sp(4, R)$ is a $3{+}1$-dimensional Minkowski spacetime. Mathematically, it is evident that this view implies restrictions on the $3+1$-dimensional space, for instance, the "quantization" of the angular momentum.

But even if we could present a more rigorous derivation of the (fundamental) relativistic EQOM of quantum mechanics — the

Dirac equation — it would not automatically imply a "derivation" of quantum mechanics as a whole. Though it is rarely explicitly mentioned, it is known (though not well-known) that the EQOM used in quantum mechanics are (taken as such) classical [80]. With respect to this question, our presentation is not new. What is new (to the knowledge of the author) is the algebraic connection between the classical (co-)symplex algebra with the Clifford algebra of the Dirac matrices.

We started with the classical Hamilton function, and were guided to interpret the fundamental variables as the quantum mechanical wave function in momentum space. Mathematically, the alleged antagonism between classical and quantum mechanics seems to evaporate in the light of what we described above. At least we could show that it is not the *mathematical structure* of the EQOM that accounts for the interpretational difficulties with quantum mechanics.

On the basis of the presented ontology of time, we found an explanation for why the quantum mechanical wave function has some features that seem to be *mysterious* or "non-classical": impossible to be directly measured and without a well-defined dimensional unit and physical meaning. However, these are exactly the properties of fundamental variables that we derived from the ontology of time and a proper logic of measurement. Supposing our ontology is "correct", it is impossible to know *what* the wave function *is*. We think that we gave an explanation why this is so: the question has no meaningful physical answer — it cannot. The consequences of the presented ontology are certainly difficult to accept: on the fundamental level, the world is apparently very different from our every day experience: loosely speaking it is "flat". The apparent $3+1$-dimensional spacetime is "in reality" a four-dimensional phase space. However, there are interpretations of quantum mechanics (and general relativity) which are way more esoteric than this.

The picture that we painted is a generalized, abstracted review of algebraic methods for a low-dimensional Γ-phase space of classical statistical mechanics based on arbitrary continuously varying fundamental variables. This is a generalization and abstraction because we did not use *a-priori* assumptions on the number of space phase points making up a "something", a piece of matter, nor did we make any *a-priori* assumptions about the type of interaction other than arguments based on symmetry considerations. It is the onto-logic of time that guided us to the idea that observables are moments of fundamental variables and the phase space coordinates themselves are not directly accessible. Here we discussed almost exclusively the second moments. However, it might be worth considering the role of fourth order and higher moments in more detail for both, the static case of "quantum systems" in their eigenstates as well as the potential role that odd moments might play in quantum jumps. There are strong arguments to assume that quantization cannot be finally understood as a linear theory. It might be enlightening to investigate the physical and algebraic meaning of higher order moments. We only shortly touched this area with Eqs. (151) and (152) with respect to single phase space points.

Following this logic, we found strong arguments for the dimensionality of energy–momentum space. Since the energy–momentum space is conjugate to spacetime, these arguments directly concern the dimensionality of the latter. Along the way we found an interpretation that describes fundamental laws of relativistic electrodynamics and quantum mechanics. We described the bridge leading to covariance matrices and statistical moments, i.e. the bridge to probability theory. Significant theories have been built on less solid grounds. Maybe the above is also a contribution to the question of whether, and in what sense, quantum mechanics deserves to be regarded as the final theoretical framework.

Acknowledgments

Mathematica® has been used for part of the symbolic calculations. Additional software has been written in "C" and compiled with the GNU©-C++ compilers on different Linux distributions. XFig 3.2.4 has been used to generate the figures, different versions of LaTeXand GNU©-emacs for editing and layout.

References

[1] Allori, V. (2015). Primitive Ontology and the Classical World, *this Volume*.

[2] Amelino-Camelia, G., Freidel, L., Kowalski-Glikman, J. and Smolin, L. (2011). Principle of relative locality. *Physical Review D*, **84**: 084010.

[3] Anna C. T. Gustavsson (2009). Constrained quantum dynamics. *Journal of Physics: Conference Series*, **174**: 012025.

[4] Bacry, H. and Boon, M. (1987). Boson algebra as a symplectic Clifford algebra. *Journal of Mathematical Physics*, **28**: 2639–2649.

[5] Baryo-Corrochano, E. (2010). What can Hamilton, Grassmann, Clifford and Hestenes tell us about perception and action systems. *Advances in Applied Clifford Algebras*, **20**: 217–240.

[6] Baumgarten, C. (2011). Use of real Dirac matrices in two-dimensional coupled linear optics. *Physical Review ST — Accelerators and Beams*, **14**: 114002.

[7] Baumgarten, C. (2012). A Symplectic Method to Generate Multivariate Normal Distributions. arXiv:1205.3601.

[8] Baumgarten, C. (2012). Geometrical method of decoupling. *Physical Review ST — Accelerators and Beams*, **15**: 124001.

[9] Baumgarten, C. (2014). A new look at linear (non-?) symplectic ion beam optics in magnets. *Nuclear Instruments and Methods in Physics Research A*, **735**: 546–551.

[10] Belen'kii, I. M. (1981). A method of regularizing the equations of motion in the central force field. *Celestial Mechanics*, **23**: 9–23.

[11] Binz, E., de Gosson, M. A. and Hiley, B. J. (2013). Clifford algebras in symplectic geometry. *Foundations of Physics*, **43**: 424–439.

[12] Blanes, S. and Budd, C. J. (2004). Explicit adaptive symplectic (easy) integrators: a scaling invariant generalisation of the Levi-Civita and KS regularisations. *Celestial Mechanics and Dynamical Astronomy*, **89**: 383–405.

[13] Briggs, J. S. and Eisfeld, A. (2012). Coherent quantum states from classical oscillator amplitudes. *Physical Review A*, **85**: 052111.

[14] Briggs, J. S. and Eisfeld, A. (2013). Quantum dynamics simulation with classical oscillators. *Physical Review A*, **88**: 062104.

[15] Brody, D. C., Gustavsson, A. C. and Hughston, L. P. (2008). Symplectic approach to quantum constraints. *Journal of Physics A: Mathematical and Theoretical*, **41**: 475301.

[16] Burić, N., Popović, D. B., Radonjić, M. and Prvanović, S. (2014). Phase space theory of quantum-classical systems with nonlinear and stochastic dynamics. *Annals of Physics*, **343**: 16–26.

[17] Buric, N. (2008). Hamiltonian quantum dynamics with separability constraints. *Annals of Physics*, **323**: 17–33.

[18] Campbell, E. T., Genoni, M. G. and Eisert, J. (2013). Continuous-variable entanglement distillation and noncommutative central limit theorems. *Physical Review A*, **87**: 042330.

[19] Cariñena, J. F., Ibort, A., Marmo, G. and Morandi, G. (2015). Geometry from Dynamics, Classical and Quantum, Springer, p. 558.

[20] Chappell, J. M., Iqbal, A. and Abbott, D. (2011). arXiv:1101. 3619v2.

[21] Cho, D.-N. and Kong O. C. W. (2014). Relativity symmetries and Lie algebra contractions. *Annals of Physics*, **351**: 275–289.

[22] Concerning complex numbers we follow the notion of D. Hestenes: *Imaginary numbers are not real* and we demand that the unit imaginary might be *convenient* but is not *required* in the formulation of a physical theory. see also: Gull, S., Lasenby, A., Doran, C. (1993). *Foundations of Physics*, **23**(9): 1175–1201.

[23] Czichos, H., Saito, T. and Smith, L. E. (eds.). (2011), (eds.): Springer Handbook of Metrology and Testing, 2nd. Ed., Springer, Berlin, Heidelberg, p. 3.

[24] Davtyan, L. S., Mardoyan, L. G., Pogosyan, G. S., Sissakian, A. N. and Ter-Antonyan, V. M. (1987). Generalised KS transformation: from five-dimensional hydrogen atom to eight-dimensional isotrope oscillator. *Journal of Physics A: Mathematical and General*, **20**: 6121–6125.

[25] de Gosson, M. A. (2012). The Symplectic Egg. arXiv:1208.5969.

[26] de Ronde, C. (2015). Representational Realism, Closed Theories and the Quantum to Classical Limit, *this Volume*, 2015.

[27] Diener, G., Weissbarth, J., Grossmann, F. and Schmidt, R. (2013). Obtaining Maxwell's equations heuristically. *American Journal of Physics*, **81**(2): 120–123.

[28] Dirac, P. A. M. (1963). A remarkable representation of the $3 + 2$ de Sitter group. *Journal of Mathematical Physics*, **4**(7), 901–909.

[29] Dragt, A. J., Neri, F. and Rangarajan, G. (1992). General moment invariants. *Physical Review A*, **45**(4): 2572–2585.

[30] Duff, M. J., arXiv:1412.2040v1.

[31] Dyson, F. J. (1990). Feynmans proof of the Maxwell equations. *American Journal of Physics*, **58**(3): 209–214.

[32] Einstein, A. (1961). Relativity: The Special and the General Theory, 15th edition, Appendix V, Crown Publishers — Bonanza Books, N.Y.

[33] Einstein, A. (1996). Autobiographical Notes. Open Court Publishing, La Salle, Illinois.

[34] Elze, H.-T. (2012). Linear dynamics of quantum-classical hybrids. *Physical Review A*, **85**: 052109.

[35] Feynman, R. P., Leighton, R. B. and Sands, M. (1964). Lecture Notes on Physics, Addison-Wesley Publ. Co., Inc., Reading, Mass.), Vol. 2, pp. 26–2 (http://feynmanlectures.caltech.edu/).

[36] Gersten, A. (1999). Euclidean special relativity. *Foundations of Physics Letters*, **12**(3): 291–298.

[37] Gibbons, G. W. (2012). in: *The Arrows of Time, Fundamental Theories of Physics*, **172**: Springer Berlin Heidelberg. See also: Igor F. Herbut (2013). Majorana mass, time reversal symmetry, and the dimension of space. *Physical Review D* **87**: 085002.

[38] Gillis, E. J. (2012). Relativity Is Not About Spacetime, arXiv:1210.3575.

[39] Goodmanson, D. M. (1996). A graphical representation of the Dirac algebra. *American Journal of Physics* **64**(7): 870–880.

[40] Grgin, E. and Petersen, A. (1974). Duality of observables and generators in classical and quantum mechanics. *Journal of Mathematical Physics*, **15**(6): 764–769.

[41] Gromov, M. (1985). Pseudo holomorphic curves in symplectic manifolds. *Inventiones mathematicae*, **82**(2): 307–347.

[42] Hamilton, W. R. (1831). The Transactions of the Royal Irish Academy, **17**: 293–423. Published by: Royal Irish Academy, Stable (http://www.jstor.org/stable/30078796).

[43] Hestenes, D. (1966). Space-Time Algebra. Gordon and Breach, New York. arXiv:0802.2728v1.

[44] Hestenes, D. (2003). *American Journal of Physics*, **71**(7): 691–741.

[45] Hiley, B. J. (2011). in: Bob Coecke (ed.), Lecture Notes in Physics Vol. 813; Springer Heidelberg Dordrecht London New York.

[46] Hoehn, P. A. and Mueller, M. P. (2016). An operational approach to spacetime symmetries: Lorentz transformations from quantum communication. arXiv:1412.8462.

[47] Horn, M. E. (2012). *Journal of Physics: Conference Series*, **380**: 012006.

[48] Horn, R. A. and Johnson, C. R. (1985). *Matrix Analysis*; Cambridge University Press.

[49] Hsu, L. and Hsu, J. P. (2012). The physical basis of natural units and truly fundamental constants. *The European Physical Journal Plus*, **127**: 11.

[50] Hurwitz, A. (1923). Über die Komposition der quadratischen Formen von beliebig vielen Variablen (in German). *Mathematische Annalen*, **88**: 1–25.

[51] Jauch, J. M. and Rohrlich, F. (1976). The Theory of Photons and Electrons; 2nd Ed., Springer.

[52] Jaynes, E. T. (1991). in: The Electron, (eds.) D. Hestenes and A. Weingartshofer. Kluwer Academic Publishers.

[53] Kaplunovsky, V. and Weinstein, M. (1985). Space-time: Arena or Illusion? *Physics Review D*, **31**: p. 1879. See also SLAC-PUB-3156.

[54] Kim, Y. S. and Noz, M. E. (1991). "Phase Space Picture of Quantum Mechanics"; World Scientific, Singapore, New Jersey, London, Hong Kong.

[55] Knuth, K. H. and Bahreyni, N. (2014). A potential foundation for emergent space-time. *Journal of Mathematical Physics*, **55**: 112501.

[56] Kobe, D. H. (1978). Derivation of Maxwell's equations from the local gauge invariance of quantum mechanics. *American Journal of Physics*, **46**(5): 342–348.

[57] Kobe, D. H. (1980). Derivation of Maxwell's equations from the gauge invariance of classical mechanics. *American Journal of Physics*, **48**(4): 348–353.

[58] Krefetz, E. (1970). A "Derivation" of Maxwell's Equations. *American Journal of Physics*, **38**(4): 513–516.

[59] Kustaanheimo, P. and Stiefel, E. (1965). Perturbation theory of Kepler motion based on spinor regularization. *Journal für die reine und angewandte Mathematik*, **218**: 204–219.

[60] Lashkari, N., McDermott, M. B. and Van Raamsdonk, M. (2014). *Journal of High Energy Physics*, **1404**: 195 and arXiv:1308.3716.

[61] Lax, P. D. (1968). Integrals of nonlinear equations of evolution and solitary waves. *Courant Institute of Mathematical Sciences* (N.Y. Univ.), Rep. NYO-1480-87; also in: *Communications on Pure and Applied Mathematics*, **21**(5): 467–490.

[62] LeBlanc L. J. (2014). Magnetic monopoles — particles carrying a single magnetic charge — have never been seen. Analogues of these entities have now been produced in an ultracold cloud of rubidium atoms. *Nature*, **505**: 627–628; and Ray, M. W., Ruokokoski, E., Kandel, S., Möttönen, M. and Hall, D. S., Observation of Dirac monopoles in a synthetic magnetic field. 657–666.

[63] Lee, D.-G. (1995). The Dirac gamma matrices as "relics" of a hidden symmetry?: As fundamental representations of the algebra $sp(4, R)$. *Journal of Mathematical Physics*, **36**: 524–530.

[64] MacFarlane, A. J. (1966). Dirac matrices and the Dirac matrix description of Lorentz transformations. *Communications in Mathematical Physics*, **2**: 133–146.

[65] Majorana, E. (1937). Teoria simmetrica dell'elettrone e del positrone (A symmetric theory of electrons and positrons). *Il Nuovo Cimento*, **5**: 171–184.

[66] Marlow, A. R. (1986). Quantum Theoretical Origin of Spacetime Structure. *International Journal of Theoretical Physics*, **25**(5): 561–571, and references therein.

[67] Mermin, N. D. (1984). Relativity without light. *American Journal of Physics*, **52**(2): 119–124.
[68] Meyer, K. R., Hall, G. R. and Offin, D., Introduction to Hamiltonian Dynamical Systems and the N-Body Problem; 2nd. Ed.
[69] Mignaco, J. A. (2001). Electromagnetic Duality, Charges, Monopoles, Topology, ... *Brazilian Journal of Physics*, **31**(2): 235–246.
[70] Nibart, G. (2006). Building the SpaceTime Structure in Theories of Physics. *AIP Conference Proceedings*, **839**: 198; doi: 10.1063/1.2216629.
[71] Ohrstrom, P. (1985). W. R. Hamilton's view of algebra as the science of pure time and his revision of this view. *Historia Mathematica*, **12**: 45–55.
[72] Okubo, S. (1991). Real representations of finite Clifford algebras. *Journal of Mathematical Physics*, **32**: 1657–1668.
[73] Pauli, W. (1936). Contributions mathematiques a la theorie des matrices de Dirac. *Annales de l'Institut Henri Poincaré*, **6**: 109–136.
[74] Penrose, R. and Rindler, W. (1984). Spinors and Spacetime, Cambridge University Press.
[75] Pirogov, Y. F. (2003). Spacetime symplectic extension. *Physics of Atomic Nuclei*, **66**(1): 135–142. From *Yadernaya Fizika*, **66**(1): 138–145.
[76] Pirogov, Y. F. (2004). Proc. of Quarks, arxiv:gr-qc/0409067; see also: www.slac.stanford.edu/econf/C0405241.
[77] Pirogov, Y. F. (2005). Space-time: emerging vs. existing. arxiv:gr-qc/0503091v1.
[78] Radovan, M. On the nature of time. arXiv:1509.01498.
[79] Ralston J. P. (2012). Quantum Theory without Planck's Constant. arXiv:1203.5557.
[80] Ralston, J. P. (1989). Berry's phase and the symplectic character of quantum time evolution. *Physical Review A*, **40**(9): 4872–4884.
[81] Richard, B. (1967). Theory of Heat, 2nd Ed., Springer Berlin, Heidelberg, New York.
[82] Scharnhorst, K. (1999). A special irreducible matrix representation of the real Clifford algebra $C(3,1)$. *Journal of Mathematical Physics*, **40**(7).
[83] Schrödinger, E. (1950). Spacetime Structure, Cambridge University Press, ISBN 0-521-31520-4.
[84] Sebastian, F. and Lombardi, O. (2015). A top-down view of the classical limit of quantum mechanics, *this Volume*.
[85] See for instance (in German) P. Schmüser: Feynman–Graphen und Eichtheorien für Experimentalphysiker; Lecture Notes in Physics No. 295, Springer 1988.

[86] See Theorem 3.2.1. in [68].

[87] Seiberg, N. (2007). Emergent Spacetime, in: *The Quantum Structure of Space and Time*, (eds.) D. Gross, M. Henneaux, and A. Sevrin, (World Scientific), pp. 163–213.

[88] Shuster, M. D. (1993). A survey of attitude representations. *The Journal of the Astronautical Sciences*, **41**(4): 439–517; ibd.: **56**(3): 359–373.

[89] Singh, S. (1986). Lorentz transformations in Mermin's relativity without light. *American Journal of Physics*, **54**(2): 183–184.

[90] Skinner, T. E. (2013). Exact mapping of the quantum states in arbitrary N-level systems to the positions of classical coupled oscillators. *Physical Review A*, **88**: 012110.

[91] Smilga, W. (2013). Emergence of space-time and gravitation. *Journal of Modern Physics*, **4**: 963–967; DOI: 10.4236/jmp.2013.47129.

[92] Sogami, I. S. (2013). Extended standard model in multi-spinor field formalism: Visible and dark sectors. *Progress of Theoretical and Experimental Physics*, 123B02. See also: Generalized Dirac equation with four orthogonal families of spin 1/2 solutions. *Progress of Theoretical and Experimental Physics*, **66**: 303–317 and New algebraic unified theory of Leptons and Quarks. *Prog. Theor. Phys.*, **78**: pp. 487–505.

[93] Steeb, W.-H. and Kunick, A. (1989). Chaos in dynamischen Systemen, Wissenschaftsverlag, B. I., Mannheim/Wien/Zürich, 2nd ed. (in German).

[94] Stueckelberg, E. C. G. (1960). Quantum theory in real Hilbert-Space. *Helvetica Physica Acta*, **33**: 727–752.

[95] Tegmark, M. (2007). The Mathematical Universe. arXiv:0704.0646.

[96] This definition holds only for a *flat spacetime*. In curved spacetime, this definition would have to be modified; in spherical trigonometry the theorem of Pythagoras also has to be replaced by the rules of John Napier.

[97] Van Raamsdonk, M. (2013). in: Merali, Z., The origins of space and time. *Nature*, **500**: 516–519.

[98] Verlinde, E. (2011). *Journal of High Energy Physics*, **1104**: 029, (arXiv:1001.0785).

[99] Videira, A. L. L., Rocha Barros, A. L. and Fernandez, N. C. (1985). Geometry as an aspect of dynamics. *Foundations of Physics*, **15**(12): 1247–1262.

[100] Wigner, E. (1960). *Communications in Pure and Applied Mathematics*, **13**(1).

[101] Zare, K. and Szebehely, V. (1975). Time transformations in the extended phase-space. *Celestial Mechanics*, **11**: 469–482.

Appendix

Appendix A: Hurwitz and Kustaanheimo–Stiefel Matrices

With a minor modification (without influence on observables), the matrix Ψ_s is proportional to the Hurwitz transformation [50]:

$$\Psi_{hw} = \frac{1}{2}\,(\psi, -\gamma_8\,\psi, \gamma_9\,\psi, -\gamma_7\,\psi)$$
$$= \frac{1}{2}\begin{pmatrix} q_1 & -p_1 & -q_2 & -p_2 \\ p_1 & q_1 & -p_2 & q_2 \\ q_2 & p_2 & q_1 & -p_1 \\ p_2 & -q_2 & p_1 & q_1 \end{pmatrix}, \tag{A1}$$

or (with a flipped sign of the last column) to the so-called Kustaanheimo–Stiefel (KS-) transformation [59]:

$$\Psi_{ks} = \frac{1}{2}\,(\psi, -\gamma_8\,\psi, \gamma_9\,\psi, \gamma_7\,\psi). \tag{A2}$$

The matrices (167) are also used in the Euler–Rodrigues formulation of attitudes and represent left- and right- multiplicative isoclinic mappings [88].

Since the KS-transformation is designed to map the Kepler- (or Coulomb-) problem to the harmonic oscillator, it is not unreasonable to speculate that the appearance of the $\frac{1}{r}$-coulomb-potential of massive (charged) particles can also be explained in this way. It has been shown elsewhere that non-bijective quadratic transformations allow to map the hydrogen atom to the harmonic oscillator [24].

Appendix B: Arbitrary constant Hamiltonians

Can the Hamiltonian equivalently be of higher order? We presumed that the Hamiltonian is a constant function of the fundamental variables ψ, i.e. a COM, we may ask if our derivation holds also for higher order functions. Given the Hamiltonian has the general form of Eq. (15):

$$\mathcal{H}(\psi) = \mathcal{H}_0 + \varepsilon^T\,\psi + \frac{1}{2}\,\psi^T\,\mathbf{A}\,\psi + \frac{1}{3}\,\mathbf{B}_{ijk}\,\psi_i\,\psi_j\,\psi_k + \\ + \frac{1}{4}\,\mathbf{C}_{ijkl}\,\psi_i\,\psi_j\,\psi_k\,\psi_l + \cdots\,, \tag{B1}$$

where all coefficients ε_i, $\mathbf{A}_{ij}$, $\mathbf{B}_{ijk}$, $\mathbf{C}_{ijkl}$ are constants. Then we still assume that not only the matrix $\mathbf{A}$ but also the higher order "tensors" are symmetric in all indices. The term $\mathcal{H}_0$ is a constant anyway and does not contribute and again we choose the zeropoint of ψ such that the linear term vanishes. Then, the time derivative of this Hamiltonian reads:

$$\begin{aligned} \dot{\mathcal{H}} &= \frac{1}{2}\,(\psi\,\mathbf{A}\,\dot{\psi} + \dot{\psi}\,\mathbf{A}\,\psi) \\ &+ \frac{1}{3}\,(\mathbf{B}_{ijk}\,\dot{\psi}_i\,\psi_j\,\psi_k + \mathbf{B}_{ijk}\,\psi_i\,\dot{\psi}_j\,\psi_k + \mathbf{B}_{ijk}\,\psi_i\,\psi_j\,\dot{\psi}_k) \\ &+ \cdots\,. \end{aligned} \tag{B2}$$

Due to the symmetry in all indices, this symplifies to:

$$\begin{aligned} \dot{\mathcal{H}} &= (\psi_k\,\mathbf{A}_{ki} + \mathbf{B}_{jki}\,\psi_j\,\psi_k + \cdots)\,\dot{\psi}_i \\ &= (\nabla_\psi \mathcal{H})_i\,\dot{\psi}_i \end{aligned} \tag{B3}$$

with the gradient given by

$$\begin{aligned} (\nabla_\psi \mathcal{H})_i &= (\mathbf{A}_{ik} + \mathbf{B}_{ijk}\,\psi_j + \mathbf{C}_{ijlk}\,\psi_j\,\psi_l + \cdots)\,\psi_k, \\ \gamma_0\,(\nabla_\psi \mathcal{H}) &= \mathbf{F}(\psi)\,\psi, \end{aligned} \tag{B4}$$

where $\mathbf{F}(\psi)$ is the product of γ_0 with a *symmetric* matrix $\tilde{\mathbf{A}}$ — and therefore a symplex:

$$\tilde{\mathbf{A}}_{ik} = \mathbf{A}_{ik} + \mathbf{B}_{ijk}\,\psi_j + \mathbf{C}_{ijlk}\,\psi_j\,\psi_l + \cdots\,, \tag{B5}$$

so that the solution still is:

$$\dot{\psi} = \gamma_0\,(\nabla_\psi \mathcal{H}). \tag{B6}$$

We look at the second moments:

$$\begin{aligned} \frac{d}{d\tau}\,(\psi\,\psi^T) &= \dot\psi\,\psi^T + \psi\,\dot\psi^T \\ &= \gamma_0\,(\nabla_\psi\mathcal{H})\,\psi^T + \psi\,(\nabla_\psi\mathcal{H})^T\,\gamma_0^T \\ &= \mathbf{F}(\psi)\,\psi\,\psi^T + \psi\,\psi^T\,\mathbf{F}^T(\psi). \end{aligned} \tag{B7}$$

Multiplication from the right with γ_0 and replacing $\mathbf{F}^T(\psi)$ with $\gamma_0\,\mathbf{F}(\psi)\,\gamma_0$ again yields:

$$\tfrac{d}{d\tau}\,(\psi\,\psi^T\gamma_0) = \mathbf{F}(\psi)\,(\psi\,\psi^T\,\gamma_0) + (\psi\,\psi^T\,\gamma_0)\,\mathbf{F}(\psi)\,\gamma_0^2, \tag{B8}$$

so that with $\gamma_0^2 = -\mathbf{1}$ and $\mathbf{S} \equiv \psi\psi^T\,\gamma_0$ we have the Lax pair:

$$\dot{\mathbf{S}} = \mathbf{F}(\psi)\,\mathbf{S} - \mathbf{S}\,\mathbf{F}(\psi). \tag{B9}$$

Note however, that the step from Eq. (B8) to Eq. (B9) is in the general case only valid for single spinors.

Appendix C: (Co-) Symplices for higher-dimensional Spacetimes

C.1. Which k-Vectors are (Co-) symplices

Given that we have a set of N pairwise anticommuting (co-) symplices $\mathbf{S}_i$ ($\mathbf{C}_i$), which can be regarded as generators of real Clifford algebras, then the question if the k-products (i.e. bi-vectors, trivectors etc.) are symplices or cosymplices, depends on a sign and can be calculated as follows:

$$\begin{aligned} (\mathbf{S}_1\,\mathbf{S}_2\,\mathbf{S}_3\,\cdots\,\mathbf{S}_k)^T &= \mathbf{S}_k^T\,\mathbf{S}_{k-1}^T\,\mathbf{S}_{k-2}^T\,\cdots\,\mathbf{S}_1^T \\ &= (-1)^{k-1}\,\gamma_0\,\mathbf{S}_k\,\mathbf{S}_{k-1}\,\mathbf{S}_{k-2}\,\cdots\,\mathbf{S}_1\,\gamma_0. \end{aligned} \tag{C1}$$

The number of permutations that is required to reverse the order of k matrices is $k\,(k-1)/2$, so that

$$(\mathbf{S}_1\,\mathbf{S}_2\,\mathbf{S}_3\,\cdots\,\mathbf{S}_k)^T = (-1)^s\,\gamma_0\,(\mathbf{S}_1\,\mathbf{S}_2\,\mathbf{S}_3\,\cdots\,\mathbf{S}_k)\,\gamma_0, \tag{C2}$$

with s given by:

$$s = k - 1 + k\,(k-1)/2 = \frac{k^2 + k - 2}{2}. \tag{C3}$$

Table 1. Signs for products of k anti-commuting (co-) symplices according to Eqs. (C3) and (C5). The "plus" signs correspond to symplices: Products of $2, 5, 6, 9, 10, \ldots$ symplices are again symplices.

k	1	2	3	4	5	6	7	8	9	10	11	12
s	0	2	5	9	14	20	27	35	44	54	65	77
$(-1)^s$	+	+	−	−	+	+	−	−	+	+	−	−
c	1	4	8	13	19	26	34	43	53	64	76	89
$(-1)^c$	−	+	+	−	−	+	+	−	−	+	+	−

If we consider k cosymplices $\mathbf{C}_i$ instead, we obtain k more sign reversals, so that

$$(\mathbf{C}_1\,\mathbf{C}_2\,\mathbf{C}_3\,\cdots\,\mathbf{C}_k)^T = (-1)^c\,\gamma_0\,(\mathbf{C}_1\,\mathbf{C}_2\,\mathbf{C}_3\,\cdots\,\mathbf{C}_k)\,\gamma_0, \tag{C4}$$

with c given by:

$$c = 2\,k - 1 + k\,(k-1)/2 = \frac{k^2 + 3\,k - 2}{2}. \tag{C5}$$

If s and c are even (odd), respectively, then the products are (co-) symplices. Table 1 lists the resulting signs for $k = 2 \ldots 10$. Clifford algebras with more than 4 generators include penta- and hexavectors as "observables". Furthermore, one may conclude from Table 1 that in $9+1$-dimensional spacetime, the pseudoscalar is the product of all 10 generators and therefore a symplex. This allows (in principle) for a scalar field. Since spatial rotators are combinations of two spatial generators, in $9+1$-dimensional spacetime there are $\binom{9}{2} = 36$ spatial rotations. A derivation of the corresponding MWEQs for such higher-dimensional spaces (if possible or not), lies beyond the scope of this chapter, but if all bi-vector boosts have a corresponding electric field component and all bi-vector rotations a corresponding magnetic field component, then we should expect 9 "electric" and 36 "magnetic" field components in $9+1$-dimensional spacetime.

C.2. The real Pauli matrices and the structure of (co-) symplices for n DOFs

For one DOF and an appropriate treatment of the 2×2-blocks of the general case, we introduce the real Pauli matrices (RPMs) according to

$$\eta_0 = \begin{pmatrix} 0 & 1 \\ -1 & 0 \end{pmatrix} \qquad \eta_1 = \begin{pmatrix} 1 & 0 \\ 0 & -1 \end{pmatrix}$$
$$\eta_2 = \begin{pmatrix} 0 & 1 \\ 1 & 0 \end{pmatrix} \qquad \eta_3 = \mathbf{1} = \begin{pmatrix} 1 & 0 \\ 0 & 1 \end{pmatrix}. \tag{C6}$$

η_0 is the symplectic unit matrix (i.e. corresponds to γ_0 of the general case). η_0, η_1, and η_2 are symplices. If we consider general $2\,n \times 2\,n$ Hamiltonian matrix $\mathbf{F}$ composed of n^2 2×2 blocks $\mathbf{A}_{ij}$, then γ_0 has the form given in Eq. (9). The $2\,n \times 2\,n$-symplex $\mathbf{F}$ that fulfills Eq. (20) has the form

$$\mathbf{F} = \begin{pmatrix} \mathbf{D}_1 & \mathbf{A}_{12} & \mathbf{A}_{13} & \dots & \mathbf{A}_{1n} \\ -\tilde{\mathbf{A}}_{12} & \mathbf{D}_2 & \mathbf{A}_{23} & \dots & \mathbf{A}_{2n} \\ -\tilde{\mathbf{A}}_{13} & -\tilde{\mathbf{A}}_{23} & \mathbf{D}_3 & \dots & \mathbf{A}_{3n} \\ \vdots & & \vdots & & \vdots \\ -\tilde{\mathbf{A}}_{1n} & -\tilde{\mathbf{A}}_{2n} & -\tilde{\mathbf{A}}_{3n} & \dots & \mathbf{D}_n \end{pmatrix}, \tag{C7}$$

where the 2×2 matrices $\mathbf{D}_k$ on the diagonal must be 2×2-symplices and can therefore be written as

$$\mathbf{D}_k = d_0^k\,\eta_0 + d_1^k\,\eta_1 + d_2^k\,\eta_2. \tag{C8}$$

If the blocks $\mathbf{A}_{ij}$ above the diagonal have the general form

$$\mathbf{A} = \sum_{k=0}^{3} a_k\,\eta_k, \tag{C9}$$

then Eq. (20) fixes the form of the corresponding blocks below the diagonal $\tilde{\mathbf{A}}_{ij}$ to

$$\begin{aligned} -\tilde{\mathbf{A}}_{ij} &= \eta_0\,\mathbf{A}_{ij}^T\,\eta_0 \\ &= -\eta_0\left(a_{ij}^0\,\eta_0 + a_{ij}^1\,\eta_1 + a_{ij}^2\,\eta_2 + a_{ij}^3\,\eta_3\right)^T\eta_0 \\ &= \eta_0\left(-\,a_{ij}^0\,\eta_0 + a_{ij}^1\,\eta_1 + a_{ij}^2\,\eta_2 + a_{ij}^3\,\eta_3\right)\eta_0 \\ &= a_{ij}^0\,\eta_0 + a_{ij}^1\,\eta_1 + a_{ij}^2\,\eta_2 - a_{ij}^3\,\eta_3 \end{aligned} \tag{C10}$$

From this, one can count that each subblock $\mathbf{A}_{ij}$ has two antisymmetric and two symmetric coefficients. We count $n(n-1)/2$ subblocks $\mathbf{A}_{ij}$. Together with n diagonal subblocks with each having one antisymmetric coefficients, we should have

$$\nu_s^a = n\,(n-1) + n = n^2 \tag{C11}$$

antisymmetric symplices. The number of independent antisymmetric parameters is $\nu^a = \nu_s^a + \nu_c^a = n\,(2\,n-1) = 2\,n^2 - n$. Therefore, we have

$$\nu_c^a = n\,(2\,n-1) - n^2 = n^2 - n. \tag{C12}$$

antisymmetric cosymplices. Since we have as many cosymplices as we have antisymmetric matrix elements $\nu_c = \nu^a$, the number of symmetric cosymplices ν_c^s is:

$$\nu_c^s = \nu_c - \nu_c^a = 2\,n^2 - n - (n^2 - n) = n^2. \tag{C13}$$

The number ν_s^s of symmetric symplices is:

$$\nu_s^s = \nu_s - \nu_s^a = 2\,n\,(2\,n+1)/2 - n^2 = n^2 + n. \tag{C14}$$

$\tilde{\mathbf{A}}_{ij}$ is called the *symplectic conjugate* of $\mathbf{A}_{ij}$. If $\mathbf{S}$ and $\mathbf{C}$ are the symplex and cosymplex-part of a matrix $\mathbf{A} = \mathbf{C} + \mathbf{S}$, then the *symplectic conjugate* is

$$\tilde{\mathbf{A}} = -\gamma_0\,\mathbf{A}^T\,\gamma_0 = \mathbf{C} - \mathbf{S}. \tag{C15}$$

As well-known, one can quickly derive that

$$\widetilde{\mathbf{A}\mathbf{B}} = \tilde{\mathbf{B}}\,\tilde{\mathbf{A}}. \tag{C16}$$

If $\mathbf{A}$ is written “classically” as

$$\mathbf{A} = \begin{pmatrix} a_{11} & a_{12} \\ a_{21} & a_{22} \end{pmatrix} \tag{C17}$$

then $\eta_0\,\mathbf{A}^T\,\eta_0$ is given by:

$$\eta_0\,\mathbf{A}^T\,\eta_0 = \begin{pmatrix} -a_{22} & a_{12} \\ a_{21} & -a_{11} \end{pmatrix}, \tag{C18}$$

so that

$$\mathbf{A}\,(\eta_0\,\mathbf{A}^T\,\eta_0) = (a_{12}\,a_{21} - a_{11}\,a_{22})\,\mathbf{1} = -\mathrm{Det}(\mathbf{A})\,\mathbf{1}. \tag{C19}$$

A matrix $\mathbf{A}$ is symplectic, if $\mathbf{A}\,\tilde{\mathbf{A}} = \mathbf{1}$. But we can say in any case, that $\mathbf{A}\,\tilde{\mathbf{A}}$ is a co-symplex:

$$\begin{aligned}\mathbf{A}\,\tilde{\mathbf{A}} &= (\mathbf{C}+\mathbf{S})\,(\mathbf{C}-\mathbf{S})\\ &= \mathbf{C}^2 - \mathbf{S}^2 + \mathbf{S}\,\mathbf{C} - \mathbf{C}\,\mathbf{S}\end{aligned} \tag{C20}$$

since squares of (co-) symplices as well as the commutator of symplex and cosymplex are cosymplices.

A $2\,n \times 2\,n$-cosymplex $\mathbf{C}$ has according to Eq. (27) the form

$$\mathbf{C} = \begin{pmatrix} \mathbf{E}_1 & \mathbf{B}_{12} & \mathbf{B}_{13} & \dots & \mathbf{B}_{1n} \\ \tilde{\mathbf{B}}_{12} & \mathbf{E}_2 & \mathbf{B}_{23} & \dots & \mathbf{B}_{2n} \\ \tilde{\mathbf{B}}_{13} & \tilde{\mathbf{B}}_{23} & \mathbf{E}_3 & \dots & \mathbf{B}_{3n} \\ \vdots & & \vdots & & \vdots \\ \tilde{\mathbf{B}}_{1n} & \tilde{\mathbf{B}}_{2n} & \tilde{\mathbf{B}}_{3n} & \dots & \mathbf{E}_n \end{pmatrix} \tag{C21}$$

where the 2×2 matrices $\mathbf{E}_k$ on the diagonal must be 2×2-cosymplices and are hence proportional to the unit matrix

$$\mathbf{E}_k = e_k\,\mathbf{1}. \tag{C22}$$

C.3. Which k-vectors of $Cl_{N-1,1}$ are (anti-) symmetric

Possible dimensionalities for emergent spacetimes are given by Eq. (84). Since all but one generator of $\mathrm{Cl}_{N-1,1}$ are symmetric, the analysis of how many k-vectors are (anti-) symmetric, is reasonably simple. The total number of k-vectors is given by $\binom{N}{k}$, the number μ_k of k-vectors generated only with spatial elements, hence is $\binom{N-1}{k}$. For k-vectors that are generated exclusively from spatial basis vectors, one finds:

$$\begin{aligned}(\mathbf{S}_1\,\mathbf{S}_2\,\mathbf{S}_3\,\cdots\,\mathbf{S}_k)^T &= \mathbf{S}_k^T\,\mathbf{S}_{k-1}^T\,\mathbf{S}_{k-2}^T\,\cdots\,\mathbf{S}_1^T\\ &= \mathbf{S}_k\,\mathbf{S}_{k-1}\,\mathbf{S}_{k-2}\,\cdots\,\mathbf{S}_1\\ &= (-1)^a\,\mathbf{S}_1\,\mathbf{S}_2\,\mathbf{S}_3\,\cdots\,\mathbf{S}_k,\end{aligned} \tag{C23}$$

where the first step is possible as all $\mathbf{S}$ are spatial basis vectors and hence symmetric. The second step reflects the number of permutations that are required to reverse the order of k anticommuting elements:

$$a = k(k-1)/2. \tag{C24}$$

If one of the symplices equals γ_0 (i.e. is anti-symmetric), then we have

$$\begin{aligned}(\mathbf{S}_1\,\mathbf{S}_2\,\mathbf{S}_3\,\cdots\,\mathbf{S}_k)^T &= \mathbf{S}_k^T\,\mathbf{S}_{k-1}^T\,\mathbf{S}_{k-2}^T\,\cdots\,\mathbf{S}_1^T\\ &= -\mathbf{S}_k\,\mathbf{S}_{k-1}\,\mathbf{S}_{k-2}\,\cdots\,\mathbf{S}_1\\ &= (-1)^{a+1}\,\mathbf{S}_1\,\mathbf{S}_2\,\mathbf{S}_3\,\cdots\,\mathbf{S}_k.\end{aligned} \tag{C25}$$

Hence, depending on whether a is even or odd, we count $\mu_k^{s,a} = \binom{N-1}{k}$ symmetric and $\mu_k^{a,s} = \binom{N}{k} - \binom{N-1}{k}$ skew-symmetric matrices or vice versa. The result for the simplest Clifford algebras is given in Table 2.

Appendix D: The Kustaanheimo–Stiefel Transformation

During the "derivation" of the Lorentz force equation in Section 8.2, we associated elements of the matrix of second moments (or $\mathbf{S} = \Sigma\,\gamma_0$) with energy and momentum of a "particle", i.e. we assumed that:

$$\begin{aligned}\mathcal{E} &= -\bar{\psi}\,\gamma_0\psi \propto \psi_0^2+\psi_1^2+\psi_2^2+\psi_3^2,\\ P_x &= \bar{\psi}\,\gamma_1\psi \propto -\psi_0^2+\psi_1^2+\psi_2^2-\psi_3^2,\\ P_y &= \bar{\psi}\,\gamma_2\psi \propto 2\,(\psi_0\,\psi_2-\psi_1\,\psi_3),\\ P_z &= \bar{\psi}\,\gamma_3\psi \propto 2\,(\psi_0\,\psi_1+\psi_2\,\psi_3).\end{aligned} \tag{D1}$$

Table 2. Numbers $\mu_k = \mu_k^s + \mu_k^a$ of k-vectors for N-dimensional dynamically emergent spacetime Clifford-algebras $\mathrm{Cl}_{N-1,1}$. μ_k^s (μ_k^a) is the number of (anti-) symmetric k-vectors. The zero-vector is identified as the unit matrix which is (of course) symmetric. $\nu_x^y = n^2 + n$ is the number of (anti-) symmetric (co-) symplices, were $y = (a), s$ denotes (anti-) symmetry and $x = (c), s$ denotes (co-) symplices.

$N = p + q$	$4 = 3 + 1$	$10 = 9 + 1$	$12 = 11 + 1$	
2^N	16	1032	4096	
n	2	16	32	
	$\mu_k = \mu_k^s + \mu_k^a$			
$k = 0$	$1 = 1 + 0$	$1 = 1 + 0$	$1 = 1 + 0$	c
$k = 1$	$4 = 3 + 1$	$10 = 9 + 1$	$12 = 11 + 1$	s
$k = 2$	$6 = 3 + 3$	$45 = 9 + 36$	$66 = 11 + 55$	s
$k = 3$	$4 = 3 + 1$	$120 = 36 + 84$	$220 = 55 + 165$	c
$k = 4$	$1 = 0 + 1$	$210 = 126 + 84$	$495 = 330 + 165$	c
$k = 5$	—	$252 = 126 + 126$	$792 = 462 + 330$	s
$k = 6$	—	$210 = 126 + 84$	$924 = 462 + 462$	s
$k = 7$	—	$120 = 84 + 36$	$792 = 462 + 330$	c
$k = 8$	—	$45 = 9 + 36$	$495 = 165 + 330$	c
$k = 9$	—	$10 = 1 + 9$	$220 = 55 + 165$	s
$k = 10$	—	$1 = 1 + 0$	$66 = 55 + 11$	s
$k = 11$	—	—	$12 = 11 + 1$	c
$k = 12$	—	—	$1 = 0 + 1$	c
$\nu_s^s = n^2 + n$	6	272	1056	
$\nu_c^s = \nu_s^a = n^2$	4	256	1024	
$\nu_c^a = n^2 - n$	2	240	992	
$\mu^s = 2n^2 + n$	10	528	2080	
$\mu^a = 2n^2 - n$	6	496	2016	

Equation (D1) is (up to factor) practically identical to the regularization transformation of Kustaanheimo and Stiefel [59], (KST). At the same time, we introduced a scaling factor γ to transform between the *eigentime* τ and the time of an observer t — which is again a similarity to the use of the KST in celestial mechanics, where a Sundman transformation

$$\frac{dt}{d\tau} = f(\mathbf{q}, \mathbf{p}) \tag{D2}$$

is used [12, 101]. Despite these remarkable similarities, there are also significant differences, since in contrast to the KST, we do *not increase* the number of variables and we do not transform coordinates-to-coordinates. The "spinor" ψ was introduced as two canonical pairs (i.e. two coordinates and the two canonical momenta) and is used in Eq. (D1) to parameterize a four-momentum vector, while the KST uses four "fictious" coordinates to parameterize three cartesian coordinates.

Appendix E: Non-Symplectic Transformations

E.1. Simple LC-Circuit

The key concept of symplectic transformations was mentioned to be the *structure preservation*. In the following, we exemplify the meaning of *structure preservation* by giving examples for non-symplectic transformations. Consider a simple LC-circuit as shown in Fig. 1. Starting up naive, we take a Hamiltonian description using the total system energy, i.e. we associate the potential energy with the energy stored in the capacitor $\frac{C}{2}\,U^2$ and the kinetic energy with $\frac{L}{2}\,I_L^2$:

$$\mathcal{H} = \frac{C}{2}\,U^2 + \frac{L}{2}\,I_L^2 = \frac{C}{2}\,U^2 + \frac{L}{2}\,I_C^2, \tag{E1}$$

and assume that U is the canonical coordinate and I_C is the canonical momentum. Then the EQOM are:

$$\begin{aligned} \dot{U} &= \frac{\partial\mathcal{H}}{\partial I_C} = L\,I_C, \\ \dot{I}_C &= -\frac{\partial\mathcal{H}}{\partial U} = -C\,U. \end{aligned} \tag{E2}$$

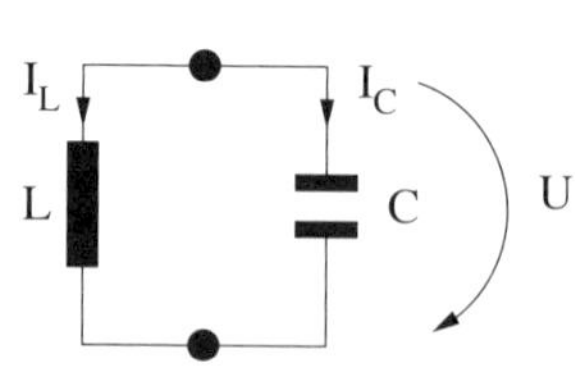

Fig. 1. Simple LC-circuit.

Quite obviously, these equations fail to describe the correct relations for capacitors and inductors, which are:

$$\begin{aligned} C\,\dot{U} &= I_C, \\ L\,\dot{I}_L &= U = -L\,\dot{I}_C. \end{aligned} \tag{E3}$$

We note that the mistake is a wrong scaling of the variables. We recall that the product of the coordinate and the corresponding conjugate momentum should have the dimension of an action, while voltage times current results in a quantity with the dimension of power. Thus, we introduce a scaling factor for the current and write:

$$\mathcal{H} = \frac{C}{2}\,U^2 + \frac{L}{2\,a^2}\,P^2, \tag{E4}$$

so that the canonical momentum is now $P = a\,I_C$ and the EQOM are:

$$\begin{aligned} \dot{U} &= \frac{\partial\mathcal{H}}{\partial P} = \frac{L}{a^2}\,P = \frac{L}{a}\,I_C, \\ \dot{P} &= -\frac{\partial\mathcal{H}}{\partial U} = -C\,U, \\ \dot{I}_C &= -\frac{C}{a}\,U. \end{aligned} \tag{E5}$$

The comparison with Eq. (E3) then yields $a = L\,C$. But note that the scaling changes the product $p_i\,q_i$ and hence is not a symplectic (structure preserving), but a non-symplectic (structure defining) transformation.

E.2. Coupled LC-circuits

If we restrict ourselves to the simplest case (i.e. two capacitors, two inductors), the structure of the only nontrivial way to couple two LC-circuits is shown in Fig. 2. The EQOM can be derived directly from the drawing using the general relations $\dot{U}_C = I_C/C$ for an ideal capacitor and $\dot{I}_L = U_L/L$ for the ideal inductor and

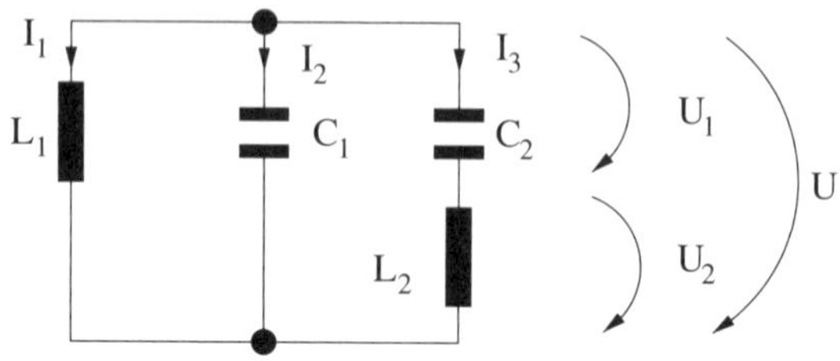

Fig. 2. Two coupled LC-circuits.

Kirchhoff's rule:

$$\begin{aligned}
\dot{I}_1 &= U/L_1, \\
\dot{U} &= -(I_1 + I_3)/C_1, \\
\dot{I}_3 &= (U - U_1)/L_2, \\
\dot{U}_1 &= I_3/C_2, \\
0 &= I_1 + I_2 + I_3, \\
U &= U_1 + U_2.
\end{aligned} \tag{E6}$$

The energy sum again yields a Hamiltonian of the diagonal form:

$$\mathcal{H} = \frac{C_1}{2}\, U^2 + \frac{L_1}{2}\, I_1^2 + \frac{C_2}{2}\, U_1^2 + \frac{L_2}{2}\, I_3^2. \tag{E7}$$

We use the voltages and currents of the Hamiltonian to define the state vector $\phi = (U, U_1, I_1, I_3)$ and we find for its derivative:

$$\begin{aligned}
\dot{\phi} &= \mathbf{F}\,\phi \\
&= \begin{pmatrix} 0 & 0 & -1/C_1 & -1/C_1 \\ 0 & 0 & 0 & 1/C_2 \\ 1/L_1 & 0 & 0 & 0 \\ 1/L_2 & -1/L_2 & 0 & 0 \end{pmatrix} \phi.
\end{aligned} \tag{E8}$$

The matrix $\mathbf{F}$ represents the structural properties of the LC-circuit. The matrix is not a symplex and cannot be transformed into a symplex by any symplectic transformation. *This does not mean* that the dynamics of the system cannot be derived from a Hamiltonian — it simply means that the transformation matrix which is required to map the system to a Hamiltonian system, is not structure *preserving* but structure *defining.*

In case of two coupled LC-circuits, it is likewise not sufficient to use scaling factors to obtain the canonical momenta and even if it was, the eigenfrequencies of the circuit cannot be guessed anymore. The square of the matrix from Eq. (E8) is given by:

$$\ddot{\phi} = \begin{pmatrix} -\frac{L_1 + L_2}{C_1\, L_1\, L_2} & \frac{1}{C_1\, L_2} & 0 & 0 \\ \frac{1}{C_2\, L_2} & -\frac{1}{C_2\, L_2} & 0 & 0 \\ 0 & 0 & -\frac{1}{C_1\, L_1} & -\frac{1}{C_1\, L_1} \\ 0 & 0 & -\frac{1}{C_1\, L_2} & -\frac{C_1 + C_2}{C_1\, C_2\, L_2} \end{pmatrix} \phi. \tag{E9}$$

Obviously, the state vector ϕ is not composed of canonical variables. The Lagrangian **L** is given by

$$\mathcal{L} = \frac{L_1}{2}\, I_1^2 + \frac{L_2}{2}\, I_3^2 - \frac{C_1}{2}\, U^2 - \frac{C_2}{2}\, U_1^2. \tag{E10}$$

Equation (E8) can be used to replace the currents by the derivatives of the voltages:

$$\begin{aligned} I_1 &= -C_1\, \dot{U} - C_2\, \dot{U}_1, \\ I_3 &= C_2\, \dot{U}_1, \\ \mathcal{L} &= \frac{L_1}{2}\, (C_1\, \dot{U} + C_2\, \dot{U}_1)^2 + \frac{L_2}{2}\, C_2^2\, \dot{U}_1^2 - \frac{C_1}{2}\, U^2 - \frac{C_2}{2}\, U_1^2, \end{aligned} \tag{E11}$$

so that for the coordinates $q_1 = U$ and $q_2 = U_1$, the canonical momenta are:

$$\begin{aligned} p_i &= \frac{\partial \mathcal{L}}{\partial \dot{q}_i}, \\ p_1 &= \frac{\partial \mathcal{L}}{\partial \dot{U}} = L_1\, C_1\, (C_1\, \dot{U} + C_2\, \dot{U}_1) \\ &= -L_1\, C_1\, I_1 \end{aligned}$$

$$\begin{aligned} p_2 = \frac{\partial \mathcal{L}}{\partial \dot{U}_1} &= L_1\,C_2\,(C_1\,\dot{U} + C_2\,\dot{U}_1) + L_2\,C_2^2\,\dot{U}_1 \\ &= -L_1\,C_2\,I_1 + L_2\,C_2\,I_3, \end{aligned} \tag{E12}$$

so that the transformation matrix from the state vector ϕ to the canonical variables ψ is given by:

$$\begin{aligned} &\psi = \mathbf{T}\,\phi \\ &\begin{pmatrix} q_1 \\ q_2 \\ p_1 \\ p_2 \end{pmatrix} = \begin{pmatrix} 1 & 0 & 0 & 0 \\ 0 & 1 & 0 & 0 \\ 0 & 0 & -L_1\,C_1 & 0 \\ 0 & 0 & -L_1\,C_2 & L_2\,C_2 \end{pmatrix} \begin{pmatrix} U \\ U_1 \\ I_1 \\ I_3 \end{pmatrix}. \end{aligned} \tag{E13}$$

The transformed matrix $\tilde{\mathbf{F}} = \mathbf{T}\,\mathbf{F}\,\mathbf{T}^{-1}$ then is:

$$\tilde{\mathbf{F}} = \begin{pmatrix} 0 & 0 & \frac{L_1+L_2}{C_1^2\,L_1\,L_2} & -\frac{1}{C_1\,C_2\,L_2} \\ 0 & 0 & -\frac{1}{C_1\,C_2\,L_2} & \frac{1}{C_2^2\,L_2} \\ -C_1 & 0 & 0 & 0 \\ 0 & -C_2 & 0 & 0 \end{pmatrix}. \tag{E14}$$

The Hamiltonian can then be expressed in the canonical coordinates as:

$$\begin{aligned} \mathcal{H} &= \frac{1}{2}\,\psi^T\,\mathbf{A}\,\psi \\ \mathbf{A} &= \begin{pmatrix} C_1 & 0 & 0 & 0 \\ 0 & C_2 & 0 & 0 \\ 0 & 0 & \frac{L_1+L_2}{C_1^2\,L_1\,L_2} & -\frac{1}{C_1\,C_2\,L_2} \\ 0 & 0 & -\frac{1}{C_1\,C_2\,L_2} & \frac{1}{C_2^2\,L_2} \end{pmatrix}, \end{aligned} \tag{E15}$$

where $\mathbf{A} = \gamma_0\,\tilde{\mathbf{F}}$. As was shown in [8], the Hamiltonian formulation (for stable oscillating systems) is always "similar" to the case of completely decoupled oscillators, i.e. the non-symplectic transformation maps the structure of Fig. 2 to two separate systems as shown in Fig. 1. The transformation $\mathbf{T}$ is not symplectic and may serve as an example for a structure *defining* transformation in contrast to symplectic structure *preserving* transformations.

E.3. The direction of time

If the matrix $\mathbf{Q}$ used in Eq. (7) is not only non-singular, but orthogonal, then

$$\mathbf{Q}^T\,\mathbf{S}\,\mathbf{Q} = \mathrm{diag}(\lambda_0\,\eta_0, \lambda_1\,\eta_0, \lambda_2\,\eta_0, \ldots, 0, 0, 0), \tag{E16}$$

with some real coefficients λ_k [48]. In this case, we would restrict to $\lambda_k = 1$ by the argument that the dynamics of the system has to be defined by the Hamilton function and not by the (otherwise arbitrary) skew-symmetric matrix $\mathbf{S}$. However, this is true only for closed systems, i.e. systems in which the matrix $\mathbf{A}$ is a function (of the constants of motion) of the variables ψ only. Since we interpreted γ_0 as the (unit vector in the) direction of time, it appears to be wise to return to the possible algebraic and physical implications of this choice for the parameters λ_k. In [8], we have demonstrated that a symplectic decoupling analysis of any oscillatory symplex $\mathbf{F}$ leads to a normal form exactly of the form on the right-hand side of Eq. (E16). The only difference is that the decoupling transformation is symplectic, while the transformation Eq. (E16) is obviously not symplectic, and not even generally orthogonal. We said that the system evolution in time is symplectic. However, the fact that the only restriction for the time-matrix $\mathbf{S}$ is its skew-symmetry calls for a deeper analysis of the possible physical implications of the non-symplectic transformation by $\mathbf{Q}$. Such a program extends the scope of this chapter. It may concern the number, type, and possible interactions and transformations between the fermions. Recall that the matrix $\mathbf{Q}$ can be used to transform $\mathbf{S}$ into all possible six already skew-symmetric matrices used to build the two types of spinors in Eq. (167). And it concerns the role that the "axial 4-vector"-components play in the transformations between these two spinor-types (or types of phase space symmetry).

E.4. Non-symplectic transformations within the Dirac algebra

In the following, we present an orthogonal (non-symplectic) transformation that enables to "rotate the time direction", i.e. to transform from γ_0 (as we used it here) to other representations of γ_0. We define the (normalized) Hadamard-matrix H_4 according to

$$\begin{aligned} H_4 &= \frac{\gamma_2 - \gamma_1 - y_4 + \gamma_5}{2} = \frac{(\gamma_{15} + \gamma_0)\,(\gamma_2 - \gamma_1)}{2} \\ &= \frac{1}{2}\begin{pmatrix} 1 & 1 & 1 & 1 \\ 1 & -1 & 1 & -1 \\ 1 & 1 & -1 & -1 \\ 1 & -1 & -1 & 1 \end{pmatrix}, \\ H_4 &= H_4^T, \\ H_4^2 &= \mathbf{1}, \\ \gamma_0\, H_4\, \gamma_0 &= H_4, \\ H_4\, \gamma_0\, H_4 &= -\gamma_0. \end{aligned} \tag{E17}$$

H_4 is a symplex and it is antisymplectic. Furthermore, we define a "shifter" matrix X according to

$$X = \frac{-\gamma_0 + \gamma_2 + \gamma_6 + \gamma_7}{2} = \begin{pmatrix} 0 & 0 & 0 & 1 \\ 1 & 0 & 0 & 0 \\ 0 & 1 & 0 & 0 \\ 0 & 0 & 1 & 0 \end{pmatrix}. \tag{E18}$$

Some properties of the X-matrix are

$$
\begin{aligned}
\gamma_0\, X\, \gamma_0 &= X^T, & X^T\, X &= \mathbf{1},\\
X^T\, \gamma_0\, X &= -\gamma_7, & X^T\, \gamma_7\, X &= -\gamma_0,\\
X^T\, \gamma_8\, X &= \gamma_{14}, & X^T\, \gamma_{14}\, X &= -\gamma_8,\\
X^T\, \gamma_9\, X &= -\gamma_{10}, & X^T\, \gamma_{10}\, X &= \gamma_9.
\end{aligned} \tag{E19}
$$

Note that X is a symplex and hence could be used as a force matrix. In this case, we have a fourth order differential equation of the form:

$$
\begin{aligned}
\psi' &= X\, \psi,\\
\psi'''' &= \psi,
\end{aligned} \tag{E20}
$$

since one finds $X^4 = \mathbf{1}$. The product of these two matrices $R_6 = H_4\, X$ is an orthogonal matrix that transforms cyclic through all possible basis systems:

$$
R_6 = H_4\, X = \frac{1}{2}\begin{pmatrix} 1 & 1 & 1 & 1\\ -1 & 1 & -1 & 1\\ 1 & -1 & -1 & 1\\ -1 & -1 & 1 & 1 \end{pmatrix}. \tag{E21}
$$

The most relevant properties are

$$
\begin{aligned}
R_6^T\, R_6 = R_6\, R_6^T &= \mathbf{1}, & R_6^T\, \gamma_0\, R_6 &= \gamma_7,\\
R_6^T\, \gamma_{10}\, R_6 &= -\gamma_8, & R_6^T\, \gamma_{14}\, R_6 &= \gamma_9,\\
R_6^T\, \gamma_7\, R_6 &= -\gamma_{14}, & R_6^T\, \gamma_8\, R_6 &= \gamma_0,\\
R_6^T\, \gamma_9\, R_6 &= \gamma_{10},
\end{aligned} \tag{E22}
$$

so that R_6 can be iteratively used to switch through all possible systems:

$$
\begin{aligned}
R_6^T\, \gamma_0\, R_6 &= \gamma_7,\\
(R_6^T)^2\, \gamma_0\, (R_6)^2 &= -\gamma_{14},\\
(R_6^T)^3\, \gamma_0\, (R_6)^3 &= -\gamma_9,
\end{aligned}
$$

$$
\begin{aligned}
(R_6^T)^4\,\gamma_0\,(R_6)^4 &= -\gamma_{10},\\
(R_6^T)^5\,\gamma_0\,(R_6)^5 &= \gamma_8,\\
(R_6^T)^6\,\gamma_0\,(R_6)^6 &= \gamma_0.
\end{aligned}
\tag{E23}
$$

Finally, one finds that $(R_6)^6 = 1$. But R_6 is neither symplectic nor is it a symplex. Expressed by the γ-matrices, R_6 is given by

$$
\begin{aligned}
4\,R_6 = 1 &+ \gamma_0 + \gamma_1 - \gamma_2 + \gamma_3 - \gamma_4 + \gamma_5 + \gamma_6\\
&+ \gamma_7 + \gamma_8 - \gamma_9 - \gamma_{10} - \gamma_{11} - \gamma_{12} - \gamma_{13} - \gamma_{14}.
\end{aligned}
\tag{E24}
$$

Another matrix with equivalent properties is given by

$$
\tilde{R}_6 = H_4\,X^T. \tag{E25}
$$

These transformations exemplifies part of the claim (i.e. the use of a specific (though arbitrary) form for γ_0) in Eq. (7).

E.5. Transformation to the conventional Dirac algebra

According to the fundamental theorem of the Dirac matrices [64], any set of Dirac matrices is similar to any other set up to a unitary transformation. However, a change in the sign of the metric tensor requires in addition a multiplication with the unit imaginary.

Let $\mathbf{U}$ be the following unitary matrix

$$
\mathbf{U} = \frac{1}{2}\begin{pmatrix} 1 & i & i & -1\\ -i & 1 & -1 & -i\\ -i & -1 & 1 & -i\\ 1 & -i & -i & -1 \end{pmatrix} \tag{E26}
$$

then it is quickly verified that

$$
\tilde{\gamma}_\mu = i\,\mathbf{U}\,\gamma_\mu\,\mathbf{U}^\dagger, \tag{E27}
$$

where $\mu \in [0\dots 3]$ and $\tilde{\gamma}_\mu$ are the conventional Dirac matrices footnote. The explicit form of the real Dirac matrices is given for instance in ([6, 8]). Using $\tilde{\gamma}_\mu$, the other matrices of the Clifford algebra are quickly constructed. However, since we multiplied by the unit imaginary, it is clear that the conventional Dirac algebra is a rep of $\mathrm{Cl}_{1,3}$.

Appendix F: Graphical Representation of Dirac Matrices

Figure 3 illustrates the geometric interpretation of the RDMs.

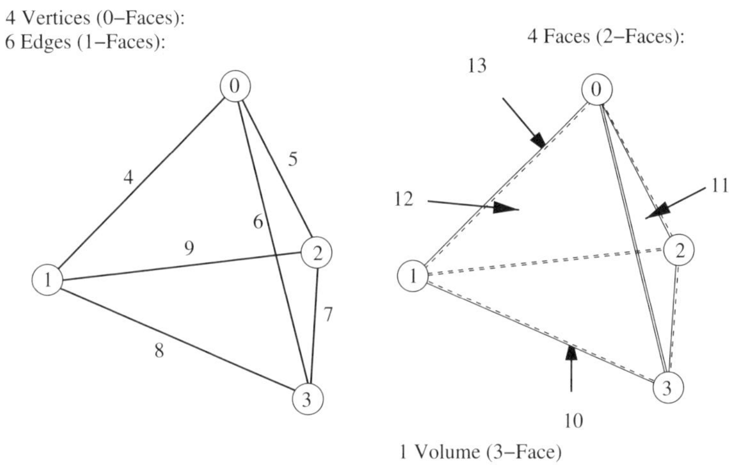

Fig. 3. The group structure of the RDMs (i.e. the Clifford algebra $Cl_{3,1}$) can be represented by a tetrahedron: The vertices represent the "basic" symplices $\gamma_0, \ldots, \gamma_3$, the edges the bi-vectors $\gamma_4, \ldots, y_9$, the surfaces the components of the axial vector $\gamma_{10}, \ldots, \gamma_{13}$ and the volume the pseudoscalar γ_{14}. Another graphical representation has been given by Goodmanson [39].

Chapter 10

The Quantum State as Spatial Displacement

Peter Holland

Green Templeton College, University of Oxford,
Oxford OX2 6HG, England
peter.holland@gtc.ox.ac.uk

We give a simple demonstration that the Schrödinger equation of wave mechanics may be recast as a self-contained second-order Newtonian law for a congruence of spacetime trajectories. This implies that a pictorial representation of the quantum state as a displacement function is implicit in the quantum description, complementary to and independent of the wave function. Quantum evolution is described by the deterministic unfolding of a continuous coordinate transformation. Introducing gauge potentials for the density and current density it is shown that the wave-mechanical and trajectory pictures are connected by a canonical transformation. The canonical trajectory theory is shown to provide an alternative basis for the quantum operator calculus and the observability of the quantum state is examined within this context. The construction illuminates some of the problems involved in connecting the quantum and classical descriptions.

1. Introduction

An unfortunate by-product of the historical debate on the interpretation of quantum mechanics is that physical ideas that may have informed the development of the subject have been marshalled into the siding of "mere philosophy". This has been the fate of the spacetime trajectory picture of quantum evolution, which is still widely assumed to be associated just with an interpretation of the theory, i.e. to depend on optional assumptions that are not inherent in the scheme of ideas that is generally accepted as constituting "quantum theory". In fact, it is an ineluctable mathematical property that the conception of a physical state based on the deterministic spacetime trajectory — comprising simultaneously well-defined position and momentum variables at each point — is implicit in the quantum description *whatever the interpretation*. More precisely, a self-contained theory of trajectories with second-order Newton-style dynamics may be obtained from the first-order Schrödinger equation by a change of variables [20, 21]. In this formulation, the quantum state is represented by the displacement function of a continuum of interacting "particles", which involves a congruence rather than a single trajectory because wave mechanics is a field theory. The characteristic features of the wave function version of state are represented by distinctive properties of the congruence. For example, the unitary evolution of the wave function corresponds to the deterministic unfolding of a continuous coordinate transformation and the single-valuedness of the wave function to non-crossing of the trajectories. Indeed, one may make the displacement function of the collective the basis of the quantum description with the wave function being regarded as a derived quantity.

The trajectory conception of state emerged following many years of studies connected with the de Broglie–Bohm theory [29] and related computational work [46] and was first clearly elaborated by the author [20, 21]. If the wave function is represented by hydrodynamic variables [34], the hydrodynamic and trajectory notions of state stand in relation to one another as the Eulerian and Lagrangian pictures of fluid dynamics [45] and we draw extensively upon the terminology and methods of that discipline. Our reference to "particles" is to be construed as referring to fluid elements but no interpretative commitment to this imagery is required (they may be considered as "elements of probability"). It is emphasized that the trajectory model is not a hidden-variable theory; it arises simply from a transformation of the independent and dependent variables employed in wave mechanics. A hidden-variable theory in this context would endow one of the trajectories in the congruence composing the state with some special property, be it a particular label or additional structure such as a corpuscle (as in the de Broglie–Bohm theory [29]), with the aim of explaining the quantum statistical predictions as the outcome of well-defined and causally connected individual events. Indeed, the trajectory representation of the state is particularly suited to describing how individuals may make up a statistical ensemble, something that is difficult to achieve using the wave function, but the additional assumptions required to do this are not part of the theory presented here. It has scarcely been noticed that much of the literature devoted to trajectory theories has been misleadingly classified as pertaining to "interpretation" when in reality many of the results relate to quantum theory itself, albeit in this unfamiliar representation. A corollary is that problems ascribed to, say, the de Broglie–Bohm theory may in fact be issues to do with quantum mechanics that would be expressed in different terms (if at all) in the usual wave function approach.

Our purpose here is to provide an introductory account of the trajectory conception of quantum dynamics and investigate some of its key features. In Section 2, we present a simple derivation of the second-order Newton-like version of Schrödinger's equation that is more transparent than the demonstration in [20]. A particular aim is to exhibit the intimate relation between the wave-mechanical and trajectory pictures by showing how they are connected by a canonical transformation when each is expressed in suitable phase space coordinates. To this end, in Section 3 we first recall the usual Hamiltonian approach to quantum mechanics. We then develop in Section 4 a novel phase space formulation of quantum mechanics in its Eulerian hydrodynamical form that is more suited to our needs by introducing potentials for the density and current density. The potentials obey a second-order field equation and exhibit a gauge freedom that may be exploited to simplify the theory. This second-order representation of the Schrödinger equation is somewhat analogous to writing Maxwell's equations in terms of the electromagnetic potentials. In Section 5, we set up the desired canonical transformation linking the potentials-based canonical theory with a canonical formulation of the trajectory model in which the potentials transform into the displacement function. A significant point is that the potentials in the latter representation are gauge invariant quantities. In this endeavor, we follow and develop a method introduced previously in classical fluid mechanics to connect the Eulerian and Lagrangian pictures [7], a procedure that is superior in several respects to the quantum canonical formulation used previously [20]. In Section 6, it is shown how the phase space trajectory theory may be used to represent the quantum operator algebra and in Section 7 we address the problem of the empirical determination of the state as it arises in the hydrodynamic formulation. Section 8 presents comments on how the trajectory view may provide the basis of a common language for the quantum and classical descriptions and how it thereby affords insight into the problems that arise in seeking to connect them.

2. Schrödinger's Equation as Newton's Law for a Continuum of Particles

2.1. Eulerian quantum hydrodynamics

We are going to show how the Schrödinger equation,

$$i\hbar\frac{\partial\psi}{\partial t} = -\frac{\hbar^2}{2m}\frac{\partial^2\psi}{\partial x_i\partial x_i} + V\psi, \tag{1}$$

may be rewritten as Newton's law for a cloud of interacting particles pursuing spacetime trajectories. To this end, we first rewrite it as two coupled real equations using the polar representation of the wave function, $\psi = \sqrt{\rho}e^{iS/\hbar}$:

$$\frac{\partial\rho}{\partial t} + \frac{\partial}{\partial x_i}\left(\rho\frac{1}{m}\frac{\partial S}{\partial x_i}\right) = 0, \tag{2}$$

$$\frac{\partial S}{\partial t} + \frac{1}{2m}\frac{\partial S}{\partial x_i}\frac{\partial S}{\partial x_i} + V_Q + V = 0, \tag{3}$$

where

$$V_Q(x) = -\frac{\hbar^2}{2m\sqrt{\rho}}\frac{\partial^2\sqrt{\rho}}{\partial x_i\partial x_i} \tag{4}$$

is the quantum potential and $i, j, k, \ldots = 1, 2, 3$. These equations are derived though multiplication and division by ψ and hence hold in non-nodal ($\psi \neq 0$) regions. The field variables ρ, S — which now represent the quantum state — inherit the continuity, boundary, and single-valuedness conditions obeyed by ψ. The latter condition is expressed by the quantization condition

$$\oint \frac{\partial S}{\partial x_i} dx_i = nh, \quad n \in \mathbb{Z}, \tag{5}$$

where the integration is over a loop fixed in space along which $\rho \neq 0$. The singularities in the phase S occur at nodes.

This way of articulating the Schrödinger equation is the basis of its Eulerian-picture hydrodynamic representation [34] where the fluid functions are expressed with respect to a fixed system of coordinates. Thus, Eq. (3) may be regarded as a Bernoulli-type equation with ρ the number density and S the velocity potential of a putative continuous "quantum fluid". Writing

$$v_i(x, t) = \frac{1}{m} \frac{\partial S(x, t)}{\partial x_i} \tag{6}$$

for the velocity field we obtain from Eq. (2) the fluid continuity equation and, differentiating Eq. (3) with respect to x, an Euler-type force law:

$$\frac{\partial \rho}{\partial t} + \frac{\partial}{\partial x_i}(\rho v_i) = 0, \tag{7}$$

$$\frac{\partial v_i}{\partial t} + v_j \frac{\partial v_i}{\partial x_j} = -\frac{1}{m} \frac{\partial}{\partial x_i}(V + V_Q). \tag{8}$$

Following Eq. (5), the circulation obeys $\oint v_i dx_i = nh/m$ so the velocity field is single-valued and irrotational except along nodal lines where it is singular:

$$\varepsilon_{ijk} \frac{\partial v_k}{\partial x_j} = \sum_{\mu} n_\mu h \int_{L_\mu} \delta(x - x_\mu)\, dx_{\mu i} \tag{9}$$

with $x_{\mu i}$ the coordinates of the μth nodal line L_μ and ε_{ijk} the antisymmetric symbol with $\varepsilon_{123} = 1$. The fluid therefore possesses quantized vortices [17, 19]. From a mathematical perspective, an advantage of the hydrodynamic formulation is that it involves only quantities ρ, v_i that are independent of the irrelevant global phase. A disadvantage is that the linearity of quantum mechanics is rendered somewhat awkwardly in these variables although they do make interference effects transparent.

If we adopt Eqs. (7) and (8) as the quantum evolution equations, they must be supplemented by the subsidiary condition (6) subject to (5). To show that we indeed obtain (1) from the hydrodynamic equations, we note that (8) with (6) inserted implies the equation

$$\frac{\partial}{\partial x_i}\left(\frac{\partial S}{\partial t}+\frac{1}{2m}\frac{\partial S}{\partial x_j}\frac{\partial S}{\partial x_j}+V_Q+V\right)=0. \tag{10}$$

It follows that the quantity in brackets is an arbitrary function of t and, absorbing this function into a redefined S, we obtain Eq. (3). Combining the latter with Eq. (7), (1) follows. Clearly, if Eq. (6) is not imposed as a subsidiary condition the flow implied by Eqs. (7) and (8) will be more general than that described by the Schrödinger equation. In fact, in the more general case the equations can still be combined into Schrödinger form but the phase, $\int mv_i(x)dx_i$, becomes path dependent and the "wave function" multivalued.

2.2. Preservation of initial conditions

An important aspect of the derivation of Eq. (1) from the hydrodynamic equations is that the subsidiary condition (6) need be assumed only at one instant. Hence, taking the latter to be $t = 0$, the condition represents a constraint just on the initial velocity field:

$$v_{0i}(x) = m^{-1}\partial_i S_0(x). \tag{11}$$

To prove this, we start by subtracting the Euler equation (8) where $v_i = m^{-1}\partial_i S$ from that equation when v_i is arbitrary:

$$\left(\frac{\partial}{\partial t}+v_j\partial_j\right)(mv_i-\partial_i S)=-\frac{1}{m}(mv_j-\partial_j S)\partial_{ij}S. \tag{12}$$

To show that $mv_i = \partial_i S$ is the unique solution to this equation for all t given (11), we use the method of characteristics. This entails passing to moving coordinates $x_i = q_i(t)$ defined by the integral curves of the velocity, obtained by solving the differential equation $v_i(x_i = q_i(t), t) = \dot{q}_i$. Then, evaluating (12) in these coordinates, we obtain

$$\frac{d}{dt}(m\dot{q}_i - \partial_i S|_{x=q(t)}) = -\frac{1}{m}(m\dot{q}_j - \partial_j S|_{x=q(t)})\partial_{ij} S|_{x=q(t)}, \tag{13}$$

where $d/dt = \partial/\partial t + \dot{q}_i \partial_i$. This relation has the form of a first-order linear ordinary differential equation $\dot{X}_i = A_{ij}(t) X_j(t)$ for which continuity of the matrix $A_{ij}(t)$ guarantees the existence and uniqueness of solutions $X_i(t)$ [16]. Then, since $m\dot{q}_i - \partial_i S = 0$ is a solution of (13), this is the unique solution for all t if it holds at $t = 0$, granted the continuity of the function $\partial_{ij} S(x = q(t))$. Returning to space coordinates, we have proved that (11) implies (6) for all t.

The preservation of gradient flow is a classical result and the details of the quantum contribution to the force in Euler's law are unimportant, other than that it contributes to an acceleration potential. There are a variety of alternative proofs, such as employing Kelvin's circulation theorem or Cauchy's vorticity theorem (see [44] for a compendium of methods). If we wish to characterize the quantum state by the ρ, v_i variables, the initial condition (11) is understood.

The theorem just proved concerns a condition for the identity of two flows corresponding to the same density and relies on Euler's force equation. In parenthesis, we remark that a complementary result holds for the density: given a flow v_i and two possible densities ρ, $|\psi|^2$ obeying the continuity equation, the initial constraint $\rho_0 = |\psi_0|^2$ ensures $\rho = |\psi|^2$ for all t. To prove this, we may again employ the characteristics but now appeal to the continuity equation. Thus, writing $f(x, t) = \rho/|\psi|^2$ we deduce from (7) that $df(x = q(t), t)/dt = 0$ or $f(x = q(t), t) = \text{const}$. Then, if $\rho_0(x) = |\psi_0(x)|^2$ for each x, we have $f = 1$ for all x and returning to space coordinates $\rho(x, t) = |\psi(x, t)|^2$ for all t.

The characteristics may also be used to show that the quantization condition (5) is temporally preserved following the flow, as we see below.

2.3. Lagrangian quantum hydrodynamics

We have shown how the spacetime trajectories defined by the integral curves of the velocity field (the trajectories employed by de Broglie and Bohm in their hidden-variable interpretation) may be invoked to demonstrate propagation properties of a quantum system. We now demonstrate that, subject to suitable initial conditions, they specify the quantum dynamics completely. To proceed, we observe that the single-valuedness of the velocity field implies that the initial position coordinates q_{0i} uniquely specify the trajectories. They therefore provide a continuously variable set of three labels to identify the curves, which we denote $q_{0i} = a_i$. The full congruence is therefore described by the displacement function $q_i(a,t)$, which is single-valued and differentiable with respect to a_i and t and the inverse mapping $a_i(q,t)$ exists and has the same properties. We may thus conceive of the system as comprising a continuum of fluid elements, or "particles", the identity of each being preserved throughout the flow and defined by the invariant a_i. This step is not merely of mathematical significance for the labeling allows us to conceive of fluid functions such as density and pressure in terms of notions not available in the Eulerian picture, namely, interparticle interactions described by the deformation matrix $\partial q_i/\partial a_l$. This description of the state, using a system of coordinates moving with the medium, corresponds to the Lagrangian picture of a fluid.

Suppose we have some means of calculating the trajectories that does not depend on first knowing the velocity field. Then, according to a result due to Euler [45], they generate the general solution for any fields ρ, v_i that satisfy the continuity equation (7). Consider the following identity obeyed by the microscopic particle density $\delta(x - q(a,t))$:

$$\frac{\partial}{\partial t}\delta(x - q(a,t)) + \frac{\partial}{\partial x_i}\left[\delta(x - q(a,t))\frac{\partial q_i(a,t)}{\partial t}\right] = 0. \qquad (14)$$

Multiplying by a function $\rho_0(a)$ and summing over all a_i, we obtain the following formulas for functions ρ, v_i obeying the continuity equation in terms of the functions $q_i(a,t)$:

$$\rho(x,t) = \int \rho_0(a)\delta(x - q(a,t))d^3a, \tag{15}$$

$$\rho(x,t)v_i(x,t) = \int \rho_0(a)\frac{\partial q_i(a,t)}{\partial t}\delta(x - q(a,t))d^3a. \tag{16}$$

As suggested by the notation, $\rho_0 = |\psi_0|^2$ is the initial value of ρ. Computing the integrals, we get the local relations

$$\rho(x,t) = J^{-1}|_{a(x,t)}\rho_0(a(x,t)), \tag{17}$$

$$v_i(x,t) = \left.\frac{\partial q_i\,(a,t)}{\partial t}\right|_{a(x,t)}, \tag{18}$$

where

$$J = \det(\partial q/\partial a) = \frac{1}{3!}\varepsilon_{ijk}\varepsilon_{lmn}\frac{\partial q_i}{\partial a_l}\frac{\partial q_j}{\partial a_m}\frac{\partial q_k}{\partial a_n}, \quad 0 < J < \infty. \tag{19}$$

The relations (17) and (18) give the desired general solution. Note that in the trajectory language, the continuity equation (7) becomes the conservation law

$$\rho(q,t)d^3q(a,t) = \rho_0(a)d^3a. \tag{20}$$

Since ρ is the particle number density, (20) states that the number of particles contained in an elementary volume is conserved by the flow. This relation prompts defining the mass of a fluid element as $m\rho d^3q$ since, assuming ρ is normalized, the total mass of the fluid is m.

To complete the trajectory representation of the fluid functions, we need a self-contained dynamical equation to determine the vector function q_i. This follows immediately from the Euler equation (8). From (18), we deduce the following relation between the accelerations in the two pictures:

$$\frac{\partial v_i}{\partial t} + v_j\frac{\partial v_i}{\partial x_j} = \left.\frac{\partial^2 q_i(a,t)}{\partial t^2}\right|_{a(x,t)}. \tag{21}$$

Substituting $x_i = q_i(a, t)$, Euler's force law then becomes in the Lagrangian picture

$$m\frac{\partial^2 q_i(a)}{\partial t^2} = -\frac{\partial}{\partial q_i}\left(V(x) + V_Q(x)\right)\big|_{x=q(a,t)}. \tag{22}$$

Here, derivatives with respect to q_i are regarded as shorthand for derivatives with respect to a_i via the formula

$$\frac{\partial}{\partial q_i} = J^{-1} J_{ij}\frac{\partial}{\partial a_j}, \tag{23}$$

where J_{ij} is the adjoint of the deformation matrix $\partial q_i/\partial a_l$ with

$$\frac{\partial q_i}{\partial a_j} J_{il} = J\delta_{lj}, \quad J_{il} = \frac{\partial J}{\partial(\partial q_i/\partial a_l)} \tag{24}$$

and ρ in V_Q is given by (17). The initial condition corresponding to (11) is $\partial q_{i0}(a)/\partial t = m^{-1}\partial S_0(a)/\partial a_i$ and the dynamics is completed by specifying $\rho_0(a)$, which appears explicitly in (22) via V_Q.

We conclude that (22) is *Schrödinger's equation in the form of Newton's second law.* The quantum state is now represented by the "displacement amplitude" $q_i(a, t)$ encoding the history of an infinite ensemble of particles whose interaction is described by the derivatives of q_i with respect to a_i (which appear up to fourth order on the right-hand side of (22)). With the appropriate initial conditions, the vector $q_i(a)$ determines the motion completely, without reference to $\psi(x)$. Complementary to the latter's unitary evolution, *quantum evolution is represented as the deterministic unfolding of a continuous coordinate transformation* $a_i \to q_i$. We remark that one can reverse the demonstration and deduce Schrödinger's equation starting from Newton's law, as shown previously [20].

As an application of this form of Schrödinger's equation, we easily derive a quantum version of Kelvin's theorem on the conservation of circulation [39, 40]:

$$\frac{\partial}{\partial t}\oint \dot{q}_i dq_i = 0. \tag{25}$$

Here the closed loop, composed of particles, remains closed during the flow due to the continuity of the function $q_i(a)$. The symmetry corresponding to this conservation law is the covariance of the theory with respect to continuous transformations of the particle label [26].

Starting from the ψ-representative of the quantum state, we may compute the q_i-representative by solving (18). Conversely, starting from the solutions q_i we may evaluate ψ using the following prescription. First of all, the initial data $\rho_0(a)$, $\dot{q}_{oi}$ fixes the initial wave function $\psi_0(a) = \sqrt{\rho_0}\exp(iS_0/\hbar)$ up to an irrelevant constant phase. To compute the wave function for all x, t up to a global phase, we first solve (22) subject to the initial conditions $q_{0i}(a) = a_i$, $\partial q_{i0}(a)/\partial t = m^{-1}\partial S_0(a)/\partial a_i$ to get the set of trajectories for all a_i, t. Next, substitute $q_i(a,t)$ in (17) to find ρ and $\partial q_i/\partial t$ in (18) to get $m^{-1}\partial S/\partial x_i$. This gives S up to an additive function of time, $f(t)$. To fix this function, apart from an additive constant, use (3). We obtain finally the following formula for the wave function as a function of the trajectory solution:

$$\psi(x,t) = \sqrt{(J^{-1}\rho_0)|_{a(x,t)}} \times \exp\left[\frac{i}{\hbar}\left(\int m\partial q_i(a,t)/\partial t|_{a(x,t)}dx_i + f(t)\right)\right]. \quad (26)$$

To summarize, we have presented three forms of the Schrödinger equation and associated concepts of state and examined their equivalence: the usual one (1) of wave mechanics where the state is represented by $\psi(x)$, the Eulerian hydrodynamic version with state $\rho(x)$, $v_i(x)$, and the Lagrangian hydrodynamic version with state $q_i(a)$, the initial velocity in the last two being subject to the gradient condition.

We remark that the single-body theory presented here readily extends to an n-body system by allowing the indices $i, j, \ldots$ to range over $3n$ values [20, 21]. The configuration space congruence may be mapped into ensembles of interlacing trajectories in three-space. Generalizing the configuration space to a Riemannian manifold, the method of representing the quantum state by Lagrangian coordinates embraces a wide variety of systems including spin 1/2 and fields [21, 22, 24] and other implications have been explored [25, 27]. A related remark is that the change of variables through which the Schrödinger equation is recast in Newtonian form is not unique. For example, relativistic considerations suggest an alternative expression for the non-relativistic velocity [28] and other models exist [23] but we shall not pursue these here.

3. Hamiltonian Formulation of Quantum Mechanics

It is well known that, using Hamiltonian methods, the Schrödinger equation may be expressed in a form that closely mirrors classical mechanics [15, 43] although the interpretations given to the symbols representing the physical state in the respective phase spaces are radically different (probability amplitude and conjugate momentum versus particle position and conjugate momentum). Here we shall examine some aspects of the usual Hamiltonian version of quantum mechanics. In the following two sections, we show how the quantum phase space variables may be chosen to resemble more closely the classical state.

We recall that in classical mechanics the state of a physical system is described by conjugate variables q_i, p_i, $i = 1, 2, 3$, obeying Hamilton's equations

$$\dot{q}_i = \frac{\partial H_{qp}}{\partial p_i}, \quad \dot{p}_i = -\frac{\partial H_{qp}}{\partial q_i}. \tag{27}$$

In the following, continuous transformations of the variables $q_i, p_i, H_{qp} \to q'_i, p'_i, H_{q'p'}$ that leave Hamilton's equation invariant — the canonical transformations — will be particularly significant. There are various necessary and sufficient signatures that a mapping is canonical. For example, if the generating function depends on the old coordinates and new momenta, $W(q, p')$, so that the transformation equations are

$$p_i = \frac{\partial W}{\partial q_i}, \quad q'_i = \frac{\partial W}{\partial p'_i}, \quad H_{q'p'} = H_{qp} + \frac{\partial W}{\partial t}, \tag{28}$$

one characterization of canonicity is the invertibility of the Hessian matrix $h_{ij} = \partial^2 W / \partial q_i \partial p'_i$. An alternative characterization is the invariance of the Poisson brackets under the transformation.

To see the close analogy between the classical and quantum Hamiltonian formalisms, it is useful to reformat Hamilton's equations in complex coordinates. Thus, introducing the coordinates $z_i = (q_i + ip_i)/\sqrt{2}$ with conjugate momenta $\pi_i = iz_i^*$, (27) become

$$\dot{z}_i = \frac{\partial H_{z\pi}}{\partial \pi_i}, \quad \dot{\pi}_i = -\frac{\partial H_{z\pi}}{\partial z_i}, \tag{29}$$

where $H_{z\pi} = H_{qp}$. This transformation from real to complex phase space is evidently canonical. Extending the range of i, Schrödinger's equation and its complex conjugate have just this form for a discrete quantum system if z_i is identified with complex normal coordinates and the Hamiltonian describes a collection of oscillators [43].

The Hamiltonian formulation of the Schrödinger equation in the position representation may be obtained by extending the discrete complex version of Hamilton's equations to a continuous system. In the usual variational approach, the Hamiltonian is written [41]

$$H_{\psi\pi}[\psi, \pi] = (i\hbar)^{-1} \int \left(\frac{\hbar^2}{2m} \frac{\partial \pi}{\partial x_i} \frac{\partial \psi}{\partial x_i} + V(x)\pi\psi \right) d^3x, \tag{30}$$

where ψ and $\pi = i\hbar\psi^*$ are conjugate variables. This is the mean value of the Hamiltonian operator in the state ψ: $H_{\psi\pi} = \langle \psi | \hat{H} | \psi \rangle$. Hamilton's equations,

$$\dot{\psi} = \frac{\delta H_{\psi\pi}}{\delta \pi}, \quad \dot{\pi} = -\frac{\delta H_{\psi\pi}}{\delta \psi}, \tag{31}$$

reproduce the Schrödinger equation (1) and its complex conjugate. Here the functional derivative is defined as follows [14]. Suppose F is a functional of some function $\phi(x)$: $F[\phi] = \int f(x, \phi, \partial\phi, \partial^2\phi, \ldots) d^3x$. Then the functional derivative of F with respect to ϕ is

$$\frac{\delta F}{\delta \phi} = \frac{\partial f}{\partial \phi} - \frac{\partial}{\partial x_i} \frac{\partial f}{\partial (\partial\phi/\partial x_i)} + \frac{\partial^2}{\partial x_i \partial x_j} \frac{\partial f}{\partial (\partial^2\phi/\partial x_i \partial x_j)} - \cdots \tag{32}$$

Using this notation, the functional Hamilton equations may be expressed in terms of Poisson brackets, the latter being defined for two functionals A, B of phase space variables $\phi(y), \gamma(y)$as follows:

$$\{A, B\}_{\phi\gamma} = \int \left(\frac{\delta A}{\delta \phi(y)} \frac{\delta B}{\delta \gamma(y)} - \frac{\delta B}{\delta \phi(y)} \frac{\delta A}{\delta \gamma(y)} \right) d^3y. \tag{33}$$

Being a Hamiltonian system, the Schrödinger equation admits canonical transformations as symmetries, which link the old phase space coordinates $\psi(x), \pi(x)$ with a new set $\psi'(y), \pi'(y)$ that obey Hamilton's equations with a transformed Hamiltonian $H_{\psi'\pi'}$. Unitary transformations $U(y, x)$ form a class of canonical transformations where the old and new coordinates are connected linearly, and likewise for the momenta:

$$\psi'(y) = \int U(y,x)\psi(x)d^3x, \quad \pi'(y) = \int U^*(y,x)\pi(x)d^3x \tag{34}$$

with

$$\begin{aligned} &\int U(y,x)U^*(y,x')d^3y = \delta(x - x'), \\ &\int U(y,x)U^*(y',x)d^3x = \delta(y - y'). \end{aligned} \tag{35}$$

Since the old and new coordinates are functionally related, we shall assume that the generating functional of the canonical transformation representing the unitary transformation is a functional of the old coordinates and new momenta:

$$W[\psi, \pi'] = \int \pi'(y)U(y,x)\psi(x)d^3yd^3x. \tag{36}$$

The equations of the canonical transformation, continuous analogues of (28), are

$$\begin{aligned} \psi'(y) &= \frac{\delta W}{\delta \pi'(y)} = \int U(y,x)\psi(x)d^3x, \\ \pi(x) &= \frac{\delta W}{\delta \psi(x)} = \int \pi'(y)U(y,x)d^3y, \end{aligned} \tag{37}$$

which, using (35), reproduce (34). The canonical character of the transformation is confirmed by evaluating the Hessian matrix,

$$\frac{\delta^2 W}{\delta \pi'(y)\delta\psi(x)} = U(y,x), \tag{38}$$

which is invertible by (35). The transformed Hamilton equations with new Hamiltonian

$$H_{\psi'\pi'} = H_{\psi\pi} + \frac{\partial W}{\partial t}, \tag{38a}$$

where $H_{\psi'\pi'} = \langle\psi'|\hat{H}'|\psi'\rangle$, $\hat{H}' = \hat{U}\hat{H}\hat{U}^\dagger + i\hbar(d\hat{U}/dt)\hat{U}^\dagger$, yield Schrödinger's equation and its complex conjugate for the new variables $\psi'(y)$. Hence, the quantum-mechanical transformation theory can be fully incorporated into the Hamiltonian language.

Non-unitary canonical transformations also arise in quantum mechanics and as a simple example, we observe that the passage to the polar representation of the wave function is a transformation of this type. Let the functions ψ, π (ρ, S) be the old (new) coordinates and momenta, respectively. Assuming the generating functional depends on the old and new coordinates ψ, ρ, we choose

$$W[\psi, \rho] = i\hbar \int \rho[\log(\psi/\sqrt{\rho}) + 1/2]d^3x. \tag{39}$$

Then, from the equations of a canonical transformation,

$$\pi = \frac{\delta W}{\delta\psi}, \quad S = -\frac{\delta W}{\delta\rho}, \tag{40}$$

we deduce the following explicit formulas for the old variables in terms of the new:

$$\psi = \sqrt{\rho}e^{iS/\hbar}, \quad \pi = i\hbar\sqrt{\rho}e^{-iS/\hbar}. \tag{41}$$

The Hessian matrix

$$h(x, x') = \frac{\delta^2 W}{\delta\rho(x')\delta\psi(x)} = \frac{i\hbar}{2\psi(x)}\delta(x - x') \tag{42}$$

has inverse $h^{-1}(x, x') = (2\psi(x)/i\hbar)\delta(x - x')$, showing that the mapping is indeed canonical. We also easily confirm the invariance of the Poisson brackets (33) in passing between the two sets of phase space coordinates:

$$\{A, B\}_{\psi\pi} = \{A, B\}_{\rho S}. \tag{43}$$

The Hamiltonian is a scalar under this time-independent transformation and its new form is

$$H_{\rho S}[\rho, S] = \int \rho \left[\frac{1}{2m}\frac{\partial S}{\partial x_i}\frac{\partial S}{\partial x_i} + U(\rho, \partial\rho) + V\right] d^3x, \qquad (44)$$

where

$$U = \frac{\hbar^2}{8m}\frac{1}{\rho^2}\frac{\partial \rho}{\partial x_i}\frac{\partial \rho}{\partial x_i} \qquad (45)$$

is the internal quantum potential energy. Hamilton's equations,

$$\dot{\rho} = \frac{\delta H_{\rho S}}{\delta S}, \quad \dot{S} = -\frac{\delta H_{\rho S}}{\delta \rho}, \qquad (46)$$

reproduce (2) and (3) where the quantum potential (4) is obtained here via the formula

$$V_Q(x) = \frac{\delta}{\delta\rho(x)}\int \rho(x')U(\rho(x'))\, d^3x'. \qquad (47)$$

4. Canonical Formulation of the Schrödinger Equation using Density and Current Density Potentials

The canonical theory just presented is economical in that no extraneous variables beyond those of physical interest (ψ, π or ρ, S) appear but it is not suitable for setting up a canonical relation with the trajectory theory. For this purpose, an enhanced phase space is required. Fortunately, there are other ways to formulate the Schrödinger equation in Hamiltonian terms. Noting that the Eulerian-picture hydrodynamic formulation of the Schrödinger equation provided a fruitful intermediary between the wave-mechanical and trajectory theories in Section 2, our strategy will be to first seek a canonical formulation of that picture of quantum hydrodynamics. Abetted by this, we shall set up a canonical transformation to the Lagrangian trajectory model (next section).

We start by writing the Hamiltonian (44) in terms of the variables ρ, v_i:

$$H_{\rho v} = \int \left(\frac{1}{2}m\rho v_i v_i + \rho U(\rho, \partial\rho) + \rho V\right) d^3x. \qquad (48)$$

For the present, we consider general vortical flows, not confined by the constraint (6). The expression (48) for the energy formally falls within the scope of classical hydrodynamics, the principal difference with typical classical fluids being the specific derivative form of the internal quantum energy U. It is a well-known feature of fluid mechanics that the Eulerian-picture variables ρ, v_i are not canonical [36]. To achieve a canonical formulation requires introducing potentials for the velocity, which results in a Clebsch-like representation: $v_i = \chi_j \partial_i \lambda_j$ (as indeed arises naturally in the ρ, S-based approach of the previous section). It is customary to employ three Clebsch parameters ($v_i = \partial_i \theta + \chi \partial_i \lambda$) but, if they are single-valued, this is not sufficient to represent the most general vector field [6, 47] (three parameters can suffice, however, if multivalued functions are admitted [31]). In our development below, an extended Clebsch-like representation (with six parameters) arises naturally. In the first instance, this is introduced as a representation of the mass current density $m\rho v_i$ rather than v_i directly.

A straightforward way to achieve the required parameterization is to make a space- and time-dependent transformation of the space coordinates to new independent variables: $x_i \to a_i = Q_i(x, t)$. Under this transformation, the dependent variables ρ (number density) and $m\rho v_i$ (mass current density = momentum density) transform according to the usual formulas of a coordinate substitution for tensor densities [42]:

$$\rho(x,t) = j(Q(x,t))\rho_0(Q(x,t)), \tag{49}$$

$$\begin{aligned} m\rho(x,t)v_i(x,t) &= \bar{P}_j(Q(x),t)j(Q(x,t))\frac{\partial Q_j(x,t)}{\partial x_i} \\ &= -P_j(x,t)\frac{\partial Q_j(x,t)}{\partial x_i}, \end{aligned} \tag{50}$$

where

$$j = \det(\partial Q/\partial x) = \frac{1}{3!}\varepsilon_{ijk}\varepsilon_{lmn}\frac{\partial Q_i}{\partial x_l}\frac{\partial Q_j}{\partial x_m}\frac{\partial Q_k}{\partial x_n}, \quad 0 < j < \infty. \tag{51}$$

Here, the transformed momentum density is denoted $\bar{P}_i$ and we have written $\bar{P}_i j$ as the function $-P_i(x)$. The minus sign in the latter definition is introduced for later convenience (in particular, so that a '+' sign appears in (58) below). Following Section 2, the mapping is assumed to be single-valued and differentiable and the inverse mapping $a_i \to x_i = q_i(a,t)$ exists and has the same properties. The transformation must obey two additional conditions. A first requirement is that the transformed density is the initial function ρ_0, expressed in the Q-coordinates, and hence has no explicit time dependence. This is necessary so that ρ obeys the continuity equation in the x-coordinates (see below). The second condition, which puts physical content into the transformation, is that the new dependent variables Q_i, P_i are canonically conjugate, i.e. they define three pairs of position and momentum variables for each space point whose temporal development is governed by Hamilton's equations. Mathematically, the new variables replace ρ, v_i as fundamental descriptors of the state and in terms of them the Hamiltonian (48) becomes

$$H_{QP}[Q,P] = \int \left(\frac{1}{2m\rho} P_i P_j \frac{\partial Q_i}{\partial x_k} \frac{\partial Q_j}{\partial x_k} + \rho U(\rho, \partial\rho) + \rho V \right) d^3x, \quad (52)$$

where ρ is shorthand for the function (49) and U is given in (45). The physical interpretation of the functions Q_i will be considered in the next section.

The functions Q_i in (49) evidently provide a "density vector potential" since, given ρ_0, we may determine the physical density ρ by differentiation. These functions also appear as potentials for the mass current (50). We thus obtain a Clebsch-type representation of the velocity v_i with six parameters Q_i and $-P_i/m\rho$.

The Clebsch parameters corresponding to a given vector field are not unique and equally viable sets χ_i, λ_i and χ_i', λ_i' are connected by a canonical transformation [11, 20, 48]. In our case, this "gauge" freedom in the parameters Q_i, P_i is restricted (but still represented by a canonical transformation) because we have the novel feature that the subset Q_i determines ρ as well as v_i. To obtain the general form of the gauge transformation, suppose that the physical fields, the left-hand sides in (49) and (50), are connected with a different set of independent canonical variables Q_i', P_i' according to the same formulas. Then the transformation linking the two canonical sets Q_i, P_i and Q'_i, P'_i that leaves the space coordinates and physical fields invariant is easily seen to be a time-independent diffeomorphism $Q'_i = f_i(Q)$ with

$$\left.\begin{aligned} &x'_i = x_i, \quad \rho' = \rho, \quad v'_i = v_i, \quad P'_i(x) = P_j(x)\frac{\partial Q_j}{\partial f_i} \\ &\rho'_0(Q'(x,t)) = \det(\partial f/\partial Q)\rho_0(Q(x,t)). \end{aligned}\right\} \tag{53}$$

That these relations define a canonical transformation may be verified by deriving them using the formulas

$$\begin{aligned} W[Q, P'] &= \int f_i(Q(x))P'_i(x)d^3x, \\ Q'_i = \frac{\delta W}{\delta P'_i}, \quad P_i &= \frac{\delta W}{\delta Q_i}, \quad H_{Q'P'} = H_{QP}. \end{aligned} \tag{54}$$

Below, we shall exploit the gauge freedom to simplify the theory.

To confirm that Q_i, P_i are suitable canonical variables, we examine Hamilton's equations:

$$\frac{\partial Q_i}{\partial t} = \frac{\delta H_{QP}}{\delta P_i} = \frac{1}{m\rho}P_k\frac{\partial Q_k}{\partial x_j}\frac{\partial Q_i}{\partial x_j} \tag{55}$$

$$\begin{aligned} \frac{\partial P_i}{\partial t} = -\frac{\delta H_{QP}}{\delta Q_i} &= j^{-1}j_{ij}\left[\frac{\partial}{\partial x_k}\left(\frac{1}{m\rho}P_mP_n\frac{\partial Q_m}{\partial x_k}\frac{\partial Q_n}{\partial x_j}\right)\right. \\ &\quad \left. - P_n\frac{\partial}{\partial x_j}\left(\frac{1}{m\rho}P_m\frac{\partial Q_m}{\partial x_k}\frac{\partial Q_n}{\partial x_k}\right) + \rho\frac{\partial}{\partial x_j}(V + V_Q)\right] \end{aligned} \tag{56}$$

where j is given by (51), j_{ij} is the adjoint of the inverse deformation matrix $\partial Q_i/\partial x_l$ with

$$\frac{\partial Q_i}{\partial x_j}j_{il} = j\delta_{lj}, \quad j_{il} = \partial j/\partial(\partial Q_i/\partial x_l), \tag{57}$$

V_Q is given in (4) and ρ in (49). These equations are unfamiliar but their physical content is easily revealed. Using the expression (50) for v_i, the first equation (55) states that the vector function Q_i is constant following the flow generated by v_i:

$$\frac{\partial Q_i}{\partial t} + v_j \frac{\partial Q_i}{\partial x_j} = 0. \tag{58}$$

This is how conservation is represented in these variables. In accordance with Q_i's role as a vector potential for the density (49), the continuity equation (7) may be derived by differentiating (58) with respect to x_j. Thus, using the formulas (51) and

$$\varepsilon_{ijk} \frac{\partial Q_i}{\partial x_p} = j \varepsilon_{prs} \frac{\partial x_r}{\partial Q_j} \frac{\partial x_s}{\partial Q_k} \tag{59}$$

we obtain

$$\frac{\partial j}{\partial t} + \frac{\partial}{\partial x_i}(j v_i) = 0. \tag{60}$$

Equation (58) ensures that $\rho_0(Q)$ is a constant of the motion and hence (7) follows from (60). With the aid of the continuity equation, it may be shown that the second Hamilton equation (56) is equivalent to the quantum Euler force law (8). Hamilton's equations therefore imply the correct hydrodynamic Eqs. (7) and (8).

To establish equivalence with Schrödinger's equation, we must fix the initial conditions $Q_{0i}(x), P_{0i}(x)$. Following (49) and (50), the latter are connected with the initial hydrodynamic functions according to

$$\begin{aligned} \rho(x, t = 0) &= j(Q_0(x)) \rho_0(Q_0(x)), \\ m\rho(x, t = 0) v_{0i}(x) &= -P_{0k}(x) \frac{\partial Q_{0k}}{\partial x_i}. \end{aligned} \tag{61}$$

These relations determine Q_{0i}, P_{0i} up to a gauge transformation (53). To simplify matters, we fix the gauge so that $Q_{0i}(x) = x_i$ since this is the condition for which $\rho(x, t = 0) = \rho_0(x)$. Invoking the initial gradient form (11) of v_i we then have, altogether,

$$Q_{0i}(x) = x_i, \quad P_{0i}(x) = -\rho_0(x) \partial_i S_0(x), \quad \rho_0 = |\psi_0|^2. \tag{62}$$

To summarize, Hamilton's equations (55) and (56) with initial conditions (62) imply the Schrödinger equation via the quantum hydrodynamic equations (7) and (8). Conversely, we may deduce

Hamilton's equations from the Schrödinger equation on substituting the relations (49) and (50) for ρ and v_i into the quantum hydrodynamic equations. We conclude that the phase space variables Q_i, P_i provide a satisfactory alternative canonical formulation of quantum propagation.

Using (50), (57) and (58), we can solve for P_i in terms of $\dot{Q}_i$:

$$P_i = m\rho g_{ij}\frac{\partial Q_j}{\partial t}, \tag{63}$$

where $g_{ij} = j^{-2}j_{ik}j_{jk}$ and $\rho = j(Q)\rho_0(Q)$. Substituting this expression in (56), we may eliminate P_i to obtain a self-governing second-order equation for the gauge potentials Q_i, which now represent the quantum state. This is the Eulerian picture equivalent of the second-order Lagrangian picture equation (22) (to which it is related via a canonical transformation, as we shall see below). This second-order version of Schrödinger's equation (which we do not give explicitly) may also be written as an Euler–Lagrange equation once the Lagrangian is obtained. The latter is found by making an inverse Legendre transformation of the Hamiltonian (52) and then substituting expression (63) for P_i:

$$\begin{aligned} L[Q, \dot{Q}, t] &= \int P_i(x)\dot{Q}_i(x)d^3x - H_{QP}[Q, P, t] \\ &= \int \rho\left(\frac{1}{2}mg_{ij}(x)\frac{\partial Q_i(x)}{\partial t}\frac{\partial Q_j(x)}{\partial t}\right. \\ &\quad \left. - \frac{\hbar^2}{8m}\frac{\partial \log\rho}{\partial x_i}\frac{\partial \log\rho}{\partial x_i} - V(x)\right)d^3x \end{aligned} \tag{64}$$

with $g_{ij} = j^{-2}j_{ik}j_{jk}$ and $\rho = j(Q)\rho_0(Q)$.

We note that to compute Q_i from a known velocity field v_i, we may use (58). This is the Eulerian equivalent of using the Lagrangian equation (18) to solve for q_i given v_i.

The relation between the first-order (ψ or ρ, v_i) and second-order (Q_i) formulations of Schrödinger's equation is analogous to that between the first-order (electric + magnetic fields) and second-order (vector potential) versions of Maxwell's equations, each formulation being self-contained with respect to its dependent variables with the two sets of initial conditions being chosen to ensure compatibility. As with electromagnetism, the quantum potentials-based approach exhibits a gauge symmetry and provides a variational basis for the theory. The quantum approach parts company with electromagnetism in one key respect, however, in the quantum case, it is possible to formulate the theory using gauge-invariant potentials. These potentials are just the displacement functions q_i, as we see next.

5. Canonical Transformation to the Spatial Trajectory Formulation

With the canonical formulation of the last section at hand, it is straightforward to transform the Schrödinger equation into Newtonian form by performing a canonical transformation $Q_i(x), P_i(x) \to q_i(a), p_i(a)$ whose remit is to effect an inversion of the independent and dependent variables, $x_i \leftrightarrow a_i$. We suppose that the generating functional of the transformation depends on the new coordinates and old momenta and is time-independent: $W[q(a), P(x)]$. The transformation formulas are therefore

$$Q_i(x) = \frac{\delta W}{\delta P_i(x)}, \quad p_i(a) = \frac{\delta W}{\delta q_i(a)}, \quad H_{qp}[q, p, t] = H_{QP}[Q, P, t]. \tag{65}$$

A generating functional with the required inversion property is given by

$$W[q, P] = \int \delta(x - q(a))\, Ja_i P_i(x) d^3x\, d^3a \tag{66}$$

with J defined in (19). For

$$Q_i(x) = \frac{\delta W}{\delta P_i(x)} = \int \delta(x - q(a)) Ja_i d^3a \tag{67}$$

and using the formula $\delta(x - q(a)) = J^{-1}\delta(a - q^{-1}(x))$ gives $Q_i(x) = q_i^{-1}(x)$. The other transformation formula yields

$$p_i(a) = \frac{\delta W}{\delta q_i(a)} = -J(q(a))\frac{\partial a_j(q(a))}{\partial q_i(a)} P_j(q(a)) \tag{68}$$

where we have used the result

$$\frac{\delta J(a')}{\delta q_i(a)} = -J_{ij}(a)\frac{\partial}{\partial a_j}\delta(a-a') \tag{69}$$

and the identity $\partial J_{ij}/\partial a_j = 0$ with J_{ij} defined in (24). The explicit solution for the new phase space variables in terms of the old is thus

$$q_i(a,t) = a_i^{-1}(x,t), \tag{70}$$

$$p_i(a,t) = -J(x)P_j(x)\left.\frac{\partial Q_j(x)}{\partial x_i}\right|_{x=q(a)} = m\rho_0(a)v_i(q(a,t),t). \tag{71}$$

The Hessian for the transformation is

$$h_{ij}(x,a) = \frac{\delta W}{\delta P_j(x)\delta q_i(a)} = -\frac{\partial a_j(q(a))}{\partial q_i(a)}\delta(a-Q(x)) \tag{72}$$

with inverse

$$h_{ij}^{-1}(x,a) = -\frac{\partial q_i(a)}{\partial a_j}\delta(x-q(a)), \tag{73}$$

which confirms the canonical nature of the mapping. This is confirmed also by the invariance of the Poisson brackets:

$$\{A,B\}_{QP} = \{A,B\}_{qp}. \tag{73a}$$

The new Hamiltonian is

$$H_{qp}[q,p,t] = \int\left[\frac{1}{2m\rho_0(a)}p_i(a)p_i(a) + \rho_0(a)U(J^{-1}\rho_0) + \rho_0(a)V(q(a))\right]d^3a \tag{74}$$

and in these variables Hamilton's equations take a recognizable form:

$$\frac{\partial q_i(a)}{\partial t} = \frac{\delta H_{qp}}{\delta p_i(a)} = p_i(a)/m\rho_0(a), \tag{75}$$

$$\frac{\partial p_i(a)}{\partial t} = -\frac{\delta H_{qp}}{\delta q_i(a)} = -\rho_0(a)\frac{\partial}{\partial q_i}(V+V_Q), \tag{76}$$

where $\partial/\partial q_i$ is given in (23). Eliminating p_i from these equations, we obtain the second-order Newtonian equation (22) for q_i. To complete the canonical mapping, we state the initial conditions obeyed by the new variables corresponding to the gauge used in (62) for the old variables:

$$q_{0i}(a) = a_i, \quad p_{0i}(a) = \rho_0(a)\partial_i S_0(a), \quad \rho_0 = |\psi_0|^2. \qquad (77)$$

As with the coordinates Q_i in (49) and (50), the coordinates q_i in (17) and (18) provide a set of potentials from which the hydrodynamic functions are obtained by differentiation: $\rho = \det(\partial q/\partial a)^{-1}\rho_0$, $v_i = \partial q_i/\partial t$. The gauge transformation (53) translates here into a time-independent diffeomorphism $a'_i = f_i(a)$, or relabeling of the trajectories, with respect to which

$$\left.\begin{aligned} &q'_i(a', t) = q_i(a, t), \quad p'_i(a', t) = \det(\partial f/\partial a)p_i(a, t),\\ &\rho'_0(a') = \det(\partial f/\partial a)\rho_0(a). \end{aligned}\right\} \qquad (78)$$

As anticipated above, we see that the potentials q_i are gauge invariant functions, as are the velocity components $\dot{q}_i$. Note that if we made a different choice for $q_{0i}(a) \neq a_i$, $\rho_0(a) \neq |\psi_0(a)|^2$.

The passage to the new variables clarifies the significance of the Eulerian-picture potentials Q_i: they provide a spacetime representation of the labels of the Lagrangian-picture trajectories and their conservation expressed in (58) represents the invariance of the particle label a_i along the line of flow it characterizes.

Transforming to the q_i-coordinates, the Lagrangian (64) becomes

$$L[q, \dot{q}, t] = \int \rho_0(a)\left[\frac{1}{2}m\frac{\partial q_i(a)}{\partial t}\frac{\partial q_i(a)}{\partial t} - \frac{\hbar^2}{8m}G_{ij}\frac{\partial \log\rho}{\partial a_i}\frac{\partial \log\rho}{\partial a_j} - V(q(a))\right]d^3a, \qquad (79)$$

where $G_{ij} = J^{-2}J_{ki}J_{kj}$ and $\rho = J^{-1}(a)\rho_0(a)$. The Euler–Lagrange equation gives (22). It will be noted how the metric characterizing the deformation appears in the kinetic term in the Q-representation (g_{ij} in (64)) and in the potential term in the q-representation (G_{ij} in (79)). The canonical theory was developed previously starting from the Lagrangian (79) [20].

We conclude once again that the deterministic continuous trajectory is tacitly contained in wave mechanics rather than being an additional structure: *The trajectory model of quantum evolution is obtained by a canonical transformation of wave mechanics when the latter is formulated in terms of the hydrodynamic phase space variables, or gauge potentials,* $Q_i(x), P_i(x)$.

6. The Quantum Formalism in q, p Phase Space

We saw in Section 3 that quantum dynamics may be formulated as a set of Hamilton equations where the Hamiltonian is the mean value of the Hamiltonian operator. We show here how the mean value, regarded as a functional, may be employed more generally to represent the quantum operator calculus and use this result to interpret the latter in terms of the alternative phase space representation q, p of the quantum state.

According to the usual formalism, to any observable represented by a self-adjoint operator $\hat{A}(x', x)$ (in the position representation) we may associate a real-valued bilinear functional of the conjugate variables ψ, π via its mean value:

$$A[\psi, \pi] = \langle\psi|\hat{A}|\psi\rangle = (i\hbar)^{-1} \int \pi(x')\hat{A}(x', x)\psi(x)\, d^3x\, d^3x'. \quad (80)$$

This is the matrix representation of the operator in the quantum phase space spanned by ψ, π and we may work with the functional A instead of the operator $\hat{A}$. The explicit form of the latter is obtained by differentiation:

$$\hat{A}(x', x) = i\hbar \frac{\delta^2 A[\psi, \pi]}{\delta\pi(x')\delta\psi(x)}. \quad (81)$$

The operator algebra may be translated into the functional calculus of mean values as follows. Introducing a second operator $\hat{B}$ with associated functional B defined as in (80), the sum and product of $\hat{A}, \hat{B}$, have the following associated functionals:

$$\langle\psi|\hat{A} + \hat{B}|\psi\rangle = A[\psi, \pi] + B[\psi, \pi], \quad (82)$$

$$\langle\psi|\hat{A}\hat{B}|\psi\rangle = i\hbar \int \frac{\delta A}{\delta\psi(x)} \frac{\delta B}{\delta\pi(x)} d^3x. \quad (83)$$

Using these formulas, we may compute the mean value of an arbitrary function of operators from the mean values of its component operators. An immediate application of this result is the well known [43] theorem that the mean value of ($(i\hbar)^{-1}$ times) the commutator $[\hat{A}, \hat{B}] = \hat{A}\hat{B} - \hat{B}\hat{A}$ is the Poisson bracket of the functionals A, B in the phase space ψ, π:

$$\begin{aligned}\langle\psi|(i\hbar)^{-1}[\hat{A}, \hat{B}]|\psi\rangle &= \{A, B\}_{\psi\pi} \\ &= \int\left(\frac{\delta A}{\delta\psi(x)}\frac{\delta B}{\delta\pi(x)} - \frac{\delta B}{\delta\psi(x)}\frac{\delta A}{\delta\pi(x)}\right)d^3x. \qquad (84)\end{aligned}$$

Having represented the operator calculus in the ψ, π phase space, insight into its physical content may be obtained by passing to the q, p phase space. Following (26) and (71), we can translate (80) into these variables as follows. For the wave function we write, ignoring the phase factor $f(t)$,

$$\psi(x = q(a,t), t) = \sqrt{(J^{-1}(a,t)\rho_0(a))}\exp\left(\frac{i}{\hbar}\int\frac{p_i(a,t)}{\rho_0(a)}\frac{\partial q_i}{\partial a_j}da_j\right). \qquad (85)$$

The conjugate momentum $\pi(x = q(a,t), t)$ is given by $i\hbar$ times the complex conjugate of (85). For x_i and $\partial/\partial x_i$ in the algebraic and differential function $\hat{A}$, we write q_i and $\partial/\partial q_i = J^{-1}J_{ij}\partial/\partial a_j$, respectively, and in integrals d^3x is replaced by Jd^3a. This implies for (80) a formula of the type

$$A[q,p] = \int f(q(a), p(a), q'(a'), p'(a'), \partial q(a), \partial q'(a'), \ldots)d^3a\, d^3a'. \qquad (86)$$

We can express the operator (81) and the composition formula (83) in terms of q, p by writing the functional derivatives of the latter in terms of those for ψ, π. A basic result is that the mean value of the commutator (84) may be expressed as the Poisson brackets for the new variables when the functionals A, B depend on q, p

through ψ, π:

$$\int \left(\frac{\delta A}{\delta q_i(a)} \frac{\delta B}{\delta p_i(a)} - \frac{\delta B}{\delta q_i(a)} \frac{\delta A}{\delta p_i(a)} \right) d^3a$$
$$= \int \left(\frac{\delta A}{\delta \psi(x)} \frac{\delta B}{\delta \pi(x)} - \frac{\delta B}{\delta \psi(x)} \frac{\delta A}{\delta \pi(x)} \right) d^3x. \tag{87}$$

An efficient way to prove this result is to express the functional derivatives on the left-hand side in terms of the polar variables,

$$\begin{aligned} \frac{\delta}{\delta q_i(a)} &= \int \left(\frac{\delta S(x)}{\delta q_i(a)} \frac{\delta}{\delta S(x)} + \frac{\delta \rho(x)}{\delta q_i(a)} \frac{\delta}{\delta \rho(x)} \right) d^3x, \\ \frac{\delta}{\delta p_i(a)} &= \int \left(\frac{\delta S(x)}{\delta p_i(a)} \frac{\delta}{\delta S(x)} + \frac{\delta \rho(x)}{\delta p_i(a)} \frac{\delta}{\delta \rho(x)} \right) d^3x, \end{aligned} \tag{88}$$

and invoke the Poisson bracket equivalence (43). The result (87) is a quantum analogue of a special case of the general relation between classical Lagrangian and Eulerian (non-Poisson) brackets [38].

We conclude that we may transcribe the quantum operator calculus into the new variables via the intermediary of mean values. For the basic set of variables of interest in physics, the formula (86) yields simple and readily interpretable expressions:

position

$$\langle \hat{x}_i \rangle = \int x_i \rho(x,t)\, d^3x = \int q_i(a,t) \rho_0(a)\, d^3a \tag{89}$$

linear momentum

$$\langle \hat{p}_i \rangle = \int \partial_i S(x,t) \rho(x,t)\, d^3x = \int p_i(a,t)\, d^3a \tag{90}$$

angular momentum

$$\langle (\hat{x} \times \hat{p})_i \rangle = \int (x \times \partial S)_i \rho\, d^3x = \int (q \times p)_i\, d^3a \tag{91}$$

kinetic energy

$$\left\langle \frac{1}{2m}\hat{p}_i\hat{p}_i \right\rangle = \int \left(\frac{1}{2m}\rho\partial_i S\partial_i S + \rho U(\rho) \right) d^3x$$
$$= \int \left(\frac{p_i p_i}{2m\rho_0} + \rho_0 U(J^{-1}\rho_0) \right) d^3a. \quad (92)$$

Note that this rendering of quantum theory differs fundamentally from phase space approaches such as that of Wigner where q, p are the independent variables and the state is represented by a quasi-distribution function $f(q, p)$ (constructed from the wave function) whose evolution is governed by a quantum analogue of the classical Liouville equation. In our case, the phase space variables define the state and the independent variables specify the particles, an analogue of classical fluid theory in its Lagrangian phase space formulation. There is, however, one property the Wigner approach and ours share: neither is unique. A wide class of quasi-distribution functions may be employed to represent the quantum state, and likewise for the law of motion in the fluid model.

7. Observability of the Quantum State

We have observed that the hydrodynamic state variables ρ, v_i play a key role in connecting the ψ- and q_i-representations of the state. We now show how these functions may also be employed to derive empirical information about the state. Assuming a means to observe these functions has been found, we may deduce the ψ, q_i-versions by integration as follows:

1. $\psi(x,t)$: From ρ we deduce $|\psi|$ and from $v_i = m^{-1}\partial_i S$ we integrate to get the phase S up to an additive function of t that is fixed apart from a constant by substituting in (3).
2. $q_i(a,t)$: From v_i we get $q_i(a)$ by integrating $v_i = \dot{q}_i$ with $a_i = q_{0i}$.

In seeking empirical methods to investigate the state, we first exclude one potential avenue by noting that the velocity field v_i is not a quantum observable: there does not exist a linear Hermitian operator for which v_i is its expectation value [18, 32, 33]. On the other hand, the local density ρ and current density j_i $(=\rho v_i)$ *are* observables in the quantum-mechanical sense: there exist Hermitian operators $\hat{\rho}_x$, $\hat{j}_{x_i}$ such that $\rho(x) = \langle\hat{\rho}_x\rangle$, $j_i(x) = \langle\hat{j}_{x_i}\rangle$ for all ψ. Specifically,

$$\hat{\rho}_x = |x_i\rangle\langle x_i|,\ \hat{j}_{x_i} = \frac{1}{2m}\left[|x_i\rangle\langle x_i|\hat{p}_i + \hat{p}_i^\dagger|x_i\rangle\langle x_i|\right], \quad \text{(no sum over } i\text{)}. \tag{93}$$

The "non-observability" of v_i may be proved by applying the following result [32] (for a comment on the proof see [18]):

Theorem: Consider three nontrivial (i.e. not multiples of the identity operator) operators $\hat{A}, \hat{B}, \hat{C}$ where $\hat{A}, \hat{B}$ are linear and the mean values obey the relation $\langle\hat{A}\rangle = \langle\hat{B}\rangle\langle\hat{C}\rangle$ for all states in Hilbert space. Then $\hat{C}$ cannot be linear.

To apply the theorem, choose $\hat{A} = \hat{j}_{x_i}$, $\hat{B} = \hat{\rho}_x$ for each $i = 1, 2, 3$. Then the ratio $v_i = j_i/\rho$ is not the mean value of a linear operator. A slightly different version of the theorem with the same implication for v_i is proved in the Appendix.

Of course, as a spacetime function v_i *is* "observable" as it may be deduced statistically from measurements of quantum observables over an ensemble of identically prepared systems (same ψ_0). For example, one can perform a sequence of position and momentum measurements at each spacetime point, which yields the local velocity [29]. Such statistical methods may be used both to learn the state when it is unknown and to check the predictions it encodes when it is known through some prior state preparation process. The methods do not, however, impinge on the ontological status of the state, in particular the hydrodynamic functions. To address the ontological issue, we consider the possibility of a direct observation of the hydrodynamic functions using the protective measurement procedure.

With the aim of demonstrating an ontological aspect of the wave function, Aharonov and co-workers [1–3] (for clarifications and reviews see [10, 12]) showed how a suitably adapted adiabatic interaction described by quantum mechanics provides a scheme to measure the expectation values of operators pertaining to a system without appreciably disturbing its quantum state. These interactions are therefore called "protective measurements". Aharonov *et al.* claimed that in certain circumstances, this technique "measures the wave function" of a single system as an extended object. They infer that the procedure, in revealing a property possessed by a single system prior to the measurement, provides evidence for the ontological character of the wave function. In fact, we shall see that it is the hydrodynamic functions that are measured by this method rather than the wave function directly.

Consider two interacting systems, an object and measuring apparatus, with initial wave functions $\psi_0(x)$ and $\beta_0(y)$, respectively. We assume for simplicity that the configuration coordinates x and y are one-dimensional. Denote by $\hat{A}$ the operator pertaining to the object whose expectation value is to be measured. Then in a protective interaction the total Hamiltonian comprises free Hamiltonians for the individual, systems and an interaction term $\hat{H}_I = g(t)y\hat{A}$. The initial combined state $\Phi_0(x,y) = \psi_0(x)\beta_0(y)$ then evolves adiabatically at time t into:

$$\Phi(x,y,t) = \psi(x,t)\beta(y,t)\exp\left[-(i/\hbar)\int_0^t g(t)y\langle\hat{A}\rangle dt\right]. \tag{94}$$

Here, $g(t)$ is a function characterizing the adiabatic interaction with $\int_0^t g(t)dt = 1$, and $\psi(x,t)$ and $\beta\,(y,t)$ are the wave functions obtained under free evolution of the two systems. It will be observed that (94) is still a product state in that the variables x and y have not become entangled. In particular, the object state is undisturbed by the interaction. The state of the apparatus has, however, acquired a phase factor depending on the expectation value $\langle\hat{A}\rangle = \langle\psi(t)|\hat{A}|\psi(t)\rangle$, which implies an observable change in the apparatus momentum. Hence, information on the state $\psi(x,t)$ can be gleaned from the apparatus by measuring the change in its momentum (via a conventional measurement). The method may be extended to protectively measure several operators simultaneously by introducing corresponding additional apparatuses. Thus, for a second operator $\hat{B}$, we introduce an apparatus with coordinate z and wave function $\gamma(z)$ and the formula (94) becomes

$$\Phi(x, y, z, t) = \psi(x, t)\beta(y, t)\gamma(z, t) \times \exp\left[-(i/\hbar)\int_0^t (g_y(t)y\langle\hat{A}\rangle + g_z(t)z\langle\hat{B}\rangle)dt\right]. \quad (95)$$

Let us choose $\hat{A} = \hat{j}_{x_i}$, $\hat{B} = \hat{\rho}_x$ for each i so that $\langle\hat{A}\rangle = \rho(x, t)v_i(x, t)$, $\langle\hat{B}\rangle = \rho(x, t)$. Then the momenta of the devices y and z are shifted by the amounts $\int_0^t g_y(t)\rho(x, t)v_i(x, t)dt$ and $\int_0^t g_z(t)\rho(x, t)dt$, respectively.

We thus have a scheme to measure the time-averaged density and current density at a space point x. Aharonov *et al.* based their claim of "measuring the wave function of a single system" on these formulas, for the special case where ψ is known to be a non-degenerate energy eigenstate but is otherwise unknown. For in that case, the density and current density are time-independent and the momentum shifts are proportional to the local values. For this particular case, we may therefore measure the functions ρ, v_i for all values of their arguments and deduce ψ (up to a gauge transformation) as noted in 1 above. As already remarked, it is the hydrodynamic variables that are measured in this scheme; ψ is deduced from them.

The protective method may be applied to states other than non-degenerate energy eigenstates but there are two caveats: (a) the full Hamiltonian that functions during the protective process depends on the state [2], which implies that we must first know ψ before we can investigate it, and (b) that investigation reveals results about time averages of (the hydrodynamic) functions of ψ rather than instantaneous values. So, in the general case, the protective scheme provides a way to empirically confirm time-averaged prior information about the state.

We conclude that functions of the hydrodynamic fields associated with a single quantum system are measurable quantities, namely, their time-averaged local values for a general state that is known prior to the measuring procedure. As a special case, the fields themselves are measurable when the state is an energy eigenstate but otherwise unknown. In the latter case, the quantum state in its ψ or q_i guise may be deduced as a single entity from the empirical data, as set out above. An important consideration in relation to the q_i-state is the consistency of this result with the unfeasibility of simultaneously measuring position and momentum, which would require the ψ-state to transform into a simultaneous eigenstate of the associated operators. The latter impossibility appears to be consistent with the protective measurement scheme since q_i is a construct from its results rather than the object of investigation. For reasons discussed elsewhere [29], the simultaneous attribution of position and momentum variables to each particle a_i is also consistent with the uncertainty relations, which constitute conditions on the statistical scatter of position and momentum measurement results in accord with the quantum formalism.

8. A Common Language for Quantum and Classical Physics?

We have seen that the self-contained second-order Newtonian version of the Schrödinger equation,

$$m\frac{\partial^2 q_i(a)}{\partial t^2} = -J^{-1}J_{ij}\frac{\partial}{\partial a_j}(V(x)+V_Q(x))|_{x=q(a,t)} \tag{96}$$

subject to $\rho_0(a) = |\psi_0(a)|^2$, $\dot{q}_{oi}(a) = m^{-1}\partial S_0(a)/\partial a_i$, attributes quantum evolution to the motion of a continuous infinity of interacting particles (in the fluid sense) having simultaneously well-defined position and momentum variables. This approach therefore appears to supply the basis of a common language for quantum and classical mechanics and might be expected to provide insight into how the theories are connected. As we shall see, while the theory does assist in this aim, it also highlights several subtleties if the quest is to treat classical-like behavior as a limiting case of an enveloping quantum description, something that is almost universally regarded as desirable and feasible.

Following (96), an obvious criterion for a quantum system to behave like a classical one, at least approximately, is that the quantum contribution to the force, $\partial_i V_Q$, is negligible compared to the classical force, for this rendering of Schrödinger's equation then reduces to Newton's classical law. This correspondence principle is state-dependent, which has the advantage of encompassing and explaining the fact that procedures which seek to characterize the limit in terms of the relative values of parameters such as m and $\hbar$ often do not lead to classical behavior [29, 30]. However, the state-dependent limiting process stated in this form is not sufficient to characterize the classical domain for several reasons.

As we have noted, a congruence of trajectories is needed for a complete description of the quantum state and none is singled out for special status, beyond the selection implicit in the initial conditions for the density and velocity. In addition, in quantum mechanics the probability density ρ refers to the likely position should a measurement of location be performed and not to the probability of current presence. The limiting process envisaged above maintains both the congruence and the quantum version of probability; it does not yield the single trajectory expected for a classical particle or the classical statistical concept based on the actual presence of the particle independent of measurement. These problems are not insurmountable and may be addressed by invoking the postulates of the de Broglie–Bohm theory [29]: that one of the paths a_i in the collective representing the state supports a material corpuscle, and ρ refers to its likely current position. Then the ensemble of paths representing the quantum state may be viewed as the potential paths of the added corpuscle, only one of which is actually realized with a frequency determined by ρ.

But even if this step is taken, there remains a further problem with the Newton-based limiting protocol [30]: the correspondence limit is exceptional. As noted above, classical behavior does not generally emerge when parametrical limiting processes are applied to specific wave functions (quantum systems with no classical analogue). And even where a quantum state is found that does imply the classical domain in some limit, the classical behavior so obtained may be but a subset of that which is allowed by classical laws for that system (classical systems with no quantum analogue). A corollary is that quantum wave functions cannot generally solve typical classical statistical problems. These points may be illustrated with simple examples, such as reflection of a particle by a wall [30].

The fundamental impediment to obtaining classical motion according to the scheme based on (96) is the congruent character of the state: at each moment at most one trajectory passes through each space point. When the putative limiting procedures are taken, the single-valuedness condition remains intact and constrains each *individual* orbit so that the set of all individuals obey the non-crossing property. This is not a characteristic of generic classical particle ensembles for which there is no single-valuedness requirement. The same problem would occur, however, in classical fluid mechanics if we wanted to consider circumstances in which the internal fluid forces are negligible relative to external body forces. In fact, we expect this would not generally be possible in a real fluid because it would destroy the mechanism that maintains its physical integrity. Is demanding a comparable reduction in the effectiveness of the internal forces in the quantum case likewise too stringent a requirement?

We make three observations about these issues. The first is that our analysis pertains to pure states and it has been suggested that the difficulties might be forestalled if the limiting process is treated using mixed states [9].

The second remark is that the issues raised may be artefacts of the position representation that go away in a different formulation of quantum mechanics (for an alternative approach see [13]).

The third point is that our considerations refer to closed systems. In practice, unless special conditions are available to isolate a system, macroscopic bodies are generally in continual interaction with their environment. It has been suggested in connection with the so-called "decoherence" program (for a review see [4]) that the "classicality" of a system may be contingent on these background interactions. This mechanism may also be invoked in the context of the de Broglie–Bohm theory as was first suggested in connection with a specific problem: that of obtaining the classical motion of a planet using the ambient stellar light as the environmental entity whose action causes potentially interfering classical segments of the planet wavefront to "decohere" [5] (another component of the problem, showing how a resultant Keplerian orbit is obtained for the planet, was not addressed in that work). To see what is involved, suppose we have a set of wave functions ψ_μ each of which is "classical" in the sense of negligible associated quantum force. Then, in a region where the functions overlap, the total wave function is given by $\sum_\mu \psi_\mu$ and the particle motion in the region of superposition is generally non-classical since the associated finite quantum force will act to preserve the congruent character of the trajectories. To disrupt this interference effect, suppose the system interacts with another system having many degrees of freedom and wave function ϕ. Then, for a suitable interaction Hamiltonian, each wave ψ_μ will couple to ϕ in a different way with a resultant total wave function $\sum_\mu \psi_\mu \phi_\mu$. For a sufficiently complex external system, each summand $\psi_\mu \phi_\mu$ occupies a distinct region of the configuration space and the de Broglie–Bohm configuration point will lie in and be guided by just one summand. Thus, the subsystem of interest will be guided by just one of the waves ψ_μ (while the others remain finite) and its classicality is ensured.

To explain classical behavior in general, the decoherence process would have to supply an interaction Hamiltonian that spontaneously, constantly, and almost universally acts in nature to bring about persistently non-overlapping configuration space packets. This means that classical mechanics, insofar as it can be discerned as a special case, emerges as a sub-dynamics in the total configuration space of the trajectory theory corresponding to the system-plus-environment. In this way, the decoherence and trajectory theories may be mutually supportive in that each supplies elements the other lacks but further analysis is needed (see for example [35]).

9. Conclusion

We have sought to show that Newton's trajectory law involving a particular type of interaction potential (the quantum potential) and subject to suitable initial conditions is just Schrödinger's equation expressed in different variables. In particular, this formulation may be obtained by a canonical transformation of wave mechanics. The displacement concept of state may thus be regarded as implicit in the quantum description. In the case of classical continuum physics, the transformation we have described — between the field-theoretic Eulerian version and the particle-like Lagrangian version — is unexceptional since the subject starts from a Newtonian analysis of physical systems as extended to continua [45]. The state of the system naturally has complementary forms. In the quantum case, the analogous result acquires a more potent significance since the theory is supposed to depart so radically from classical notions that it is felt that a trajectory formulation, even if possible, must be at best a metaphysical supplement of no import to the theory itself. To be sure, the physical significance of the quantum trajectories, such as what may travel along them, may be an issue of interpretation but their definition and employment as an alternative and independent characterization of the quantum state is a matter of mathematical transformation.

According to these results, the notion of "wave-particle duality" has some currency but it refers to different ways of viewing a single process, not to mutually exclusive experimental contexts as has been asserted historically. Our thesis developed elsewhere [21, 25, 27] is that this duality is not specific to quantum theory (or to classical continuum theories) but is an aspect of generic field theories that may be expected to admit Lagrangian trajectory formulations.

Is this alternative version of quantum mechanics useful? It is already well established that a trajectory outlook brings computational benefits to quantum theory [46]. It also allows us to examine questions that may be difficult to formulate in the ψ-description, such as criteria for chaos (e.g. [8]), time of transit (e.g. [37]) and, as we have seen, classical-like behavior. This is a field where a great deal of foundational and numerical work has been done but it still lacks significant theorems. It also lacks empirical support although, as we have seen, a scheme may be conceived to measure the hydrodynamic functions with which the trajectory variables are intimately connected. Perhaps the observation that, in the end, the trajectory is just a different way of doing quantum mechanics may attract further interest.

References

[1] Aharonov, Y. and Vaidman, L. (1993). Measurement of the Schrödinger wave of a single particle. *Physics Letters A*, **178**: 38.

[2] Aharonov, Y., Anandan, J. and Vaidman, L. (1993). Meaning of the wave function. *Physical Review A*, **47**: 4616.

[3] Anandan, J. (1993). Protective measurement and quantum reality. *Foundations of Physics Letters*, **6**: 503.

[4] Bacciagaluppi, G. (2012). The role of decoherence in quantum mechanics. *Stanford Encyclopedia of Philosophy*, (http://plato.stanford.edu/entries/qm-decoherence/).

[5] Bohm, D. and Hiley, B. J. (1993). *The Undivided Universe*, Routledge, London.

[6] Bretherton, F. P. (1970). A note on Hamilton's principle for perfect fluids. *Journal of Fluid Mechanics*, **44**: 19.

[7] Broer, L. J. F. and Kobussen, J. A. (1974). Conversion from material to local coordinates as a canonical transformation. *Journal of Applied Sciences Research*, **29**: 419.

[8] Contopoulos, G. Delis, N. and Efthymiopoulos, C. (2012). Order in de Broglie–Bohm quantum mechanics. *Journal of Physics A: Mathematical and Theoretical*, **45**: 165301.

[9] Cushing, J. T. (2000). Bohmian insights into quantum chaos. *Philosophy of Science Association*, **67**: S430.

[10] Dass, N. D. H. and Qureshi, T. (1999). Critique of protective measurements. *Physical Review A*, **59**: 2590.

[11] Eckart, C. (1960). Variation principles of hydrodynamics. *Physics of Fluids*, **3**: 421.

[12] Gao, S. (ed.) (2015). *Protective Measurement and Quantum Reality*, Cambridge University Press, Cambridge.

[13] Garcia de Polavieja, G. (1996). A causal quantum theory in phase space. *Physics Letters A*, **220**: 303.

[14] Greiner, W. and Reinhardt, J. (1996). *Field Quantization*, Springer-Verlag, Berlin. Sec. 2.3.

[15] Heslot, A. (1985). Quantum mechanics as a classical theory. *Physical Review D*, **31**: 1341.

[16] Hirsch, M. W., Smale, S. and Devaney, R. L. (2004). *Differential Equations, Dynamical Systems, and an Introduction to Chaos*, 2nd edn., Elsevier, Amsterdam, p. 150.

[17] Hirschfelder, J. O. (1977). The angular momentum, creation, and significance of quantized vortices. *Journal of Chemical Physics*, **67**: 5477.

[18] Hirschfelder, J. O. (1978). Quantum mechanical equations of change. *Journal of Chemical Physics*, **68**: 5151.

[19] Hirschfelder, J. O., Goebel, C. J. and Bruch, L. W. (1974). Quantized vortices around wave function nodes. II *Journal of Chemical Physics*, **61**: 5456.

[20] Holland, P. (2005). Computing the wave function from trajectories: particle and wave pictures in quantum mechanics and their relation. *Annals of Physics (NY)*, **315**: 505.

[21] Holland, P. (2005). Hydrodynamic construction of the electromagnetic field. *Proceedings of The Royal Society A*, **461**: 3659.

[22] Holland, P. (2006). Hidden variables as computational tools: the construction of a relativistic spinor field. *Foundations of Physics*, **36**: 1.

[23] Holland, P. (2009). Schrödinger dynamics as a two-phase conserved flow: an alternative trajectory construction of quantum propagation. *Journal of Physics A: Mathematical and Theoretical*, **42**: 075307.

[24] Holland, P. (2010). Quantum field dynamics from trajectories, in *Quantum Trajectories* (ed.) Chattaraj, P., Taylor & Francis/CRC, Boca Raton.

[25] Holland, P. (2012). Hydrodynamics, particle relabelling and relativity. *International Journal of Theoretical Physics*, **51**: 667.

[26] Holland, P. (2013). Symmetries and conservation laws in the Lagrangian picture of quantum hydrodynamics, in *Concepts and Methods in Modern Theoretical Chemistry: Statistical Mechanics*, (eds.) Ghosh, S. K. and Chattaraj, P. K., Taylor & Francis/CRC, Boca Raton.

[27] Holland, P. (2014). On systems having Poincaré and Galileo symmetry. *Annals of Physics (NY)*, **351**: 935.

[28] Holland, P. and Philippidis, C. (2003). Implications of Lorentz covariance for the guidance formula in two-slit quantum interference. *Physical Review A*, **67**: 062105.

[29] Holland, P. R. (1993). *The Quantum Theory of Motion*, Cambridge University Press: Cambridge.

[30] Holland, P. R. (1996). Is quantum mechanics universal? in *Bohmian Mechanics and Quantum Theory: An Appraisal*, (eds.) Cushing, J. T., Fine, A. and Goldstein, S., Kluwer, Dordrecht, p. 99.

[31] Jackiw, R., Nair, V. P., Pi, S.-Y. and Polychronakos, A. P. (2004). Perfect fluid theory and its extensions. *Journal of Physics A: Mathematical and General*, **37**: R327.

[32] Kan, K. K. and Griffin, J. J. (1977). Single-particle Schrödinger fluid. I. Formulation. *Physical Review C*, **15**: 1126.

[33] London, F. (1945). Planck's constant and low temperature transfer. *Reviews of Modern Physics*, **17**: 310.

[34] Madelung, E. (1926). Quantentheorie in hydrodynamischer form. *Z. Phys.*, **40**: 322.
[35] Matzkin, A. (2009). Bohmian mechanics, the quantum-classical correspondence and the classical limit: the case of the square billiard. *Foundations of Physics*, **39**: 903.
[36] Morrison, P. J. (1998). Hamiltonian description of the ideal fluid. *Reviews of Modern Physics*, **70**: 467.
[37] Muga, J. G. and Leavens, C. R. (2000). Arrival time in quantum mechanics. *Physics Reports*, **338**: 353.
[38] Salmon, R. (1988). Hamiltonian fluid mechanics. *Annual Review of Fluid Mechanics*, **20**: 225.
[39] Schönberg, M. (1954). On the hydrodynamical model of the quantum mechanics. *Nuovo Cimento*, **12**: 103.
[40] Schönberg, M. (1955). Vortex motions of the Madelung fluid. *Nuovo Cimento*, **1**: 544.
[41] Schiff, L. (1968). *Quantum Mechanics*, 3rd edn. McGraw-Hill, London.
[42] Schouten, J. A. (1959). *Tensor Analysis for Physicists*, 2nd edn., Clarendon Press, Oxford.
[43] Strocchi, F. (1966). Complex coordinates and quantum mechanics. *Reviews of Modern Physics*, **38**: 36.
[44] Truesdell, C. (1954). *The Kinematics of Vorticity*, Indiana University Press, Bloomington.
[45] Truesdell, C. and Toupin, R. A. (1960). The classical field theories, in *Principles of Classical Mechanics and Field Theory*, Handbuch der Physik, Band III/I, (ed.) Flügge, S., Springer-Verlag: Berlin.
[46] Wyatt, R. E. (2005). *Quantum Dynamics with Trajectories*, Springer: New York.
[47] Yoshida, Z. (2009). Clebsch parameterization: Basic properties and remarks on its applications. *Journal of Mathematical Physics*, **50**: 113101.
[48] Ziman, J. M. (1953). Quantum hydrodynamics and the theory of liquid helium. *Proceedings of the Royal Society of London A*, **219**: 257.

Appendix A: Proof that the Velocity Field is Not a Quantum Observable

Theorem: Consider three operators $\hat{A}, \hat{B}, \hat{C}$ where $\hat{A}, \hat{B}$ are linear and the mean values ($A_\psi = \langle\psi|\hat{A}|\psi\rangle$ etc.) obey the relation

$$A_\psi = B_\psi C_\psi \tag{A.1}$$

for all Hilbert space states $|\psi\rangle$ with $A_\psi, B_\psi \neq 0$ for at least one state. Then $\hat{C}$ cannot be linear.

Proof. We may use (A.1) and the linearity of $\hat{A}, \hat{B}$ to compute $A_{2\psi}$ in two ways:

$$A_{2\psi} = 4A_\psi = 4B_\psi C_\psi \tag{A.2}$$

$$A_{2\psi} = B_{2\psi} C_{2\psi} = 4B_\psi C_{2\psi}. \tag{A.3}$$

Equating the right-hand sides of these relations, we have $C_{2\psi} = C_\psi$. If $\hat{C}$ is linear, this implies $C_\psi = 0$, which contradicts (A.1) if $|\psi\rangle$ is a state for which $A_\psi, B_\psi \neq 0$. Hence $\hat{C}$ cannot be linear. □

As noted in the text, to apply the theorem put $\hat{A} = \hat{j}_{x_i}$, $\hat{B} = \hat{\rho}_x$ for each $i = 1, 2, 3$. Then the velocity component $v_i = j_i/\rho$ is not the mean value of a linear operator. Note that this result holds whatever definition is adopted for the density and current density fields, so long as these are expressible as mean values of linear operators. In particular, it applies to alternative trajectory theories such as that derived from the non-relativistic limit of the Dirac theory where the current density necessarily differs from that used here [28]. The latter example provides a salutary lesson that a conserved current derived from a wave equation may contain a trivial (identically conserved) component that is physically nontrivial.

Chapter 11

Symmetry and Natural Quantum Structures for Three-Particles in One-Dimension

N. L. Harshman

Department of Physics
American University
4400 Massachusetts Ave. NW
Washington, DC 20016-8058, USA
harshman@american.edu

How symmetries induce natural and useful quantum structures is explored by looking at models of three interacting particles in one-dimension. Models with three interacting particles manifest the nearly full spectrum of possibilities for dynamical systems. The chapter demonstrates that the related but distinct notions of integrability, separability, and solvability identify useful collective observables for Hamiltonians with sufficient symmetry. In turn, these observables induce tensor product structures on the Hilbert space that may be especially useful for storing and processing quantum information and may provide some deeper guidance to interpretation and phenomenology of quantum few-body physics.

1. Introduction

This chapter presents a collection of examples of how quantum structures are relevant to the study of three particles in one dimension. The collection is curated around a theme: symmetry selects preferred quantum structures, and preferred quantum structures instantiate observation, information, control, and interpretation "naturally".

The ironic quotes around "naturally" indicate that this may be an idiosyncratic (and possibly recursive) definition. Whatever "naturally" means, physicists know that some representations, coordinate systems, reference frames, etc. are more natural than others for studying specific systems. While underlying physics must be independent of representation, a judicious choice of representation simplifies analysis by eliminating irrelevant variables, indicating conserved quantities, consolidating dimensional variables into unified scales, allowing approximation methods to converge more rapidly, and so on. Often finding the "better" representation requires a transformation from observables associated with reductive constituents to observables associated with the emergent collective. So a thread running through the collection is how symmetry identifies collective observables that induce physically-meaningful quantum structures for the three-body system.

Why three bodies? Three-body systems have been at the frontier of dynamical analysis since Newton, and their study is the first baby step from two-body to many-body systems. The three-body problem is perennially fertile; new periodic solutions to the even classical three-body gravitational system (in three dimensions) have been discovered recently [28]. This chapter considers models with three non-relativistic particles in one dimension. Even in one dimension, the general three-body problem is not integrable and does not have exact and entire solutions unless some additional symmetries provide extra integrals of the motion [9]. When additional symmetry is not present, some classical, non-integrable three body-systems have soft chaos [14]: they stay close to integrable models where perturbation theory works well, at least for certain initial conditions. Other models have hard chaos and deterministic prediction is impractical for realistic uncertainties and any initial conditions.

Diagnosing how integrability affects the emergence of "natural" collective quantum structures is a central (and unresolved) goal of this inquiry. Integrability means there are as many independent integrals of the motion as degrees of freedom, and these integrals of the motion are observables in involution (defined by the Poisson bracket) with the Hamiltonian and with each other. When a system is classically integrable, the integrals of motion certainly provide a "natural" way to describe the system in terms of action-angle variables. The complete set of integrals are determined by initial conditions and define a manifold in phase space that is essentially a multi-dimensional torus. The trajectory lies on the manifold and it is trivially expressed in terms of action-angle coordinates, although the canonical map back to position and momentum coordinates may be highly nontrivial. Note that integrable is not the same thing as separable in the classical sense, i.e. explicitly finding a set of coordinates that separate the Hamilton–Jacobi equation so that the equations of motion can be solved by quadrature. Separability is used in this chapter both in the sense of separable coordinates for solving the three-body Schrödinger equation and separable from the quantum information perspective of Hilbert space tensor product structures and entanglement. Clarifying that connection is another goal.

There is still debate about how best to carry over the definition of integrability to quantum systems [3, 6, 22, 31]. Some aspects of integrability transition neatly from classical mechanics: integrals of the motion are realized as operators on the Hilbert space that commute with the Hamiltonian and each other. However, sticking points include how to extend to discrete degrees of freedom and finite Hilbert spaces, how to define independence of integral operators, and how to deal with indistinguishable particles. For certain three-body scenarios, where integrability is clear, a reasonable hypothesis is that the set of commuting observables associated to integrals of motion induce tensor product structures that are "natural". For example, when quantum structures are induced by observables that commute with the Hamiltonian, information measures based on those structures, like entanglement, are invariant in time. And sometimes these structures indicate preferred collective degrees of freedom that are compatible with symmetrization of identical particles. Using a broad definition of information, by choosing quantum structures that align with integrability, we find a representation for the system in which for "typical" states, the embodied information can be shifted to the observables.

Another motivation for this study is a system of experimental interest: identical fermions or bosons (with and without spin or other internal components) in an one-dimensional trap with tunable two-body interactions interaction. This model has been probed experimentally using ultracold atoms in effectively one-dimensional optical traps. The two-body interactions can be tuned by Feshbach and confinement-induced resonances and are well-approximated by delta-function contact potential [25]. See [7] and [13] for reviews of theory and experiment for one-dimensional bosonic and fermionic systems, respectively, and [17, 18] for an overview of symmetries of the few-body model and recent theoretical references. One interesting feature of this model is that the strength of the contact interaction is described by a single parameter, and as this parameter varies from zero to infinity, the model interpolates between two integrable cases, the non-interacting limit and the so-called unitary limit. Although conclusive statements are not yet forthcoming, this system demonstrates how "natural" quantum structures fade and re-emerge as symmetry and integrability is tuned between two solvable models.

This introduction has argued that one-dimensional, quantum three-body models are a minimally-complicated but experimentally-relevant system to investigate open questions about the relationships among integrability, separability, and symmetry. The rest of the chapter is a series of essays on these relationships and how they induce quantum structures. Thinking about quantum structures, and in particular the relativity of entanglement [2, 15, 20, 32, 33], has proved useful for a variety of purposes, including developing decoherence-free subspaces for error-correcting codes [29], understanding the dynamics of open systems [1], and classifying entanglement structures in two-body systems [16]. The goal here is to see whether thinking about quantum structures can possibly explain the emergence of "natural" observables for storing and processing quantum information. Since the models under consideration are relevant to ultracold atomic gases, one of the possible working materials for proposed quantum information devices like computers and simulators, this inquiry into quantum structures has practical ramifications. A hypothesis is that systems near solvable models will work best in such applications, and so the first section reviews some examples of solvable models used throughout the chapter.

2. Solvable Models

The family of models under consideration have an Hamiltonian $\hat{H}$ expressed in terms of the particle position and momentum operators $\hat{X}_i$ and $\hat{P}_i$ like

$$\hat{H} = \sum_i \left(\frac{1}{2m} \hat{P}_i^2 + V^1(\hat{X}_i) \right) + \sum_{\langle i,j \rangle} V^2(|\hat{X}_i - \hat{X}_j|), \tag{1}$$

where the second sum is over all pairs $\langle i, j \rangle$. The one-particle external potential $V^1(x)$ $(x \in \mathbb{R})$ and the two-particle interaction potential $V^2(r)$ $(r \in \mathbb{R}^+)$ are assumed to be sufficiently well-behaved functions so that when the position observables $\hat{X}_i$ are inserted into them, the resulting operators are still self-adjoint on the Hilbert space. Notice that I am only considering spin-independent Hamiltonians in this chapter. Although not dynamically coupled to spatial degrees of freedom, spin degrees of freedom play a role due to symmetrization of identical indistinguishable particles.

For certain choices of the functions V^1 and V^2, the Hamiltonian (1) is solvable. Solvable models are like lamps illuminating the landscape of dynamical systems. They are touchstones for analysis because (as the joke goes [4]) if you are looking for your lost keys, under the lampposts is where there is enough light for you to search effectively.

2.1. Solvability for one particle

First, note that all one-particle, one-dimensional systems are integrable. The one-particle Hamiltonian

$$\hat{h} = \frac{1}{2m} \hat{P}^2 + V^1(\hat{X})$$

certainly commutes with itself, so there are as many integrals of motion (one) as there are degrees of freedom (also one). Classically, the equation of motion can be solved by quadrature for the classically allowed spectrum of energies. Quantum mechanically, the energy levels of bound states (the only kind of states considered here) are discretized by a quantum number $n \in \{0, 1, 2, \ldots\} \equiv \mathbb{N}$ that labels energies ϵ_n in the one-particle spectrum $\{\epsilon_0, \epsilon_1, \epsilon_2, \ldots\} = \sigma^1$. In principle, each ϵ_n is the solution of an integral equation that depends on the functional form of V^1, so in addition to being integrable, all one-dimensional, one-particle systems in an external potential are also solvable.

When the spectrum σ^1 is known, one can always define a function $\epsilon(z)$ by a finite series such that $\epsilon(n) = \epsilon_n$ for all $n \leq n_{max}$, with n_{max} some arbitrary cutoff. When the function $\epsilon(n)$ is a finite algebraic expression in terms of the quantum number n for all $n \in \mathbb{N}$, the system is algebraically solvable. Algebraically solvable one-dimensional potentials include the infinite square well, harmonic oscillator, Pöschl–Teller, and Morse potentials.

2.2. Non-interacting three-particle models

When there are no interactions $V^2 \equiv 0$, every $\hat{H}$ with form (1) is integrable because $\hat{H} = \hat{h}_1 + \hat{h}_2 + \hat{h}_3$ decomposes into a sum of three independent one-particle, one-dimensional Hamiltonians $\hat{h}_i$. The total non-interacting spectrum σ_0 is all possible sums of three elements of σ^1, and therefore all energies $E \in \sigma_0$ are

- non-degenerate when $E = 3\epsilon_j$;
- three-fold degenerate when $E = 2\epsilon_j + \epsilon_k$ $(j \neq k)$;
- six-fold degenerate when $E = \epsilon_j + \epsilon_k + \epsilon_l$ $(j \neq k \neq l)$.

Harmonic traps $V^1(x_i) = m\omega x_i^2/2$ are an example of a three-particle system that is superintegrable, i.e. the Hamiltonian has more independent integrals of motion than the three degrees of freedom [10]. The three-particle model is isomorphic to one particle in an isotropic harmonic potential in three-dimensions with nine independent invariant operators [19, 23]: the three single particle Hamiltonians $\hat{h}_i$, the three "angular momentum"-like operators $\hat{Q}_1\hat{P}_2 - \hat{P}_1\hat{Q}_2$, and the Demkov operators [21] $\hat{P}_1\hat{P}_2 + m^2\omega^2\hat{X}_1\hat{X}_2$.

2.3. Interacting three-particle models

When interactions are included, here are three notable solvable cases:

- Models with harmonic interactions $V^2(|x_i - x_j|) = \gamma(x_i - x_j)^2$ are solvable when the external field is quadratic $V^1(x_i) = Ax_i^2 + Bx_i + C$. In particular, for harmonic traps ($A > 0$) and homogeneous potentials ($A = B = 0$), the model is algebraically solvable. This solvability is a consequence of the separability of the center-of-mass and relative degrees of freedom for quadratic potentials; any other external potential function $V^1(x)$ does not preserve this separation. Further, for harmonic traps, the interacting system is superintegrable because an angular-momentum-like operator remains invariant.
- Models with the contact interaction $V^2(|x_i - x_j|) = \gamma\delta(x_i - x_j)$ are solvable in the unitary limit $\gamma \to \infty$, and are algebraically solvable whenever the one-particle potential $V^1(x)$ is algebraically solvable. Further, for the case of a homogeneous potential and infinitely hard-wall boundary conditions, the model is solvable for any value of γ by the technique of Bethe's ansatz [26]. This case is an example where integrability does not follow from some choice of separable coordinates.
- The Calogero–Moser model with the inverse-square potential $V^2(|x_i - x_j|) = \gamma/(x_i - x_j)^2$ and the harmonic trap potential $V^1(x_i) = m\omega^2 x_i^2/2$ is algebraically solvable [4]. There are also several other related solvable models when $V^2(|x_i - x_j|)$ is a reciprocal sinusoidal or hyperbolic function [5].

Ideally, there would be an unified method to characterize separability, integrability, solvability, and algebraic solvability in terms of symmetries. And certainly, much work has been done here, but no complete, coherent picture has emerged. However, for integrable and solvable systems, the integrals give "natural" observables that induce "natural" quantum structures on the Hilbert space that are useful for analyzing and extracting entanglement, as the following sections explore.

3. Separability in Configuration Space

Let us identify one chain of separability, a more refined notion than integrability, that selects a preferred tensor product structure on the Hilbert space.

1. Identify a natural Hamiltonian that is totally separable. In other words, there is a canonical coordinate transformation that separates the Hamilton–Jacobi equation into a sum of action-angle contributions.
2. The quantum system is also separable in that coordinate system, i.e. the Hamiltonian becomes a wave equation that is a separable partial differential equation. For each spatial degree of freedom, there is a countably-infinite tower of harmonics with energies bounded from below. Products of these harmonics are energy eigenstates with a phase that simply rotates in time.
3. There are specific observables associated with the separable coordinate system. These observables are constructed from the generators of the symmetries implied by the separability of the system. Sometimes the generators for the algebra are inherited from an underlying model with more symmetry. The algebra of observables can be partitioned further into commuting subalgebras for each independent degree of freedom.
4. These subalgebras induce a partitioning of the Hilbert space. In other words, the tensor products of the representations of the subalgebras are the same as the representation of the entire system. Since these observables commute with the Hamiltonian, entanglement with respect to this tensor product structure is invariant with respect to time.

This sounds like a plausible method for inducing a “natural” tensor product structure for classically separable systems, but the devil is in the details. This section teases out a few of those details, and for the limited set of models considered here, arrives at a heuristic classification.

One advantage of working with three identical particles in one dimension is that the three-dimensional configuration space has familiar geometrical structures, coordinates, and transformations. One realization of the spatial Hilbert space $\mathcal{K}$ is the space of Lebesgue-square-integrable functions $L^2(\mathcal{X})$ on the configuration space $\mathcal{X} \sim \mathbb{R}^3$. The manifold $\mathcal{X}$ has many different possible coordinate systems. The simplest case is where each coordinate is just the particle position $\xi_i = x_i \in (-\infty, \infty)$ with respect to a common origin in $\mathcal{X}$, corresponding to the decomposition $\mathbb{R}^3 = \mathbb{R} \times \mathbb{R} \times \mathbb{R}$. This is just one of an equivalence class of rectangular coordinate systems for $\mathcal{X}$ related by translation, rotation, and orthogonal scaling.

There are a host of other coordinate systems, but in physics the most interesting from the perspective of quantum structures and solvability for the three-body, one-dimension problem are the orthonormal coordinate systems that separate the Schrödinger equation in three dimensions. Delightfully, these have already been entirely classified. Depending on the functional form of the potential $V(\mathbf{x})$, the time-independent Schrödinger equation

$$-\frac{\hbar^2}{2m}\left(\frac{\partial^2}{\partial x_1^2} + \frac{\partial^2}{\partial x_2^2} + \frac{\partial^2}{\partial x_3^2}\right)\Psi(\mathbf{x}) + (V(\mathbf{x}) - E)\,\Psi(\mathbf{x}) = 0 \qquad (2)$$

can be solved by separation of variables in 11 different orthonormal coordinate systems [11, 24]: rectangular, cylindrical, elliptical cylindrical, parabolic cylindrical, spherical, conical, parabolic, prolate spheroidal, oblate spheroidal, ellipsoidal, and paraboloidal. All of these coordinate systems can be unified into a single framework, i.e. they are all degenerations of ellipsoidal coordinates. For each coordinate system, the transformed potential $\tilde{V}(\xi)$ in the separable coordinates $\xi = \{\xi_1, \xi_2, \xi_3\}$ must have a particular form in order for (2) to separate. For example, in spherical coordinates $\{r, \theta, \phi\}$, the potential must have the form

$$\tilde{V}(\xi) = v_1(r) + \frac{v_2(\theta)}{r^2} + \frac{v_3(\phi)}{r^2 \sin^2\theta}.$$

For our case of three identical particles interacting via Galilean-invariant interactions, the potential $V(\mathbf{x})$ is constrained to the form

$$\begin{aligned} V(\mathbf{x}) = V^1(x_1) + V^1(x_2) + V^1(x_3) + V^2(|x_1 - x_2|) \\ + V^2(|x_2 - x_3|) + V^2(|x_3 - x_1|), \end{aligned} \qquad (3)$$

where $V^1(x_i)$ is the external trap potential and $V^2(|x_i - x_j|)$ is the Galilean invariant two-body interaction between particles i and j. The specific form (3) eliminates most of the 11 possibilities for separable potentials. When there are nontrivial interactions V^2, only rectangular and cylindrical remain, and then only for harmonic traps. (Harmonic traps with no interactions are separable in eight of the 11 coordinate systems [21])

Even when the model Hamiltonian is separable, there are still distinctions between different kinds of separability depending on the nature of the separation constants [24]. This has consequences for the quantum structures that can be defined, and I categorize these into three standards of separability: gold, silver, and bronze.

3.1. Gold separability

The most straightforward example of separability of the Schrödinger equation leading to separable degrees of freedom is rectangular coordinates, when some linear transformation of the particle coordinates $\mathbf{q} = R\mathbf{x} + \mathbf{b}$ separates the potential $V(\mathbf{x})$ into a sum of independent functions like

$$\tilde{V}(\mathbf{q}) = v_1(q_1) + v_2(q_2) + v_3(q_3).$$

When there are no two-body interactions, every three-body system is certainly of this form with $R = I_3$ and $q_i = x_i$, but there are interacting examples, too. Quadratic potentials $V^1(x_i) = Ax_i^2 + Bx_i + C$ with harmonic two-body interactions $V^2(|x_i - x_j|) = \gamma(x_i - x_j)^2$ are separable in rectangular Jacobi rectangular coordinates:

$$\begin{aligned} q_1 &= \frac{1}{\sqrt{3}}(x_1 + x_2 + x_3), \\ q_2 &= \frac{1}{\sqrt{2}}(x_1 - x_2), \\ q_3 &= \frac{1}{\sqrt{6}}(x_1 + x_2 - 2x_3). \end{aligned} \tag{4}$$

Note that any orthogonal combination $q_2' = \cos\vartheta q_2 + \sin\vartheta q_3$ and $q_3' = -\sin\vartheta q_2 + \cos\vartheta q_3$ (where $\vartheta \in [0, 2\pi)$) also separates the quadratic trap with harmonic interactions.

Why is this the gold standard? Because systems separable in rectangular coordinates correspond most directly to our notions of separable degrees of freedom and independent modes, and therewith to a "naturally" separable Hilbert space. The Hamiltonian breaks into a sum of three terms $H = h_1 + h_2 + h_3$ where

$$h_i = -\frac{\hbar^2}{2m}\frac{\partial^2}{\partial q_i^2} + v_i(q_i).$$

The energy $E = \epsilon_1(\eta_1) + \epsilon_2(\eta_2) + \epsilon_3(\eta_3)$ can be expressed as a sum of functions $\epsilon_i(\eta_i)$, each of which is a function of only one of the three separation constants $\{\eta_1, \eta_2, \eta_3\}$. The energy eigenstates are products of functions X_i which each depend on single separation constant η_i as

$$X_1(q_1; \eta_1) X_2(q_2; \eta_2) X_3(q_3; \eta_3).$$

Denote the spectrum for each separation constant η_i by σ_i. These spectra σ_i are independent for rectangular coordinates and so the general state $|\Phi\rangle \in \mathcal{K}$ is expressed as an independent triple sum over a basis labeled by the separation constants (i.e. quantum numbers) η_i

$$|\Phi\rangle = \sum_{\eta_1 \in \sigma_1} \sum_{\eta_2 \in \sigma_2} \sum_{\eta_3 \in \sigma_3} c_{\eta_1 \eta_2 \eta_3} |\eta_1 \eta_2 \eta_3\rangle.$$

Put another way, each subspace can be treated mathematically like a totally independent subsystem. The spatial Hilbert space is the tensor product

$$\mathcal{K} = \mathcal{K}_1 \otimes \mathcal{K}_2 \otimes \mathcal{K}_3, \tag{5}$$

where each $\mathcal{K}_i$ is realized by square-summable sequences $\ell^2(\eta_i)$ labeled by the quantum number η_i. The abstract Hamiltonian $\hat{H}$ (as opposed to its representation H on $\mathcal{X}$) is the sum of three local operators

$$\hat{H} = \hat{h}_1 \otimes \hat{\mathbb{I}} \otimes \hat{\mathbb{I}} + \hat{\mathbb{I}} \otimes \hat{h}_2 \otimes \hat{\mathbb{I}} + \hat{\mathbb{I}} \otimes \hat{\mathbb{I}} \otimes \hat{h}_3. \tag{6}$$

The set of observables $\{\hat{h}_1, \hat{h}_2, \hat{h}_3\}$ is complete and commuting, so by Zanardi's theorem [15, 32, 33], these subalgebras of observables induce the tensor product structure (5) on $\mathcal{K}$. This relatively simple example of separability in rectangular coordinates provides a useful contrast with the cases below.

3.2. Silver and bronze separability

Unlike gold separability, for general separable coordinate systems $\xi = \{\xi_1, \xi_2, \xi_3\}$, the different form of the Hamiltonian cannot be written as a sum of three local sub-Hamiltonians. Although the eigenstates are separable functions of the coordinates, they are not separable with respect to the separation constants η_i as

$$X_1(\xi_1; \eta_1, \eta_2, \eta_3) X_2(\xi_2; \eta_1, \eta_2, \eta_3) X_3(\xi_3; \eta_1, \eta_2, \eta_3).$$

What distinguishes silver and bronze is whether the spectra σ_i of separation constants are independent (silver) or not (bronze).

As an example, consider two cases of silver separability in Jacobi cylindrical coordinates $\{\rho, \phi, z\}$: the harmonic trap with either harmonic interactions

$$\begin{aligned} H &= -\frac{\hbar^2}{2m}\nabla^2 + \frac{1}{2}m\omega^2\mathbf{x}^2 + \gamma\left((x_1 - x_2)^2 + (x_2 - x_3)^2 + (x_3 - x_1)^2\right) \\ &= -\frac{\hbar^2}{2m}\nabla^2 + \frac{1}{2}m\omega^2 q_1^2 + \left(\frac{1}{2}m\omega^2 + 3\gamma\right)\rho^2 \end{aligned} \tag{7}$$

or the Calogero–Moser Hamiltonian with inverse quadratic interactions

$$\begin{aligned} H &= -\frac{\hbar^2}{2m}\nabla^2 + \frac{1}{2}m\omega^2\mathbf{x}^2 + \frac{\gamma}{(x_1 - x_2)^2} + \frac{\gamma}{(x_2 - x_3)^2} + \frac{\gamma}{(x_3 - x_1)^2} \\ &= -\frac{\hbar^2}{2m}\nabla^2 + \frac{1}{2}m\omega^2(q_1^2 + \rho^2) + \frac{2\gamma}{\rho^2} \\ &\quad \times \left(\frac{1}{\cos^2\phi} + \frac{1}{\cos^2(\phi - 2\pi/3)} + \frac{1}{\cos^2(\phi - 4\pi/3)}\right) \\ &= -\frac{\hbar^2}{2m}\nabla^2 + \frac{1}{2}m\omega^2(q_1^2 + \rho^2) + \frac{18\gamma}{\rho^2 \sec^2(3\phi)}, \end{aligned} \tag{8}$$

where $\rho^2 = q_2^2 + q_3^2$ and $\tan\phi = q_3/q_2$ in terms of Jacobi rectangular coordinates (4).

In both of these cases, the potential in Jacobi cylindrical coordinates satisfies

$$\tilde{V}(\rho, \phi, z) = v_\rho(\rho) + \frac{v_\phi(\phi)}{\rho^2} + v_z(z).$$

However, note that the configuration space Hamiltonian cannot be written as a sum of three differential operators that each only depend on a single variable. The second term ties together the separation constants $\eta_\rho \equiv \nu$ and $\eta_\phi \equiv \mu$ (but not $\eta_z \equiv \eta$) and the general energy eigenstate has the form

$$R_{\nu\mu}(\rho)\Phi_\mu(\phi)Z_\eta(z).$$

In principle, one could imagine defining a tensor product structure like

$$\mathcal{K}_\rho \otimes \mathcal{K}_\phi \otimes \mathcal{K}_z \sim L^2(\mathbb{R}^+) \otimes L^2(\mathbb{S}_1) \otimes L^2(\mathbb{R}),$$

but the Hamiltonian would not have a decomposition like (6). Therefore, this is not "gold" separability.

Although the separation constants are not independent, their spectra are, and that is what makes both of these cases "silver" separable in this classification scheme. In terms of the separation constants, the energy for the harmonic interaction Hamiltonian (7) is

$$E_{\eta\nu\mu} = \hbar\omega(\eta + 1/2) + \hbar\sqrt{\omega^2 + 4m\gamma}(2\nu + |\mu| + 1)$$

with $\nu \in \sigma_\nu \mathbb{N}$, $\eta \in \sigma_\eta = \mathbb{N}$, and $\mu \in \sum_\mu = \mathbb{Z}$. For the Calogero–Moser Hamiltonian (8) the energy is

$$E_{\eta\nu\mu} = \hbar\omega\left[\eta + 2\nu + |\mu| + 3/2\left(1 + \sqrt{1 + 2m^2\gamma}\right)\right]$$

again with $\nu \in \sigma_\nu = \mathbb{N}$ and $\eta \in \sigma_\eta = \mathbb{N}$, but this time only for $\mu/3 \in \mathbb{Z}$, i.e. σ_μ contains all integer multiples of 3. Because the spectra of separation constants are not dependent, for both cases, the states can still be expressed in terms of energy eigenstates like

$$|\Phi\rangle = \sum_{\nu\in\sigma_\nu} \sum_{\mu\in\sigma_\mu} \sum_{\eta\in\sigma_\eta} c_{\nu\mu\eta}|\nu\mu\eta\rangle,$$

where the order of the sum does not matter.

An example of "bronze" separability is the harmonic, non-interacting model solved in spherical coordinates $\{r, \theta, \phi\}$ with the usual separation constants $\{n, l, m\}$. The energy eigenstates have the form

$$R(r; n, l)\Theta(\theta; l, m)\Phi(\phi; m),$$

and that is definitely not gold separability because of the multiple separation constant dependence of the harmonics. Further, the spectrum of possible m depends on l, so the triple sum

$$|\Phi\rangle = \sum_{n \in \mathbb{N}} \sum_{l \in \mathbb{N}} \sum_{m=-l}^{l} c_{nlm} |nlm\rangle$$

cannot be rearranged arbitrarily.[a]

Why do these distinctions among different types of separability matter? A quantum structure is useful from the point of view of control and information when there are experimentally-accessible observables that are complete and local with respect to that tensor product structure. For silver and bronze separability, the Hamiltonian does not decompose into a sum of local subsystem Hamiltonians on a partitioned Hilbert space, so the virtual subsystems induced by such a partition are less "natural". However, at least for the silver separability, there are complete and local operators that characterize the spectra and basis for each partition, making this case somewhat more natural. For example, it seems more physically reasonable to imagine controllably entangling the radial and polar degrees of freedom for the two cases in cylindrical coordinates than controlling entanglement between the m-degree of freedom and the l-degree of freedom in spherical coordinates. Future work will try to make this fuzzy notion of "reasonable" and "natural" well-defined, and the key will likely be better understanding of the relationship between separability and symmetry.

[a] For the familiar Coulomb potential (which has no equivalent one-dimensional three-body model), σ_l also depends on n and so the separability is even 'bronzer' than the isotropic harmonic trap.

4. Types of Symmetries

Here I classify some different types of symmetries that generate structures in the Hilbert space via their representations. Of most concern are configuration space and phase space symmetry transformation groups that leave the Hamiltonian invariant, but dynamical symmetries are briefly mentioned at the end because of their connection to separability.

4.1. Configuration space symmetries

Repeating for convenience, in configuration space $\mathbf{x} = \{x_1, x_2, x_3\} \in \mathcal{X}$ the configuration space Hamiltonian H has the form

$$H = -\frac{\hbar^2}{2m}\nabla^2 + \sum_i V^1(x_i) + \sum_{\langle i,j\rangle} V^2(|x_i - x_j|). \tag{9}$$

Here I restrict attention to cases where $V^1(x_i)$ is a one-body trapping potential such that $\lim_{x\to\infty} V^1(\pm x) \to \infty$ and $V^2(|x_i - x_j|)$ is a Galilean-invariant two-body interaction potential.

A configuration space symmetry is a group of transformations of $\mathcal{X}$. These transformations do not have to be linear, but of present interest are subgroups of orthogonal transformations $O \in O(3)$. An orthogonal transformation O on configuration space induces a unitary transformation $\hat{U}(O)$ on the Hilbert space by

$$\psi(O\mathbf{x}) = \langle O\mathbf{x}|\psi\rangle = \langle \mathbf{x}|\hat{U}^\dagger(O)|\psi\rangle = (\hat{U}(O)\psi)(\mathbf{x}). \tag{10}$$

When there is a complete basis for the Hilbert space, this method can be used to explicitly calculate the representation (although this is usually not the most practical method).

For otherwise arbitrary functions V^1 and V^2, the only symmetry that the Hamiltonian H necessarily has in configuration space is the geometrical realization of particle permutation symmetry P_3, which is isomorphic to the abstract symmetric group on three objects S_3. The six elements of P_3 can be broken into the following:

- The identity, which maps $\{123\}$ into $\{123\}$ and acts as the identity matrix I_3 on $\mathcal{X}$.

- The pairwise exchanges $\{213\}$, $\{132\}$, and $\{321\}$ which act as reflections in $\mathcal{X}$ across the planes $x_1 = x_2$, $x_2 = x_3$, and $x_1 = x_3$, respectively.
- The 3-cycles $\{231\}$ and $\{312\}$ that act as rotations by $\pm 2\pi/3$ around the line $x_1 = x_2 = x_3$.

This set of six transformations is isomorphic to the three-dimensional point group denoted C_{3v} in Schönflies notation or [3] in Coxeter notation.

Notice that all six of these transformations are orthogonal transformations $O \in O(3)$ that leave invariant the Jacobi cylindrical coordinates q_1 and ρ. The quantity ρ is expressed in particle coordinates like

$$\rho^2 = \sum_{\langle i,j \rangle} (x_i - x_j)^2 = 2x_1^2 + 2x_2^2 + 2x_3^2 - 2x_1x_2 - 2x_2x_3 - 2x_1x_3$$

and it quantifies the scale of the relative particle distribution, i.e. the Euclidean distance in $\mathcal{X}$ from the line of three-particle coincidence $x_1 = x_2 = x_3$.

The Hamiltonian H can have additional configuration space symmetries if the trap is parity symmetric $V^1(-x + 2b) = V^1(x)$ about some point b. Denote by Π_i the configuration space operator realizing this reflection $x_i \to -x_i + 2b$. In general, a single particle inversion Π_i does not leave the Hamiltonian invariant because of the interaction terms.[b] However, the total inversion $\Pi = \Pi_1\Pi_2\Pi_3$ does leave H invariant, and it commutes with all elements of P_3. Putting this together, the minimal configuration space symmetry of H is $P_3 \times O(1) \sim S_3 \times Z_2$, where the inversion subgroup is denoted $O(1)$. This group is isomorphic to the point group $D_{3d} \sim [[3]]$.

For certain traps and potentials, there is more than the minimal configuration space symmetry for three interacting particles P_3 (no parity) or $P_3 \times O(1)$ (parity). Clearly there is more than minimal

[b]In the absence of interactions, each single particle parity Π_i is a symmetry of H, but the Π_i do not commute with particle permutations. The group structure that describes that combination is the wreath product (a kind of semi-direct product) of the symmetric group with the reflection group $S_3 \wr Z_2$.

symmetry in a harmonic well with any interaction potential:

$$\begin{aligned} H &= -\frac{\hbar^2}{2m}\nabla^2 + \frac{m\omega^2}{2}(x_1^2 + x_2^2 + x_3^2) + V^2(|x_1 - x_2|) \\ &\quad + V^2(|x_2 - x_3|) + V^2(|x_1 - x_3|) \\ &= -\frac{\hbar^2}{2m}\nabla^2 + \frac{m\omega^2}{2}(\rho^2 + q_1^2) \\ &\quad + V^2(\rho|\cos\phi|) + V^2(\rho|\cos(\phi - 2\pi/3)|) \\ &\quad + V^2(\rho|\cos(\phi - 4\pi/3)|). \end{aligned} \tag{11}$$

In Jacobi cylindrical coordinates, total inversion takes the form $\Pi \cdot \{q_1, \rho, \phi\} = \{-q_1, \rho, \phi + \pi\}$. As mentioned above, the rotations and reflection in P_3 leave q_1 and ρ invariant but transform ϕ. For example, the particle exchange $\{213\}$ maps ϕ into $\pi - \phi$ and the 3-cycle $\{231\}$ maps ϕ to $\phi + 2\pi/3$. But now there is an additional independent symmetry: the relative and center-of-mass degrees of freedom can be independently inverted. In other words, relative parity $\Pi_r \cdot \{q_1, \rho, \phi\} = \{q_1, \rho, \phi + \pi\}$ is a good quantum number for Hamiltonians with harmonic traps. This means that for three particles in a harmonic trap with arbitrary two-particle interactions, the configuration space symmetry group is $P_3 \times O(1) \times O(1)$ and is isomorphic to $D_{6h} \sim [3, 2]$. This additional symmetry is an example of an emergent symmetry, i.e. a symmetry of the model than cannot be generated by products of one-particle operators like Π_i and permutations in P_3.

A group of nonlinear symmetry transformations in configuration space are relevant for the Calogero–Moser Hamiltonian (8) for any interaction strength or for unitary limit $\gamma \to \infty$ of the contact interaction Hamiltonian in any external trap

$$H = -\frac{\hbar^2}{2m}\nabla^2 + \sum_i V^1(x_i) + \gamma\left(\delta(x_1 - x_2) + \delta(x_2 - x_3) + \delta(x_1 - x_3)\right). \tag{12}$$

For these models, position wave functions with finite energy must vanish on the coincidence manifold $\mathcal{V}$ defined as the union of the planes $x_1 = x_2$, $x_2 = x_3$, and $x_1 = x_3$. The effective configuration space is therefore $\mathcal{X}' = \mathcal{X} - \mathcal{V}$. The manifold $\mathcal{V}$ divides $\mathcal{X}$ into six equivalent sectors $\mathcal{X}_{ijk}$, each with fixed particle order $x_i > x_j > x_k$. The Hilbert space of finite-energy states $\mathcal{K}' \subset \mathcal{K}$ is realized by

$$L^2(\mathcal{X}') \sim L^2(\mathcal{X}_{ijk}) \otimes \mathbb{C}^6. \tag{13}$$

In other words, every eigenstate of (8) or (12) is six-fold degenerate for distinguishable particles. Sector permutations act as orthogonal transformations of the degeneracy space, and the solvability of those two models can be understood as a consequence of this symmetry. A permutation of the six sectors $\mathcal{X}_{ijk}$ is a nonlinear transformation of $\mathcal{X}'$ because of the discontinuity across the coincidence planes. For the contact interaction, the energy eigenfunctions in each sector $\mathcal{X}_{ijk}$ are just the fermionic (totally-antisymmetric) solutions to the non-interacting problem with the same trapping potential V^1 restricted to the sector, a result usually attributed to Girardeau [12].

4.2. Phase space symmetries

Any orthogonal transformation O of configuration space $\boldsymbol{q} = O\mathbf{x}$ induces an orthogonal, canonical transformation in six-dimensional phase space $\{\boldsymbol{q}, \boldsymbol{k}\} = \{O\mathbf{x}, O\mathbf{p}\}$ that is also a symmetry transformation of H. There is always at least one additional symmetry in phase space beyond configuration space, namely time translation T_t generated by H itself. Classically, this is just the Hamiltonian flow in time along the trajectories through phase space. The quantum mechanical consequence for a completely bound system is the discretization of energy.

So for an asymmetric trap with no additional symmetries, the phase space symmetry group is $T_t \times P_3$ and for a symmetric trap it is $T_t \times P_3 \times O(1)$. As before, for a harmonic trap like (11), there is an additional, emergent symmetry due to the separable center-of-mass degree of freedom. Any rotation in the q_1-k_1 hyperplane, where k_1 is the momentum conjugate to q_1, is a phase space symmetry. This is the standard U(1) symmetry of the one-dimensional harmonic oscillator. The group U(1) contains the relative parity operation, so the total phase space symmetry for three-harmonically trapped particles is $T_t \times P_3 \times O(1) \times U(1)$.

4.3. Dynamic symmetries

More generally, the group of all unitary operators that commute with the Hamiltonian is the kinematic symmetry group of the Hamiltonian. Only kinematic symmetry groups associated to transformations of the configuration space or phase space have been considered; the case of accidental kinematic symmetries is briefly discussed in the next section. Sometimes there are transformations of configuration space, phase space, or the Hilbert space that do not commute with the Hamiltonian but map its energy spectrum onto itself in a regular way. These are examples of dynamic symmetries, also known as spectrum-generating symmetries. In a Lie-algebraic dynamic symmetry, the Hamiltonian is not invariant under a group of transformations of the phase space, but the generators of that transformation combined with the Hamiltonian form a Lie algebra of operators defined on the Hilbert space. One example of a Lie-algebraic dynamic symmetry is the "hidden" SO(2, 1) symmetry of the harmonic oscillator [27, 30]. First consider the one-particle case and define the operators

$$\hat{W}^{\pm} = \frac{1}{4\sqrt{2}}\left(\frac{1}{m\hbar\omega}\hat{P}^2 - \frac{m\omega}{\hbar}\hat{X}^2 + \frac{i}{\hbar}\hat{X}\hat{P} + \frac{i}{\hbar}\hat{P}\hat{X}\right). \tag{14}$$

These operators have the commutation relation with the one-particle harmonic Hamiltonian $\hat{h}$

$$[\hat{h}, \hat{W}^{\pm}] = \pm 2\hbar\omega\hat{W}^{\pm},$$

showing that they act like ladder operators connecting states with an energy difference of $2\hbar\omega$, i.e. two rungs apart. The commutator between the ladder operators is

$$[\hat{W}^{+}, \hat{W}^{-}] = -\frac{1}{2\hbar\omega}\hat{h}$$

where the minus sign indicates that this an SO(2, 1)-like Lie algebra and not an SO(3)-like Lie algebra.

This dynamic SO(2, 1) symmetry can be extended to three non-interacting particles in a harmonic trap in several ways and it is related to the separability of radial coordinates (which have a length scale) and angular coordinates (which do not) [30]. This dynamic symmetry-induced separability can also be extended to interacting Hamiltonians that separate in Jacobi cylindrical coordinates like harmonic trap/harmonic interaction Hamiltonian (7), the Calogero–Moser Hamiltonian (8), and the unitary limit of the contact interaction Hamiltonian (12).

For an example of a different kind of dynamic symmetry relevant for few-body systems called state permutation symmetry [8], see [17, 18].

5. Symmetry and Structures

There is a natural tensor product structure for systems with three particles with spin in one dimension:

$$\mathcal{H} = \mathcal{H}_1 \otimes \mathcal{H}_2 \otimes \mathcal{H}_3. \tag{15}$$

This tensor product structure can be induced by the one-particle subalgebras of observables. Further, since the one-particle spin operators commute with the spatial operators, they form complete, commuting subalgebras and each particle's Hilbert space can be factored into a spatial part $\mathcal{K}_i$ and a spin part $\mathcal{S}_i$ like

$$\mathcal{H}_i = \mathcal{K}_i \otimes \mathcal{S}_i \sim \mathrm{L}^2(\mathbb{R}) \otimes \mathbb{C}^J, \tag{16}$$

where J is the number of spin (or internal) discrete levels for each particle.

This chapter only considers spin-independent Hamiltonians, so the spaces $\mathcal{S}_i$ are only relevant for symmetrization of identical particles. Therefore, the six-fold tensor product structure implied by the composition of (15) and (16) can be repartitioned into the dynamically-invariant tensor product structure

$$\mathcal{H} = \mathcal{K} \otimes \mathcal{S} \sim \mathrm{L}^2(\mathbb{R}^3) \otimes \mathbb{C}^{J^3}, \tag{17}$$

where

$$\mathcal{K} = \bigotimes_i \mathcal{K}_i \quad \text{and} \quad \mathcal{S} = \bigotimes_i \mathcal{S}_i.$$

The one-parameter group of time translation T_t with elements $t \in \mathbb{R}$ is represented by the unitary operator $\hat{U}(t)$ on $\mathcal{H}$ which factors on the tensor product (17)

$$\hat{U}(t) = \hat{U}_{\mathcal{K}}(t) \otimes \hat{\mathbb{I}},$$

i.e. it is local with respect to the structure (17). One way to think about this is that the spin-independent Hamiltonian $\hat{H}$ has $\mathrm{U}(J^3)$ symmetry since it commutes with any unitary operator acting only on $\mathcal{S}$. The dynamical-invariance of the tensor product structure (17) means that the amount of cntanglement between the spatial degrees of freedom and spin degrees of freedom is an invariant in time. For example, any pure initial state $|\Psi(0)\rangle$ can be decomposed like

$$|\Psi(0)\rangle = \sum_j \lambda_j |\kappa_j(0)\rangle \otimes |\sigma_j\rangle \tag{18}$$

with orthogonal vectors $|\kappa_j(0)\rangle \in \mathcal{K}$ and $|\sigma_j\rangle \in \mathcal{S}$. Although the spatial states $|\kappa_j(t)\rangle$ evolve in time, the coefficients λ_j that measure the entanglement between spin and spatial degrees of freedom do not.

The connections among the spin-invariance of the Hamiltonian, commuting subalgebras of observables, the induced "natural" tensor product structure, and the consequences for entanglement dynamics are reasonably straightforward. The rest of this section develops some more formal notions of how symmetry representations induce quantum structures and then gives a few examples.

5.1. Irreducible representation decompositions

Symmetries induce quantum structures through their representation on the Hilbert space. Denote a kinematic symmetry group by G. In the chapter, only finite and compact groups appear as kinematic groups, so all irreducible representations (irreps) of G can be realized by unitary finite-dimensional matrices acting on an irrep space $\mathcal{M}^\mu$. Denote these irreps by D^μ where μ labels the irrep. The dimension of the irrep is $d(\mu)$ and the irrep space is $\mathcal{M}^\mu \sim \mathbb{C}^{d(\mu)}$. The Hilbert space can be decomposed into subspaces called irrep towers $\mathcal{H}^\mu$

$$\mathcal{H} = \bigoplus_{\{\mu\}} \mathcal{H}^\mu. \tag{19}$$

The set of irreps $\{\mu\}$ for a finite group is finite, and for a general compact group it is countable. Each irrep tower $\mathcal{H}^\mu$ can be decomposed into a direct sum of equivalent irrep spaces $\mathcal{M}_i^\mu \sim \mathbb{C}^{d(\mu)}$ like

$$\mathcal{H}^\mu = \bigoplus_i \mathcal{M}_i^\mu. \tag{20}$$

This decomposition of the tower $\mathcal{H}^\mu$ is formally isomorphic to the direct product of a single irrep space $\mathcal{M}^\mu$ with a space $\mathcal{A}^\mu$ that represents the degeneracy

$$\bigoplus_i \mathcal{M}_i^\mu = \mathcal{M}^\mu \otimes \mathcal{A}^\mu. \tag{21}$$

The irrep degeneracy space $\mathcal{A}^\mu$ can be finite or infinite dimensional. The tensor product decomposition (21) is "natural" when the space $\mathcal{A}^\mu$ is diagonalized by an observable $\hat{A}_\mu$ (or set of observables) that is defined on the tower $\mathcal{H}^\mu$, that commutes with G, and that distinguishes different appearances of the same irrep $\mathcal{M}^\mu$. Such an observable $\hat{A}_\mu$ induces a tensor product structure that "separates" the irrep tower $\mathcal{H}^\mu$.

Better from the point of view of "natural" separability is when the same observable $\hat{A}$ is defined over all $\mathcal{H}$ and it separates all the towers. Then the space $\mathcal{A}$ is the same for all towers and can be factored out of the decomposition like

$$\mathcal{H} = \left(\bigoplus_{\{\mu\}} \mathcal{M}^\mu \right) \otimes \mathcal{A}. \tag{22}$$

Depending on the operator (or set of operators) $\hat{A}$, the space $\mathcal{A}$ also may carry a representation of another symmetry group, say G$'$. If so, then the process can be repeated: $\mathcal{A}$ can be decomposed into G$'$-towers, perhaps inducing another separable factor.

5.2. Example of symmetry-induced structure I: Time translation

Let us bring this back to the case of three trapped particles. The most important symmetry group has already been mentioned, time-translation symmetry T_t. The representation on the Hilbert space is of course just $\hat{U}(t) = \exp(-i\hat{H}t/\hbar)$. The group $\mathrm{T}_t \sim \mathbb{R}$ is abelian and has one-dimensional irreps $\mathcal{M}^E$ characterized by a real number, the energy E. The Hilbert space is therefore decomposable into irrep towers

$$\mathcal{H} = \bigoplus_{E\in\sigma} \mathcal{H}^E, \tag{23}$$

where σ is the spectrum of eigenvalues of $\hat{H}$ on $\mathcal{H}$. If there were no other symmetries, then the space $\mathcal{H}^E$ is one-dimensional and isomorphic to the T_t-irrep $\mathcal{H}^E \sim \mathcal{M}^E$. More generally, each energy level has a degeneracy $d(E)$ and is decomposed as

$$\mathcal{H}^E = \bigoplus_{i=1}^{d(E)} \mathcal{M}^E = \mathcal{A}^E \sim \mathbb{C}^{d(E)}.$$

Note that any operator that commutes with the Hamiltonian decomposes into a block diagonal form, acting only on the spaces $\mathcal{A}^E$. The goal of spectroscopy (from the limited perspective of a group theorist) is therefore to find the symmetry group G of the Hamiltonian and its representation on the whole Hilbert space $\mathcal{H}$ such that every degenerate energy subspace $\mathcal{A}^E$ is an irrep space $\mathcal{M}^\mu$ of G. When that is possible, there is a unique association from every $E \in \sigma$ onto some irrep μ of the group G. Then $\mathcal{H}$ has the alternate decomposition

$$\mathcal{H} = \bigoplus_{\{\mu\}} \mathcal{H}^\mu \tag{24a}$$

and

$$\mathcal{H}^\mu = \bigoplus_{E\in\sigma|_\mu} \mathcal{M}^\mu_E = \mathcal{M}^\mu \otimes \mathcal{A}^E, \tag{24b}$$

where the labels for the copies of $\mathcal{M}^{\mu}_{E}$ into which the tower $\mathcal{M}^{\mu}$ decomposes are in fact just the energies E in the spectrum $\sigma|_{\mu}$ restricted to only μ-types irreps. For three trapped particles, the spectrum $\sigma|_{\mu}$ is discrete, bounded from below, and countable, and $\mathcal{A}^{E} \sim \ell^{2}(E)$, the square-summable sequences. Whether this kind of partition of the Hilbert space is useful depends very much on the interplay between the spectrum of energies and spectrum of irreps of G.

Again, this decomposition (24) applies in the ideal case where the maximal symmetry group of $\hat{H}$ has been identified. If there are energy levels with degeneracies $d(E)$ that do not correspond to irreps dimensions $d(\mu)$, then there are two possibilities: (1) a global symmetry is missing from G, where by "global symmetry" I mean symmetry that is induced on the Hilbert space from a symmetry transformation defined on the whole configuration space or phase space; or (2) a truly "accidental" symmetry, i.e. a conspiracy of parameters that leads to certain energy levels lining up by "accident". The canonical example of this accidental symmetry would be the Pythagorean degeneracy of certain non-interacting three-particle energy levels when the confining potential is a perfect, infinite cube.

5.3. Example of symmetry-induced structure II: Particle permutations

To make this more concrete, at a minimum the symmetry group of the Hamiltonian must have the subgroup $P_3 \sim S_3$. This group has three irreps:

- The one-dimensional totally symmetric irrep is denoted [3] or $\yng(3)$. On $\mathcal{M}^{[3]}$, every element $p \in P_3$ is represented by 1.
- The one-dimensional totally antisymmetric irrep is denoted $[1^3]$ or $\yng(1,1,1)$. On $\mathcal{M}^{[1^3]}$, odd permutations are represented by -1 and even permutations by 1.
- The two-dimensional irrep with mixed symmetry is denoted [21] or $\yng(2,1)$. On $\mathcal{M}^{[21]}$, elements p can be represented by orthogonal 2×2 matrices.

Thus, the Hilbert space can be decomposed like

$$\mathcal{H} = \mathcal{H}^{[3]} \oplus \mathcal{H}^{[21]} \oplus \mathcal{H}^{[1^3]} \tag{25}$$

and the states of bosonic identical particles are elements of $\mathcal{H}^{[3]}$ with spectrum $\sigma|_{[3]}$ and fermionic particles are $\mathcal{H}^{[1^3]}$ with spectrum $\sigma|_{[1^3]}$. Consider the case when the particles have no spin or internal components ($J = 1$) and so $\mathcal{H} = \mathcal{K}$ and $\mathcal{S}$ is trivial. If there are no other symmetries, then each one of these spaces $\mathcal{H}^{\mu}$ is a tower of S_3 irreps $\mathcal{M}^{\mu}$ with different energies. For example, since $\mathcal{M}^{[3]}$ is trivial, the bosonic sector is

$$\mathcal{H}^{[3]} = \bigoplus_{E \in \sigma|_{[3]}} \mathcal{M}_E^{[3]}$$

if there are no other symmetries. In other words, $\mathcal{H}^{[3]}$ is isomorphic to the space square-summable series $\ell^2(E)$ on the index $E \in \sigma|_{[3]}$. If there are additional symmetries, then there may be degenerate bosonic level and the decomposition is

$$\mathcal{H}^{[3]} = \bigoplus_{E \in \sigma|_{[3]}} \mathcal{M}_E^{[3]} \otimes \mathcal{A}^E. \tag{26}$$

Then the goal is to find observables that diagonalize the finite-dimensional space $\mathcal{A}^{[3],E}$. Ideally, these spaces (and the observables that diagonalize them) are the same for every E so an $\dim(\mathcal{A})$-dimensional degeneracy space $\mathcal{A}^E \equiv \mathcal{A}$ factors out of the sum (26) so the Hilbert space can be realized by $\ell^2(E) \otimes \mathbb{C}^{\dim(\mathcal{A})}$. This seems to be possible whenever the three particle model is integrable, although a proof has not yet been found.

6. Conclusion

The reader who has made it this far deserves two confessions. First, obviously this research is still a work in progress. The goal of finding how symmetry selects preferred quantum structure has not been met and a unifying framework for identifying emergent collective observables has not been found. Nor has a novel practical protocol for manipulating robust and error-protected quantum information and entanglement been discovered. However, pursuing the relationships among integrability, solvability, and separability has revealed that not all degrees of freedom are created equal, and that quantum structures induced by some symmetries and observables are more natural than others.

Second, the reader deserves to know a "secret" motivation for this work. Each month, it seems another physicist adds their voice to the "it from bit" chorus, and I admit I find their songs compelling. Perhaps quantum information theory will simultaneously open doorways to the next generation of technology, provide new techniques to resolve difficult quantum problems in everything from superconductivity to nuclear structure, and resolve long-standing interpretational questions. That would certainly be exciting! But I still do not understand how to define information in quantum theory without conceding some sort of meta-theoretical status to a privileged observer. And so, this essay on three particles in one dimension is part of a continuing inquiry into how "nature" can select "natural" observables that turn data into information, and eventually, into physical meaning.

References

[1] Arsenijević, M., Jeknić-Dugić, J., Todorović, D. and Dugić, M. (2014). Entanglement relativity in the foundations of the open quantum systems theory. arXiv:1412.7829.

[2] Barnum, H., Knill, E., Ortiz, G. and Viola, L. (2003). Generalizations of entanglement based on coherent states and convex sets. *Physical Review A*, **68**: 032308; Barnum, H., Knill, E., Ortiz, G., Somma, R. and Viola, L. (2004). A subsystem-independent generalization of entanglement. *Physical Review Letters*, **92**: 107902.

[3] Braak, D. (2011). Integrability of the Rabi model. *Physical Review Letters*, **107**: 100401.

[4] Calogero, F. (1971). Solution of the one-dimensional n-body problems with quadratic and/or inversely quadratic pair potentials. *Journal of Mathematical Physics*, **12**: 419.

[5] Calogero, F. (2008). Calogero-Moser system. *Scholarpedia*, **3**(8): 7216.

[6] Caux, J.-S. and Mossel, J. (2011). Remarks on the notion of quantum integrability. *Journal of Statistical Mechanics*, P02023.

[7] Cazalilla, M. A., Citro, R., Giamarchi, T., Orignac, E. and Rigol, M. (2011). One dimensional bosons: From condensed matter systems to ultracold gases. *Reviews of Modern Physics*, **83**: 1405.

[8] Chen, J.-Q., Ping, J. J. and Wang, F. (2002). *Group Representation Theory for Physicists*, 2nd Ed., Chap. 3, (World Scientific, Singapore).

[9] Chencinier, A. (2007). Three body problem. *Scholarpedia*, **2**(10): 2111.

[10] Evans, N. W. (1990). Superintegrability in classical mechanics. *Physical Review A*, **41**: 5666.

[11] Eisenhart, L. P. (1948). Enumeration of potentials for which one-particle Schroedinger equations are separable. *Physical Review*, **74**: 87.

[12] Girardeau, M. (1960). Relationship between systems of impenetrable bosons and fermions in one dimension. *Journal of Mathematical Physics*, **1**: 516–523.

[13] Guan, X.-W., Batchelor, M. T. and Lee, C. (2013). Fermi gases in one dimension: From Bethe Ansatz to experiments. *Reviews of Modern Physics*, **85**: 1633.

[14] Gutzwiller, M. (1990). *Classical and Quantum Chaos*, Springer, Berlin.

[15] Harshman, N. L. and Ranade, K. S. (2011). Observables can be tailored to change the entanglement of any pure state. *Physical Review A*, **84**: 012303.

[16] Harshman, N. L. (2012). Observables and entanglement in the two-body system. *AIP Conference Proceedings*, **1508**: 386.

[17] Harshman, N. L. (2016). One-dimensional traps, two-body interactions, few-body symmetries: I. One, two, and three particles. *Few-Body Systems*, **57**: 11.

[18] Harshman, N. L. (2016). One-dimensional traps, two-body interactions, few-body symmetries: II. N Particles. *Few-Body Systems*, **57**: 45.

[19] Jauch, J. M. and Hill, E. L. (1940). On the problem of degeneracy in quantum mechanics. *Physical Review*, **57**: 641.

[20] Jeknić-Dugić, J., Arsenijević, M. and Dugić, M. (2013). *Quantum Structures: A View of the Quantum World*, Lambert Academic Publishing.

[21] Kalnins, E. G., Williams, G. C., Miller Jr. W. and Pogosyan, G. S. (1999). Superintegrability in three-dimensional Euclidean space. *Journal of Mathematical Physics*, **40**: 708.

[22] Larson, J. (2013). Integrability versus quantum thermalization. *Journal of Physics B*, **46**: 224016.

[23] Louck, J. D. (1965). Group theory of harmonic oscillators in n-dimensional space. *Journal of Mathematical Physics*, **6**: 1786.

[24] Morse, P. M. and Feshbach, H. (1958). *Methods of Theoretical Physics*, Part 1, Chap. 8, McGraw-Hill, New York.

[25] Olshanii, M. (1998). Atomic scattering in the presence of an external confinement and a gas of impenetrable bosons. *Physical Review Letters*, **81**: 938.

[26] Oelkers, N., Batchelor, M. T., Bortz, M. and Guan, X.-W. (2006). Bethe ansatz study of one-dimensional Bose and Fermi gases with periodic and hard wall boundary conditions. *Journal of Physics A*, **39**: 1073.

[27] Pitaevskii, L. P. and Rosch, A. (1997). Breathing modes and hidden symmetry of trapped atoms in two dimensions. *Physical Review A*, **55**: R853.

[28] Šuvakov, M. and Dmitrašinović, V. (2013). Three classes of Newtonian three-body planar periodic orbits. *Physics Review Letters*, **110**: 114301.

[29] Viola, L., Fortunato, E. M., Pravia, M. A., Knill, E., Laflamme, R. and Cory, D. G. (2001). Experimental realization of noiseless subsystems for quantum information processing. *Science*, **293**: 2059.

[30] Werner, F. and Castin, Y. (2006). Unitary gas in an isotropic harmonic trap: Symmetry properties and applications. *Physical Review A*, **74**: 053604.

[31] Weigert, S. (1992). The problem of quantum integrability. *Physica D*, **56**: 107.

[32] Zanardi, P. (2001). Virtual quantum subsystems. *Physical Review Letters*, **87**: 077901.

[33] Zanardi, P., Lidar, D. A. and Lloyd, S. (2004). Quantum tensor product structures are observable induced. *Physical Review Letters*, **92**: 060402.

Chapter 12

Quantum to Classical Transitions via Weak Measurements and Post-Selection

Eliahu Cohen

H.H. Wills Physics Laboratory, University of Bristol, Tyndall Avenue, Bristol, BS8 1TL, UK

Yakir Aharonov

School of Physics and Astronomy, Tel Aviv University, Tel Aviv 6997801, Israel

Schmid College of Science, Chapman University, Orange, CA 92866, USA

Alongside its immense empirical success, the quantum mechanical account of physical systems imposes a myriad of divergences from our thoroughly ingrained classical ways of thinking. These divergences, while striking, would have been acceptable if only a continuous transition to the classical domain was at hand. Strangely, this is not quite the case. The difficulties involved in reconciling the quantum with the classical have given rise to different interpretations, each with its own shortcomings. Traditionally, the two domains are sewed together by invoking an *ad hoc* theory of measurement, which has been incorporated in the axiomatic foundations of quantum theory. This work will incorporate a few related tools for addressing the above conceptual difficulties: deterministic operators, weak measurements, and post-selection. Weak Measurement, based on a very weak von Neumann coupling, is a unique kind of quantum measurement with numerous theoretical and practical applications. In contrast to other measurement techniques, it allows to gather a small amount of information regarding the quantum system, with only a negligible probability of collapsing it onto an eigenstate of the measured observable. A single weak measurement yields an almost random outcome, but when performed repeatedly over a large ensemble, the averaged outcome becomes increasingly robust and accurate. Importantly, a long sequence of weak measurements can be thought of as a single projective measurement. We claim in this work that classical variables appearing in the macro-world, such as center of mass, moment of inertia, pressure, and average forces, result from a multitude of quantum weak measurements performed in the micro-world. Here again, the quantum outcomes are highly uncertain, but the law of large numbers obliges their convergence to the definite quantities we know from our everyday lives. By augmenting this description with a final boundary condition and employing the notion of "classical robustness under time-reversal", we will draw a quantitative borderline between the classical and quantum regimes. We will conclude by analyzing the role of macroscopic systems in amplifying and recording quantum outcomes.

1. Introduction

The vague border between the classical and quantum realms gives rise to the well-known measurement problem. The problem is best understood by considering the unique properties of the quantum state space, which is boosted in size compared to the classical phase space, in order to accommodate distinctly non-classical entangled states and states of superposition. The former entails nonlocal correlations; the latter, stemming from the mutual incompatibility of conjugate observables, implies that the quantum reality cannot be accounted for in classical terms of definite physical properties.

However, a superposition is never observed directly — a measurement will yield one definite value of a physical property even when the state is not an eigenstate of the measured observable (recall Schrödinger's famous cat, which is always found to be either dead or alive, but not both). In such cases, nothing in the quantum description dictates the exact result of a measurement. Textbook QM supplements the unitary evolution of Schrödinger equation (SE) with a second dynamical law, which spells a non-unitary break in the evolution upon measurement, a collapse, instantaneously changing the state of the system to an eigenstate of the measured observable. Accordingly, the result will only be determined probabilistically, where the probability is given by the square amplitude of the eigenstate term, a postulate known as the Born rule. This is in stark contrast to classical mechanics, which only exhibits probabilities stemming from ignorance about the exact phase space state of the system, while remaining fully deterministic and local at the fundamental level of the physical laws.

While the collapse postulate makes QM perfectly operational, it introduces ambiguity into the theory. Given that any macroscopic object is just an aggregate of microscopic objects, as suggested by the lack of criterion for otherwise distinguishing them, it is not clear why the SE should not suffice for the full dynamical description of any process in nature. And if it were all encompassing, QM would have been deterministic, and not probabilistic, as the collapse and the Born rule maintain. Attempts to give satisfactory explanations to this predicament lead to discussions about the completeness of the quantum description, and different interpretations. The different approaches range from collapse theories such as the Ghirardi–Rimini–Weber (GRW) Spontaneous Localization Model [19] and successors thereof, through deterministic variable theories such as Bohmian mechanics [15], the objective general-relativistic collapse suggested by Penrose [23], and all the way to the relative state interpretation by Everett [18], which assumes nothing other than the standard axioms. In the latter, the different branches of a superposition are said to represent different co-existing states of reality, where the observation of a certain outcome is attributed to the specific state of the observer that is correlated to it in the superposition. Each of the superposition terms constitutes a "branching world", and is part of a Universal deterministically evolving wave function. Hence, it is also known as the "Many-Worlds Interpretation" (MWI) [32].

The novel approach we shall present tries to tackle the difficulties without resorting to the usual notion of collapse. It will inherit the advantages of the MWI without assuming multiple realities. This approach suggests that a complete description of the physical state has to include two state-vectors, forming the "two-state". The states evolve independently by the same unitary dynamical law (and same Hamiltonian), but in opposite temporal directions (where the forward direction is defined according to the direction of entropy increase in the observed Universe). Whenever macroscopic objects, i.e. many-particle systems, are entangled with a microscopic system, as in a measurement, the setup/environment selects a preferred basis, while the unknown backward-evolving state selects a definite outcome from the known forward-evolving state, giving rise to a single definite physical reality. Thus, the probabilistic nature of quantum events can be thought of as stemming from our ignorance of the backward-evolving state, reintroducing the classical concept of probability as a measure of knowledge.

The decoherence program [26, 36] has been successful in reducing, locally, the unobserved coherent superposition of macrostates into a mixture of effectively classical states, pointer states. The damping of the interference terms in the pointer states basis is attributed to the near orthogonally of environmental states entangled with them. By tracing out the environmental degrees of freedom, one may unveil the mixed state in which the system and apparatus are given. However, the trace operation is a purely mathematical procedure, which indicates no reduction of the global state to a single definite measurement outcome. Using the backward-evolving state of the two-state vector formalism (TSVF), we shall demonstrate how a selection of a single outcome may be achieved.

We will focus our attention on the boundary conditions posed in each realm. In classical mechanics, initial conditions of position and velocity for every particle fully determine the time evolution of the system. Therefore, trying to impose a final condition would either lead to redundancy or inconsistency with the initial conditions. This situation is markedly different in the realm of quantum mechanics. Because of the uncertainty principle, an initial state-vector does not determine, in general, the outcome of a future measurement. However, adding another constraint, namely, the final (backward-evolving) state-vector, results in a more complete description of the quantum system in between these two boundary conditions, that has bearings on the determination of measurement outcomes. The usefulness of the backward-evolving state-vector was demonstrated in the works of Aharonov *et al.*

The emergence of specific macrostates seems non-unitary from a local perspective, and constitutes an effective "collapse", a term which will be used here to denote macroscopic amplification of microscopic events, complemented by a reduction via the final state. We will show that a specific final state can be assigned so as to enable macroscopic time-reversal or "classical robustness under time-reversal", i.e. reconstruction of macroscopic events in a single branch, even though "collapses" have occurred. An essential ingredient in understanding the quantum-to-classical transition is the robustness of the macrostates comprising the measuring apparatus, which serves to amplify the microstate of the measured system and communicate it to the observer. The robustness guarantees that the result of the measurement is insensitive to further interactions with the environment. Indeed, microscopic time-reversal within a single branch is an impossible task because evolution was not unitary. Macroscopic time-reversal, which is the one related to our everyday experience, is possible, although nontrivial.

A measurement generally yields a new outcome state of the quantum system and the measuring device. This state may be treated as an effective boundary condition for both future, and past events. We suggest it is not the case that a new boundary condition is independently generated at each measurement event by some unclear mechanism. Rather, the final boundary condition of the Universe includes the appropriate final boundary conditions for the measuring devices which would evolve backward in time to select a specific measurement outcome. In the following sections, we shall demonstrate how this boundary condition arises at the time of measurement due to a two-time decoherence effect. Indeed, we will see that in the pointer basis (determined by decoherence), the outcome of the measurement can only be the single classical state corresponding to the final boundary condition. We thus suggest a particular final boundary condition for the Universe, in which each classical system (measuring device) has, at the time of measurement, a final boundary condition equal to one of its possible classical states (evolved to the final time).

A further requirement is that the final state in the pointer basis will induce, backwards in time, an appropriate distribution of outcomes so as to recover the empirical quantum mechanical probabilities for large ensembles, given by the Born rule. The determination of the measurement statistics by the correspondence between the two states may lead one to conclude that within the framework of TSVF, the Born rule is a coincidental state of affairs rather than a law of nature. That is, that the Born rule is a product of an empirically verifiable, yet contingent, relation between the initial and final boundary conditions, one that needs to be postulated for the sake of deducing the rule. This specific relation is contingent in that it leaves open the possibility of a different relation which will lead to a modified version of the Born rule, while the rest of physics remains where it stands. This, however, is inaccurate. It can be shown that this specific law follows, in the infinite N limit, from the compatibility of quantum mechanics with classical-like properties of macroscopic objects [3]. Under the assumption that for macroscopically large samples, the results of physical experiments are stable against small perturbations, a final state pertaining to the Born rule is the most likely final state, for any ensemble.

The work is structured as follows: Section 2 introduces von Neumann's measurement scheme. It will be used for performing strong (projective), as well as weak measurements throughout the work. The description of quantum reality is then augmented with a final boundary state (Section 3) and deterministic operators (Section 4). In Section 5, averages taken over large ensembles are shown to be deterministic operators representing the cumulative result of a set of weak measurements. Section 6, which is the heart of the work, presents the two-time decoherence scheme and the micro–macro quantitative boundary required for attaining time-reversal symmetry. Section 7 concludes the work.

2. von Neumann Interaction — Projective and Weak

Throughout this work, the von Neumann scheme [33] has a key role in realizing quantum measurements. It was traditionally used for describing projective measurements and the decoherence process [16, 25, 35]. We shall briefly review these arguments and then discuss an important limiting case where the coupling between the measured system and measuring device is much smaller than the quantum uncertainty of the latter. This case is known as *weak measurement.*

Let S denote our system to be measured, prepared at some state $|\psi\rangle$. Suppose A is the Hermitian operator we wish to measure on the system S, having n eigenvectors $|a_i\rangle$ such that $A|a_i\rangle = a_i|a_i\rangle$. When expressed in the eigenbasis of A, the system's wave function takes the form:

$$|\psi\rangle = \sum_i \alpha_i|x_i\rangle. \tag{1}$$

Let $|\phi_d\rangle$ denote the wave function of the measurement device (also called a "pointer"). When represented in the position basis, it will be written as:

$$|\phi\rangle = |\phi_d\rangle = \int_q \phi(q)|q\rangle dq, \tag{2}$$

where q is the position variable of the measuring pointer. Let Q_d be the position operator such that $Q_d|q\rangle = q|q\rangle$ (here, we use Q_d to distinguish the operator Q_d from its eigenvector $|q\rangle$ and eigenvalue q, the subscript d is used for denoting the measuring device). It is assumed that initially $\phi(q)$ behaves normally around 0 with some variance σ^2:

$$\phi(q) = (2\pi\sigma^2)^{-\frac{1}{4}}e^{-q^2/4\sigma^2}. \tag{3}$$

The measuring device $|\phi_d\rangle$ is later examined and the shift in the pointer's position is measured.
Consider the interaction Hamiltonian H_{int} ([10, 28]):

$$H = H_{\text{int}} = g(t)A \otimes P_d. \tag{4}$$

Here $g(t)$ is a coupling impulse function satisfying:

$$\int_0^T g(t)dt = g, \tag{5}$$

where T is the coupling time, g is the coupling strength and P_d is the operator conjugate to Q_d such that $[Q_d, P_d] = i\hbar$.

We shall start the measurement process with the vector:

$$|\psi\rangle \otimes |\phi(q)\rangle, \tag{6}$$

in the product space of the two systems. Then we apply the following time evolution based on the interaction Hamiltonian above

$$e^{-i\int Hdt/\hbar}|\psi\rangle \otimes |\phi(q)\rangle. \tag{7}$$

It is easy to see that on each of the vectors $|a_i\rangle \otimes |\phi(q)\rangle$ the Hamiltonian H takes Q_d to $Q_d + ga_i$, (Heisenberg evolution):

$$Q_d(T) - Q_d(0) = \int_0^T dt \frac{\partial Q_d}{\partial t} = \int_0^T \frac{i}{\hbar}[H, Q_d]dt = ga_i. \tag{8}$$

The corresponding transformation of the coordinates of the wave function is:

$$e^{-i\int Hdt/\hbar}|\psi\rangle \otimes |\phi\rangle = \sum_i \alpha_i |a_i\rangle \otimes |\phi(q - ga_i)\rangle. \tag{9}$$

In case $ga_i \gg \sigma$ (i.e. the coupling strength is much larger than the pointer's uncertainty), then $|\phi(q - ga_i)\rangle$ and $|\phi(q - ga_j)\rangle$ are almost orthogonal for $i \neq j$. Hence, the different possible measurement outcomes are projected into distinct states of the measuring device. By this, the first stage of the measurement (also known as pre-measurement) is over. In experiments, however, we do not observe our measured system entangled with various pointer states — we see only one outcome. Therefore, a second, non-unitary amplification stage is needed. Once amplified due to coupling with the environment (which is considered within this work to be part of the macroscopic measurement device), we would see only one eigenstate out of the initial superposition. This reflects the collapse on a single measurement outcome a_i.

Alternatively, when $ga_i \ll \sigma$, the functions $|\phi(q - ga_i)\rangle$ are highly overlapping and hence the measurement result is inconclusive. In this case, known as a weak measurement, the von Neumann interaction does not end in a collapse onto an eigenstate of the measured observable, nor in an unambiguous information about the measured system. We gather a minute amount of information regarding the system at the price of slightly changing the measured state. When repeating the weak measurements over a large ensemble containing N particles, the outcomes accumulate like independent identically distributed normal variables, hence the relative error drops like $\frac{1}{\sqrt{N}}$.

2.1. Weak measurements

Originally developed to test the prediction of the TSVF, Weak measurement [6] has already been proven to be very helpful in several experimental tasks [21, 22, 27, 34], as well as in revealing fundamental concepts [8, 9, 11, 14, 30, 31] (which are admittedly under controversy now). Tasks traditionally believed to be self-contradictory by nature such as determining a particle's state *between* two measurements prove to be perfectly possible with the aid of this technique. Within the framework of the TSVF, weak measurements reveal new and sometime puzzling phenomena. For a general discussion on weak measurements see [10, 17, 28].

3. The Final State

We would like to add now an important ingredient to our analysis — a final boundary state. The idea that a complete description of a quantum system at a given time must take into account two boundary conditions rather than one is known from the TSVF. The TSVF is a time-symmetric formulation of standard quantum mechanics, which posits, in addition to the usual state vector, a second state vector evolving from the future towards the past. This approach has its roots in the works of Aharonov, Bergman, and Lebowitz [7], but it has since been extensively developed [5, 24], and has led to the discovery of numerous peculiar phenomena [3].

The TSVF provides an extremely useful platform for analyzing experiments involving pre- and post-selected ensembles. Post-selection is permitted in quantum mechanics due to the effective indeterminacy of measurement, which entails that the state of a system at one time and its Hamiltonian only partially determine measurement outcomes at later times. *Weak measurements* enable us to explore the state of the system at intermediate times without disturbing it. The two-state $\langle\phi|\ |\psi\rangle$ created by both boundary conditions allows to define for any operator the *weak value*

$$\langle A\rangle_w = \frac{\langle\phi|A|\psi\rangle}{\langle\phi|\psi\rangle}. \tag{10}$$

This weak value naturally appears as the pointer's shift when we perform a weak measurement on a pre-/post-selected ensemble $\langle\psi_f|\ |\psi_i\rangle$

$$\begin{aligned}\langle\psi_f|e^{-i\int Hdt/\hbar}|\psi_i\rangle\otimes|\phi\rangle &\approx \langle\psi_f|1-igA\otimes P_d|\psi_i\rangle\otimes|\phi\rangle\\ &= \langle\psi_f|\psi_i\rangle\,(1-ig\langle A\rangle_w P_d)\,|\phi\rangle\\ &\approx \langle\psi_f|\psi_i\rangle e^{-ig\langle A\rangle_w P_d}|\phi\rangle\\ &= \langle\psi_f|\psi_i\rangle|\phi(q-g\langle A\rangle_w P_d)\rangle. \end{aligned} \tag{11}$$

The power to explore the pre- and post-selected system by employing weak measurements motivates a literal reading of the formalism, i.e. as more than just a mathematical analytic tool. It motivates a view according to which future and past are equally important in determining the quantum state at intermediate times, and hence equally *real.* Accordingly, in order to fully specify a system, one should not only pre-select, but also post-select a certain state using a projective measurement.

4. Deterministic Operators

Describing quantum mechanics through operators within the Heisenberg representation allows to identify *deterministic* operators. We find these operators vital for understanding the "classical-quantum interplay". Although operating on *quantum* states, the outcome is deterministic, as if we have performed a classical experiment with *classical* observables. Identifying the set of all deterministic operators with respect to given state, amounts to providing a complete description of the quantum system (equivalent to that of Schrödinger) yet somewhat more compact and predictable [13].

In the Schrödinger representation, a quantum system is fully described by vector in a Hilbert space. Its time evolution is dictated by the Hamiltonian and calculated according to the SE. Observables are usually described by time-independent operators. In the Heisenberg representation however, a physical system can be described by a closed set (under addition and multiplication) of *deterministic operators*, evolving in time according to the Heisenberg equation, whereas the state does not change in time. Deterministic operators are Hermitian "eigenoperators", i.e. Hermitian operators for which the system's state is an eigenstate:

$$\{A_i \text{ such that } A_i|\psi\rangle = a_i|\psi\rangle, a_i \in \Re\}. \tag{12}$$

It is easy to show that for describing a particle in an n-dimensional Hilbert space, a set of $(n-1)^2+1$ deterministic operators, whose eigenvectors span the relevant sub-space, is required [29]. The physical significance of these operators stems from the possibility to measure all of them without disturbing the particle, i.e. without inducing collapses. Therefore, they can also be measured successively without mutual disturbance

$$[A_i, A_j]|\psi\rangle = 0, \tag{13}$$

for any i, j.

The mathematical equivalence between the Schrödinger and Heisenberg representations assures that there is a one-to-one correspondence between the wave function and the set of deterministic operators describing the same physical system. However, the wave function also expresses non-deterministic properties, such as positions in a delocalized system. We do not consider these to be real properties of the single particle, maintaining that they are exhausted by the set of deterministic properties. If, for instance, the position operator is not deterministic, the question "where is the particle?" bears no meaning. The non-deterministic operators do not represent properties intrinsic to the particle. They only reflect probabilistic properties at the ensemble level. It should be noted though, that an ensemble possesses a set of deterministic operators larger than that of the single particle. Importantly, as we shall see in the next section, the average value of any one-particle operator is deterministic. The wave function itself is another deterministic operator of an ensemble of particles [13].

Within a time-symmetric formalism adding a final state amounts to adding a second set of deterministic operators on top of the one dictated by the initial state, thereby enlarging the assortment of system properties. The properties expressed by this two-fold set are the ones which we believe to constitute the primitive ontology of quantum mechanics.

5. Averages as Deterministic Operators

We claim in this section that most classical experiments we perform are based on weak measurements. In fact, any physical quantity averaged over a large ensemble of particles can be understood as a result of weak coupling to each of them. This follows from the following statement: *The state of a large ensemble is nearly an eigenstate of the "average" operator* [4]. In this sense, averages are *deterministic* operators, thus having a classical nature.

The proof is as follows. We shall utilize the well-known relation

$$A|\Psi\rangle = \bar{A}|\Psi\rangle + \Delta A|\Psi_\perp\rangle, \tag{14}$$

where $\Delta A = [(\bar{A^2}) - (\bar{A})^2]^{1/2}$ and $|\Psi_\perp\rangle$ is some state orthogonal to $|\Psi\rangle$. For an ensemble of identical particles, we thus obtain:

$$\begin{aligned}\frac{1}{N}\sum_{i=1}^{N} A_i \prod_{i=1}^{N} |\Psi\rangle_i &= \frac{1}{N}\sum_{i=1}^{N}(\bar{A}|\Psi\rangle_i + \Delta A|\Psi_\perp\rangle_i)\prod_{k\neq i}^{N}|\Psi\rangle_k \\ &= \bar{A}\prod_{i=1}^{N}|\Psi\rangle_i + \frac{\Delta A}{N}\sum_{i=1}^{N}|\Psi_\perp\rangle_i \prod_{k\neq i}^{N}|\Psi\rangle_k,\end{aligned} \tag{15}$$

where A_i is the observable of interest applied to the i-th particle. Due to mutual orthogonality, the last term's norm is $O(\frac{1}{\sqrt{N}})$ and hence can be neglected. Therefore, the product of N identical states is an eigenstate of any average operator at the limit of $N \to \infty$.

Moreover, the procedure of obtaining it is a weak measurement based on the interaction Hamiltonian $H = g(t)\frac{1}{N}\sum_{i=1}^{N} A_i$ creating a very weak coupling (scales like the inverse of the ensemble's size) to each particle.

A few generalizations of the above procedure for obtaining the average value can simply follow:

(i) If instead of a product of N identical states, we have a slight fluctuations $|\delta\psi\rangle_i$ in each single particle state, then as long as these fluctuations are random, we would be able to neglect them at the limit of large N.

(ii) If rather than a product of N identical states, we have N_1 particles prepared in some state $|\chi\rangle_1$, N_2 particles prepared in $|\chi\rangle_2, \ldots, N_m$ particles prepared in $|\chi\rangle_m$, such that $\sum_{i=1}^{m} N_m = N$ and $\forall m\ N_m \gg 1$ the proof can be repeated when again the uncertainty term is negligible. The average would now be $\bar{A} = \frac{\sum_{i=1}^{m} N_i \bar{A}_i}{N}$, an eigenstate of A, achieved through weak measurements.

(iii) If instead of a single pointer we have a sequence of k pointers, where $N/k \gg 1$ we would still be able to utilize the law of large numbers for each of them and arrive at the previous result.

Another classical property of the average operators is their almost exact commutativity [2]. If, for instance, we have an ensemble of spins prepared in the same state then the average spin along the x-direction is defined by $\bar{S_x} = \frac{\hbar}{2}\frac{\sum_{i=1}^{N}\sigma_x^i}{N}$, and

$$\lim_{N\to\infty}[S_x, S_y] = i\hbar \lim_{N\to\infty}\frac{\bar{S_z}}{N} = 0, \tag{16}$$

implying commutativity of the average spin operators along the x and y directions (recall the notion of deterministic operators presented in the previous section).

We are then led to the conclusion that our everyday classical experience, which is based on averaged properties of large ensembles, may essentially originate from the theory of weak measurements. Any macroscopic event, perceived by us as deterministic, can be decomposed into a large number of weak measurements, each of which is uncertain.

6. Time-Reversal Symmetry and the Macroscopic Threshold

We shall see now that a post-selected state of the universe may account for the apparent collapses we see in nature. Furthermore, and strictly related to the above discussion, it will be shown that a time-reversal symmetry sets a lower bound on the number of particles comprising a macroscopic object.

Assume that the world is symmetrically described by an initial, as well as a final, fine-tuned boundary condition. We further assume the existence of a thermodynamic arrow of time set in the direction of entropy increase, hence the final state will characterize a highly entangled world with high entropy. Macroscopic objects (to be quantitatively defined below) are comprised of at least $N \gg 1$ microscopic elements, that can be coupled to other, external, microscopic elements. As discussed in Section 2, quantum measurements will take the form of the von Neumann scheme (assumed in this section to be carried out using a qubit rather than a continuous pointer) and will be followed by a macroscopic amplification (i.e. a macroscopic record of the microscopic result encoded in the state of at least N microscopic particles, or "environment" in the language of decoherence). As a result of the measurement, a "collapse" of the measured microscopic degrees of freedom might seem to occur from the experimentalist's point of view. This apparent collapse, resulting from a partial overlap with a specific post-selected state, obeys time-reversal symmetry under several conditions.

To illustrate the suggested scheme, we shall discuss first the simplest case where only one measurement is performed and the macroscopic world (including measuring devices) does not collapse. In the next subsection, we will analyze the important case in which the distinction between microscopic and macroscopic world depends only on the number of microscopic elements comprising the objects in question. That is, part of the macroscopic measuring device will be assumed to "collapse", but nevertheless, our macroscopic world will be shown to maintain its robustness under time-reversal. For further details and generalizations, we refer the reader to [12].

6.1. A single ideal measurement

Let our system be initially described by

$$|\Psi(t_0)\rangle = (\alpha|1\rangle + \beta|2\rangle)|\text{READY}\rangle|\epsilon_0\rangle, \tag{17}$$

where $\alpha|1\rangle + \beta|2\rangle$ is the state of a microscopic particle, $|\text{READY}\rangle$ is the pointer state of the measuring device and $|\epsilon_0\rangle$ is the state of the environment (for simplicity of notation, Normalizations are omitted hereinafter). Following the von Neumann scheme, we create at time $t = t_1$ a coupling between the particle state and the pointer state, establishing a one-to-one correspondence between them. We will denote the orthogonal pointer states by "I" and "II". The pointer will shift to $|I\rangle$ in case the particle is in $|1\rangle$ and to $|II\rangle$ in case the particle is in $|2\rangle$:

$$|\Psi(t_1)\rangle = (\alpha|1\rangle|I\rangle + \beta|2\rangle|II\rangle)|\epsilon_0\rangle. \tag{18}$$

Then, in the course of a short time t_d, the preferred pointer state is selected and amplified by a multi-particle environment in the process of decoherence. Since the pointer state basis is favored by system-environment interactions, it is not prone to further entanglement and decoherence. Therefore, it enables us to read off the result of the measurement from the environment in which it is encoded in a unitary fashion. The reading of a specific result does not correspond to just one specific state of the apparatus/environment but rather to a subset of states taken from a very large state-space, where distinct readings correspond to orthogonal states. Physically, these may be spatially separated blotches on a photo-detector, or concentration of molecules in a corner of a chamber. We represent these distinct environmental subsets as ϵ_1 and ϵ_2, and the dynamical process is thus

$$|\Psi(t_1 + t_d)\rangle = \alpha|1\rangle|I\rangle|\epsilon_1\rangle + \beta|2\rangle|II\rangle|\epsilon_2\rangle. \tag{19}$$

This is a macroscopic amplification of the microscopic measurement, which results in what we call "measurement" of the particle. After this point, the particle may continue to interact with other objects (microscopic or macroscopic).

Now comes the crucial part. Let the backward-evolving state at $t = t_f$ contain only a single term out of the preferred pointer basis

$$\langle \Phi\left(t_f\right)| = \langle \phi|\langle I|\langle \epsilon_1|, \tag{20}$$

where $\langle \phi|$ is a final state of the microscopic particle. Within the TSVF, our system will be described by the two-state:

$$\langle \phi|\langle I|\langle \epsilon_1|\ (\alpha|1\rangle|I\rangle|\epsilon_1\rangle + \beta|2\rangle|II\rangle|\epsilon_2\rangle), \tag{21}$$

for $t_1 + t_d < t < t_2$. This is essentially a future choice of $|I\rangle$, which may serve as a reason for the initial outcome represented by the microscopic state $|1\rangle$. The approximate orthogonality of $|\epsilon_1\rangle$ and $|\epsilon_2\rangle$ assures that after reducing the density matrix to include only the observable degrees of freedom, within the interval $t_1 + t_d < t < t_2$, only the first term in Eq. 21 will contribute, accounting for the macroscopic result we witness. In the most general case, the backward environment-pointer state need not be exactly identical to the corresponding term in the forward-state, as long as the measure of its projection on it is exponentially (in the number of particles) larger than the measure of its projection on the non-corresponding term(s) of the forward state.

Regarding weak values (Eq. 10), if any were measured during the intermediate times, they would have been determined by the specific selection made by the final state. Moreover, any interaction with this pre- and post-selected ensemble would reflect this final state in the form of the "weak potential" [9].

The effective boundary condition for the past of the backward-evolving state determines the observed measurement outcome by a backward decoherence process. That is, just the same as the backward-state sets the boundary for the future of the forward-evolving state, the forward-evolving state sets the boundary for the past of the backward-state. Together with the regular decoherence, this amounts to a symmetric two-time decoherence process [1, 20], allowing for a generalization to multiple-time measurements. This subject is formalized in the next subsection, where we present a detailed description of two consecutive measurements.

The important conclusion we should bear in mind is related to the time-reversed process. Starting from the state $|\Psi(t_f)\rangle$ as described by Eq. 20 and going backwards in time, we are able to reconstruct the pointer reading $|I\rangle$ although the measured microscopic particle has changed its state. This relates to the concept of "macroscopic robustness under time-reversal" on which we elaborate within the next subsection.

We note that in cases where the free Hamiltonian is nonzero, we will have to apply the forward time-evolution operator on the final boundary condition, which would then cancel upon backward time-evolution to the present state. This clearly does not change the results, and therefore we preferred to discuss a zero Hamiltonian.

6.2. When macroscopic objects also collapse

We are now in position to address the issue of time reversibility of the dynamical equations governing the macrostates. While this property is most naturally present in the MWI, as long as the macroscopic objects stay intact, it may also exist in a single branch, even if its history includes non-unitary events. This is due to the fact that the result of a measurement performed on a microscopic state is stably stored within the macroscopic objects, as we have seen in the last chapter, and can theoretically be extracted. Therefore, while the measured microstate may change non-unitarily from our local perspective, our measurement reading may not. This possibility is what we will refer to as "classical robustness under time-reversal". As well as being a landmark of classical physics, time reversibility is vital in order to draw valid conclusions about the early Universe from our current observations.

Let the system not be completely isolated, and allow external quantum disturbances which interfere with the evolution in an indeterministic and thus irreversible way (from the single-branch perspective). It is generally accepted that the pointer states selected by the environment are immune to decoherence, and are naturally stable [26, 36]. Problems start when the (macroscopic) measurement devices begin to disintegrate to their microscopic constituents, which may couple to other macroscopic objects and effectively "collapse". These collapses seem fatal from the time-reversal perspective, as time-reversed evolution would obviously give rise to initial states very different from the original one. To tackle this, we demand that subtle environmental interactions, mildly altering the macrostate, will not stray too far from the subset of states indicating the perceived measurement result, compared to the orthogonal result.

Considering the free evolution of the measuring device and applying it backwards from the final and slightly altered state, the state at the time of measurement will still project heavily onto the same sub-space, indicating the same reading. This may be regarded as macroscopic physics having time-symmetric dynamics. While it might be the case that we do not reconstruct the starting microscopic configuration, being macroscopic objects, this should not upset us as long as our experience remains the same, i.e. as long as macroscopic readings, depending on the macrostate of their N micro-particles, do not change when backward evolution is applied. This will be shown to be the case when several assumptions are made regarding the macroscopic objects and the rate of collapse.

To derive the limit between microscopic and macroscopic regimes, we will assume now that the amplification mechanism consists of at least $N \gg 1$ particles belonging to the environment or measuring device, from which only $n \ll N$ particles may later be measured and collapsed without rendering the dynamics irreversible. By "measured", we do not necessarily mean that an observer entangled them with a device designated for measurement. Rather, we mean that they may get entangled with some other degrees of freedom and decohere. We believe that it is reasonable to assume that $n \ll N$ always, because measuring N (which is typically, 10^{23}) particles and recording their state is practically impossible. Equations 17–19 still have the same form, but the measurement of the environment at some $t = t_2$ leads to

$$|\epsilon_i(N)\rangle \quad \rightarrow \quad \prod_{j=1}^{N-n} |C_i^{(j)}\rangle|\epsilon_i(N-n)\rangle, \tag{22}$$

for $i = 1, 2$ representing encodings of two orthogonal microstates. The environment states are N-particles states at first, and later contain only $N - n$ particles, while their other n components "collapse", for simplicity of calculation, to the product state $\prod_{j=1}^{n} |C_i^{(j)}\rangle$. The trivial point, although essential, is that

$$\langle C_1^{(j)} | \epsilon_1^{(j)} \rangle = \gamma_1^{(j)} \neq 0, \tag{23}$$

for every $j = 1, 2, \ldots, n$, where $\epsilon_1^{(j)}$ is the j-th environment state before the collapse, i.e. collapse can never reach an orthogonal state. For later purposes, let us also assume

$$\langle C_2^{(j)} | \epsilon_1^{(j)} \rangle = \gamma_2^{(j)} \neq 0. \tag{24}$$

It is not necessarily different from 0, but as will be demonstrated below, this is the more interesting case. We would like to show that the final boundary state of Eq. (20) still has much higher probability to meet $|\epsilon_1\rangle$ than $|\epsilon_2\rangle$, and hence the pointer reading is determined again by the specific boundary condition, despite the collapse of the classical apparatus. Indeed, under the assumption of ending the evolution in the following final boundary condition:

$$\langle \Phi(t_f) | = \langle \phi | \langle I | \langle \epsilon_1 |, \tag{25}$$

we can define the "robustness ratio" as a ratio of probabilities: The probability to reach backwards in time the "right" state $|I\rangle$ divided by the probability to reach the ("wrong") $|II\rangle$ state. This ratio ranges from zero to infinity suggesting low (values smaller than 1) or high agreement (values greater than 1) with our classical experience in retrospect. In our case, it is

$$\begin{aligned} \frac{Pr(\text{Right})}{Pr(\text{Wrong})} &= \frac{\prod_{j=1}^{n} \gamma_1^{(j)}}{|\langle \epsilon_1(N-n) | \epsilon_2(N-n) \rangle|^2 \prod_{j=1}^{n} \gamma_2^{(j)}} \\ &\simeq |\langle \epsilon_1(N-n) | \epsilon_2(N-n) \rangle|^{-2}. \end{aligned} \tag{26}$$

Hence for a sufficiently large ratio of N/n "classical robustness" is attained — the result of Eq. (26) is exponentially high. If one is not ready yet to accept that below some minimal probabilistic threshold events do not occur in our universe (and hence, only the outcome chosen by the boundary state occurs), we can choose a better tuned boundary condition allowing this ratio to diverge (i.e. at least one of the $\gamma_2^{(j)}$ factors is zero).

The significance of the above result is the following: even though from the perspective of the single branch a non-unitary evolution has occurred, there exists a final boundary state which can reproduce with high certainty the desired macroscopic reality when evolved backwards in time. This "robustness ratio" can be used also for the definition of macroscopic objects, i.e. defining the border between classical and quantum regimes. We thus understand the significance of macroscopic objects in storing information for extended times until the final boundary condition arrives and information about past events can be released.

We see that at the cost of having to introduce a final boundary condition, we are able to reclaim determinism (in the two-state sense) and ensure macroscopic time-reversal. It was already assumed, that subsequent to the measurement interaction, decoherence causes an effectively irreversible branching of the superposition into isolated terms. Therefore, no inconsistencies can arise from the existence of a special final boundary condition of the form described before, which simply causes the selection of a single specific branch from the many-worlds picture. In this view, the measurement process does not increase the measure of irreversibility beyond that of regular thermodynamics. Additionally, accounting for the apparent collapse, TSVF does not suggest a microscopic quantum mechanical arrow of time. It does however assume asymmetric initial and final boundary conditions.

According to the TSVF, any post-selected state which is not orthogonal to the pre-selected state is permissible. However, in our model, we have discerned a special boundary condition which accounts for the experimental result, i.e. for the single outcomes which actually occurred in measurements that were actually performed, as well as for the Born rule statistics. This choice is justified on several grounds:

(i) It unites the two dynamical processes of textbook quantum mechanics — the SE and collapse — under one heading. In doing so, it renders QM deterministic and local on a global level, and above all, rids of the ambiguity involved in the approach of "unitary time evolution+non-unitary collapse".

(ii) It allows for robustness under time-reversal of macroscopically large systems.

(iii) It is a natural framework to understand weak values and weak reality.

(iv) As explained, it is not the case that we could have chosen any sort of boundary condition and still maintain classicality on the macro-level. Only states pertaining to the Born rule allow for that. So empirical observations other than the Born rule by itself (e.g. stability under perturbations) can be seen as supporting evidence for a backwards-evolving state with these properties, if any at all.

We find these reasons enough to postulate such a boundary condition. Moreover, the initial state of our universe can also be regarded as "unique", and therefore, we would like to perceive these two boundary conditions as reasonable, constructive, and even necessary for explaining our current observations, rather than artificial. It should also be stressed that in spite of this "uniqueness", the final state has high thermodynamical entropy and also high entanglement entropy, since it encodes all the measurement outcomes of microscopic objects.

It should be noted that n cannot grow to be N, i.e. there is always a "macroscopic core" to every macroscopic object which contained initially N or more particles. It can be shown that $\frac{dn}{dt} \leq 0$ and also that $\frac{dn}{dt} \rightarrow 0$ for long enough times, assuming, for example, an exponential decay of the form:

$$N(t) = N(0)\text{exp}(-t/T), \tag{27}$$

where T is some constant determining the lifetime of macroscopic objects. Also, on a cosmological scale (inflation of the universe), it can be shown that after long time, measurements become less and less frequent (macroscopic objects which can perform measurements are simply no longer available). That means there is more than one mechanism responsible for a finite number (and even much smaller than N) of collapses at any finite or infinite time of our system's evolution.

The MWI was invoked in order to eliminate the apparent collapse from the unitary description of QM. Within the MWI, the dynamics of the universe is both symmetric and unitary. We have now shown that these valuable properties can be attained even at the level of a single branch, i.e. without the need of many-worlds, when discussing macroscopic objects under suitable boundary conditions. Despite the seemingly non-unitary evolution of microscopic particles at the single branch, macroscopic events can be restored from the final boundary condition backwards in time due to the encoding of their many degrees of freedom in the final state.

7. Conclusions

Three complementary approaches for understanding the transition from quantum to classical physics were discussed: deterministic operators, weak measurements, and post-selection.

We began with the observation that deterministic operators capture in some sense the gist of classical determinism even when applied to quantum systems. We then showed that in quite general cases, the quantum average is such a deterministic operator. As an offshoot of this approach, we have seen that a measurement of a classical average is in fact a quantum weak measurement of each particle in the ensemble. Finally, the addition of post-selection and the requirement for classical robustness under time-reversal then helped us defining the border line between classical and quantum regimes.

We hope that in future works, we will be able to strengthen the relations between the three approaches and combine them into a coherent, fresh perspective on the quantum ontology.

Acknowledgments

We would like to thank Tomer Landsberger for many helpful discussions. E.C. was supported by ERC AdG NLST. Y.A. acknowledges support from Israel Science Foundation Grant No. 1311/14, ICORE Excellence Center "Circle of Light" and the German–Israeli Project Cooperation (DIP).

References

[1] Aharonov, Y. and Gruss, E. (2005). Two time interpretation of quantum mechanics, arXiv:quant-ph/05072692.

[2] Aharonov, Y. and Reznik, B. (2002). How macroscopic properties dictate microscopic probabilities. *Physical Review A*, **65**: 052116.

[3] Aharonov, Y. and Rohrlich, D. (2005). Quantum paradoxes: quantum theory for the perplexed, Weinheim: Wiley-VCH, Chapters 16–18.

[4] Aharonov, Y. and Vaidman, L. (1990). Properties of a quantum system during the time interval between two measurements. *Physics Review A*, **41**: 11.

[5] Aharonov, Y. and Vaidman, L. (2002). The two-state vector formalism of quantum mechanics, in *Time in Quantum Mechanics*, J.G. Muga *et al.* (eds.), *Springer*, 369–412.

[6] Aharonov, Y., Albert, D. Z. and Vaidman, L. (1988). How the result of a measurement of a component of a spin 1/2 particle can turn out to be 100?, *Physics Review Letters*, **60**: 1351–1354.

[7] Aharonov, Y., Bergmann, P. G. and Lebowitz, J. L. (1964). Time symmetry in the quantum process of measurement. *Physics Review*, **134**: 1410–1416.

[8] Aharonov, Y., Botero, A., Popescu, S., Reznik, B. and Tollaksen, J. (2002). Revisiting Hardy's paradox: counterfactual statements, real measurements, entanglement and weak values. *Physics Letters A*, **301**: 130–138.

[9] Aharonov, Y., Cohen, E. and Ben-Moshe, S. (2014). Unusual interactions of pre- and post-selected particles, *EPJ Web Conf.*, **70**: 00053.

[10] Aharonov, Y., Cohen, E. and Elitzur, A. C. (2014). Foundations and applications of weak quantum measurements. *Physics Review A*, **89**: 052105.

[11] Aharonov, Y., Cohen, E., Grossman, D. and Elitzur, A. C. (2013). Can weak measurement lend empirical support to quantum retrocausality. *EPJ Web of Conferences*, **58**: 01105.

[12] Aharonov, Y., Cohen, E., Gruss, E. and Landsberger, T. (2014). Measurement and collapse within the two-state-vector formalism. *Quantum Studies: Mathematics and Foundations*, **1**, 133–146.
[13] Aharonov, Y., Landsberger, T. and Cohen, E. (2015). A nonlocal ontology underlying the time-symmetric Heisenberg representation, arXiv:1510.03084.
[14] Aharonov, Y., Popescu, S., Rohrlich, D. and Skrzypczyk, P. (2013). Quantum Cheshire cats. *New Journal of Physics*, **15**: 113015.
[15] Bohm, D. (1952). A Suggested interpretation of the quantum theory in terms of hidden variables I and II. *Physics Review*, **85**: 166–193.
[16] Brune, M., Hagley, E., Dreyer, J., Maitre, X., Maali, A., Wunderlich, C., Raimond, J. M. and Haroche, S. (1996). Observing the progressive decoherence of the "meter" in a quantum measurement. *Physics Review Letters*, **77**: 4887.
[17] Dressel, J., Malik, M., Miatto, F. M., Jordan, A. N. and Boyd, R. W. (2014). Colloquium: Understanding quantum weak values: Basics and applications. *Reviews of Modern Physics*, **86**: 307–316.
[18] Everett, H. (1957). "Relative state" formulation of quantum mechanics. *Review of Modern Physics*, **29**: 454.
[19] Ghirardi, G. C., Rimini, A. and Weber, T. (1986). Unified dynamics for microscopic and macroscopic systems, *Physics Review D*, **34**: 470.
[20] Gruss, E. (2000). M.Sc. Thesis at The Hebrew University of Jerusalem, arXiv:quant-ph/0006070.
[21] Hosten, O. and Kwiat, P. (2008). Observation of the spin Hall effect of light via weak measurements. *Science*, **319**: 787.
[22] Jordan, A. N., Tollaksen, J., Troupe, J. E., Dressel, J. and Aharonov, Y. (2015). Heisenberg scaling with weak measurement: A quantum state discrimination point of view. *Quantum Studies: Mathematics and Foundations*, **2**: 5–15.
[23] Penrose, R. (1996). On gravity's role in quantum state reduction. *General Relativity and Gravitation*, **28**: 581.
[24] Reznik, B. and Aharonov, Y. (1995). Time-symmetric formulation of quantum mechanics. *Physics Review A*, **52**: 2538–2550.
[25] Schlosshauer, M. (2005). Decoherence, the measurement problem, and interpretations of quantum mechanics. *Review of Modern Physics*, **76**: 1267.
[26] Schlosshauer, M. (2007). Decoherence and the quantum-to-classical transition. pp. 13–112, 329–357. Springer, Heidelberg, Berlin.
[27] Starling, D. J., Dixon, P. B., Jordan, A. N. and Howell, J. C. (2009). Optimizing the signal-to-noise ratio of a beam-deflection measurement with interferometric weak values. *Physics Review A*, **80**: 041803.

[28] Tamir, B. and Cohen, E. (2013). Introduction to weak measurements and weak values. *Quanta*, **2**: 7–17.
[29] Tollaksen, J. and Aharonov, Y. (2009). The deterministic set of operators, quantum interference phenomena, and quantum reality. *Journal of Physics: Conference Series*, **196**: 012006.
[30] Tollaksen, J., Aharonov, Y., Casher, A., Kaufherr, T. and Nussinov, S. (2010). Quantum interference experiments, modular variables and weak measurements. *New Journal of Physics*, **12**: 013023.
[31] Vaidman, L. (2013). Past of a quantum particle. *Physics Review A*, **87**: 052104.
[32] Vaidman, L. (2014). Many-worlds interpretation of quantum mechanics, in *The Stanford Encyclopedia of Philosophy*, Edward N. Zalta (ed.), (http://plato.stanford.edu/entries/qm-manyworlds/).
[33] von Neumann, J. (1955). Mathematical foundations of quantum mechanics, trans. R.T. Beyer, Princeton: Princeton U. Press, Chapters V–VI.
[34] Xu, X. Y., Kedem, Y., Sun, K., Vaidman, L., Li, C. F. and Guo, G. C. (2013). Phase estimation with weak measurement using a white light source. *Physics Review Letters*, **111**: 033604.
[35] Zurek, W. H. (2003). Decoherence and the transition from quantum to classical-Revisited, quant-ph/0306072.
[36] Zurek, W. H. (2003). Decoherence, einselection, and the quantum origins of the classical. *Reviews of Modern Physics*, **75**: 715.

Chapter 13

Bound States as Fundamental Quantum Structures

R. E. Kastner

Foundations of Physics Group
University of Maryland, College Park, MD, USA

Bound states arise in many interactions among elementary field states, and are represented by poles in the scattering matrix. The emergent nature of bound states suggests that they play an perhaps under-appreciated role in specifying the ontologically relevant degrees of freedom pertaining to composite systems. The basics of this ontology are presented, and it is discussed in light of an example of M. Arsenijević, J. Jeknić-Dugić and M. Dugić.

1. Bound States: A Definition

Suppose we are given two or more interacting degrees of freedom, at least of which can be modeled as a field mode of an applicable quantum field. One can apply scattering theory to these degrees of freedom, and for attractive interactions certain final states will be bound states, which establish a new composite entity.

The paradigmatic example is the hydrogen atom. One can treat the proton as the generator of a potential well and the electron as an incoming wave function. A scattering matrix can be defined, which assigns amplitudes to various outgoing states for the particles. For suitable incoming momenta of the electron, some of those states will correspond to the situation in which the scattered electron state has imaginary energies; these will result in an exponentially decaying wave function that traps the electron in the potential well due to the proton. These are the bound states, which result in the creation of the hydrogen atom as a new composite entity. The bound states (and resonances which are metastable bound states) correspond to poles in the scattering matrix for the interacting quanta.[a] Thus, for purpose of this discussion, we define the bound state as a final scattering state characterized by a pole in the scattering matrix describing an encounter between two or more degrees of freedom.[b]

2. Bound States as Emitters and Absorbers

A bound system such as an hydrogen atom is of course subject to many different energy levels, corresponding to the different possible bound states for the electron. An excited state is subject to decay to a lower state, upon which the electron emits a photon. In the transactional interpretation (TI) of Cramer [2], at least one additional absorbing bound system is required as a condition for a real photon to be emitted from an excited atom. This is because TI is based on a direct-action theory, in which radiation occurs as a result of a mutual direct interaction between emitter and the absorber. The classical version of the direct-action theory was developed by Wheeler and Feynman [4] and the quantum relativistic version by Davies [3].

TI defines the usual quantum state $|\Psi\rangle$ as an "offer wave" (OW), and it defines the advanced response $\langle a|$ of an absorber A as a 'confirmation wave' (CW). In general, many absorbers $A, B, C, \ldots$. respond to an OW, where each absorber responds to the component of the OW that reaches it. The OW component reaching an absorber X would be $\langle x|\Psi\rangle|x\rangle$, and it would respond with the adjoint (advanced) form $\langle x|\langle\Psi|x\rangle$. The product of these two amplitudes corresponds to the final amplitude of the "echo" of the CW from X at the locus

[a]See [6], p. 431 for a careful discussion of this.
[b]Technically, we need to add that the potential modeling the interaction falls off faster than any exponential as separation of the degrees of freedom approaches infinity.

of the emitter and corresponds to the Born Rule. Meanwhile, the sum of the weighted outer products (projection operators) based on all CW responses corresponds to the mixed state identified by von Neumann as resulting from the non-unitary process of measurement. Thus, TI provides a physical explanation for both the Born Rule and the measurement transition from a pure to a mixed state.

Due to energy conservation, a free charged particle can neither emit nor absorb a photon; so bound states are the true emitters and absorbers. The present author has proposed a relativistic extension of TI, called the possibilist transactional interpretation, or "PTI" [5]. PTI regards bound states as important and fundamental quantum structures. According to PTI, spacetime is a discrete manifold that is emergent from a quantum level of physical possibilities, which are the quantum degrees of freedom (including bound states). It is transactions between an emitting bound state and an absorbing bound state that generate spacetime events and their connections: the emitter and absorber define the endpoints, and the transferred quantum of conserved quantities (energy, momentum, spin, etc.) defines their connection. Thus, when an electron in an atom emits a photon, the entire atom is actualized as a spacetime object, and similarly, the entire absorbing atom is actualized as a spacetime object as well. This is because, in each case, the entire atom undergoes a well-defined physical change of state as a result of the transaction; it is not just the component electron that is affected.

3. A Physical Criterion for Structurally Significant Degrees of Freedom

In this volume, Arsenijević *et al.* [1] discuss the challenge of deciding on physically relevant decomposition of composite systems into subsystems. Analyzing the example of an electrically neutral atom in a Stern–Gerlach apparatus, they point out that if the action of the field on the atom is indeed correctly modeled by the usual Hamiltonian interaction term inducing entanglement between the atom's center of mass and its spin, the external magnetic field of the S–G acts on the center of mass of the atom as a whole, and not individually on the electrons and nucleus. For if the latter were the case, the internal state of the atom would be observably changed after passing through the field.

According to the criteria in Section 1, and supported by the considerations of Section 2, this is because the atomic bound state is an emergent structure. Its status as a new, non-separable entity corresponds to the fact that it is described by a pole in the scattering matrix for its constituent subsystems. Such a singularity in the theoretical description is often an indicator of an emergent process, in which one model breaks down and must be superseded by another. For example, in quantum field theory, a real quantum can be seen as an emergent structure as follows: the propagator function, defined over the complex frequency plane, describes virtual quantum propagation. A pole in the propagator describes a real quantum, which differs physically from a virtual quantum since the former is representable by a Fock state (stable field excitation) but the latter is not. Similarly, the bound state is not simply an entangled state of two or more degrees of freedom, in which the entanglement can be viewed as relative to a choice of observable. Rather, it is an emergent structure arising from a particular kind of interaction among its constituents, in which they lose their separate identity as free field excitations and form a new collective final state. This observation holds regardless of interpretation; i.e. the bound state can be understood as an emergent structure based on standard scattering theory and on the mathematical properties of the energies of the applicable final states for quanta entering a bound state.

In the PTI picture, clearly the dynamical conserved quantities are considered the fundamental "preferred" observables, while the position observable (characterizing an aspect of spacetime phenomena) is secondary and emergent. One might ask why there should be such an asymmetry between the observables. This asymmetry is naturally found at the relativistic level, since at that level there is no well-defined position observable (and time is not even an observable at the non-relativistic level). In contrast, energy, momentum, spin, etc. are all well-defined observables at the relativistic level. And it is at the relativistic level where one finds emission (creation of field quanta) and absorption (destruction of field quanta) which must always occur in any measurement context, i.e. whenever it is possible to unambiguously apply a determinate value of an observable to a system. Thus, the dynamical conserved quantities merit consideration as the naturally preferred observables, and this in turn allows for a fundamental structural criterion.

It should be noted that one can still obtain some information about the components of bound states. One can image to some extent the component degrees of freedom through probing the shape of the scattering potential they create. The composite system is used as a target, and a probe system is scattered from it.[c] But there is no quantum entanglement between the probe system and the target components: the probe does not "see them" individually at the level of their quantum states, but only collectively through the overall shape of the target potential. An early example is Rutherford's gold foil experiment, in which alpha particle were used as probes for the gold atoms.

4. Conclusion

It has been argued that bound states, as represented by poles in the scattering matrix for their constituents, constitute emergent non-separable structures that are properly characterized by degrees of freedom reflecting their wholeness, such as the center-of-mass degree of freedom used to model atoms interacting with a Stern–Gerlach field. The entanglement of the composite degrees of freedom of a bound state is of a fundamentally different nature rather than of non-bound degrees of freedom, in that it is not relative to a choice of observable, but is an intrinsic, formative property of the entire structure. The bound state's description in terms of a center-of-mass degree of freedom can thus be viewed as physically unambiguous and descriptive of the composite system in a way that the component degrees of freedom are not. This is yet another demonstration of

[c] Here we consider only elastic collisions, since inelastic ones alter or destroy the internal structure such that a collective degree of freedom would not accurately describe it anyway.

the signature of the quantum world that the "whole is greater than the sum of the parts." It is also noted that the dynamical conserved quantities are the natural "preferred observables" when the relativistic level is taken into account in investigating the origins of quantum structures.

Acknowledgment

I would like to thank Miroljub Dugić and Jasmina Jeknić-Dugić for valuable discussions.

References

[1] Arsenijević, M., Jeknic-Dugić, J., Dugić, M. A top-down versus a bottom-up hidden-variables description of the Stern–Gerlach experiment, forthcoming in Kastner, R. E., Jeknić-Dugić J., Jaroszkiewicz, G. (eds.), *Quantum Structural Studies.* Singapore: World Scientific Publishers. (not yet published). Preprint version: http://arxiv.org/abs/1601.05555.
[2] Cramer, J. G. (1986). The Transactional interpretation of quantum mechanics. *Reviews of Modern Physics*, **58**: 647–688.
[3] Davies, P. C. W. (1971, 1972). Extension of Wheeler–Feynman quantum theory to the relativistic domain I. scattering processes. *Journal of Physics A: Gen. Phys.*, **4**: 836; and Extension of Wheeler–Feynman quantum theory to the relativistic domain II. emission processes. *Journal of Physics A: Gen. Phys.*, **5**: 1025–1036.
[4] Feynman, R. P. and Wheeler, J. A. (1945, 1949). Interaction with the absorber as the mechanism of radiation, *Reviews of Modern Physics*, **17**: 157–161; and Classical electrodynamics in terms of direct interparticle action, *Reviews of Modern Physics*, **21**: 425–433.
[5] Kastner, R. E. (2012). *The Transactional Interpretation of Quantum Mechanics: The Reality of Possibility.* Cambridge: Cambridge University Press.
[6] Weinberg, S. (1995). *The Quantum Theory of Fields I.* Cambridge: Cambridge University Press.

Methodological Approaches

Chapter 14

A Top-down View of the Classical Limit of Quantum Mechanics

Sebastian Fortin* and Olimpia Lombardi†

*CONICET — University of Buenos Aires,
Buenos Aires, Argentina

†CONICET — University of Buenos Aires,
Theiss Research, Argentina

The problem of the classical limit of quantum mechanics consists in explaining how the classical realm "emerges" from the quantum domain. Although along the history of quantum mechanics the problem has been addressed from many different perspectives, at present it is recognized that the classical limit also involves some kind of physical process, which transforms quantum states in such a way that they finally can be interpreted as classical states. This process is known as *quantum decoherence*.

According to the orthodox approach, decoherence is an environment induced phenomenon: the classical world emerges from the interaction between subsystems of a quantum system. This approach is based on a *bottom-up view*, since it begins by the analysis of the subsystems of the whole system and of their interactions. In the present chapter, we propose an inverse perspective, a *top-down view*, which begins by analyzing the whole closed system and its evolution, and on this basis identifies the degrees of freedom that will behave classically. From this perspective, decoherence is not a yes-or-no process, but a phenomenon *relative* to the decomposition of the closed system selected in each case. In turn, the classical limit based on this top-down view is more general than that resulting from the traditional environment induced approach to decoherence, and dissolves some of its conceptual difficulties.

1. Introduction

Since its birth in the early 20th century, quantum mechanics raised a number of questions and problems, many of which are still a source of lively debate. The attempts to address those issues have led to a multiplicity of interpretations and theoretical developments that have enriched the scientific knowledge about the theory. Perhaps the most discussed problem in this context is the so-called *quantum measurement problem*, based on the theoretical difficulty to explain how measuring devices with classical pointers are able to produce definite results when acting on quantum systems [12, 26, 109]. Another question that has been the subject of intensive research is the problem of *the classical limit of quantum mechanics* [21, 101]. According to the correspondence principle ([23]; for a recent discussion, see [24]), there should be a limiting procedure that accounts for the classical behavior of a system in terms of the laws of quantum mechanics. The problem of the classical limit consists in explaining how the classical realm "emerges" from the quantum domain. The two problems just mentioned carry the burden of facing the issue, already pointed out by Bohr, of the "cut" between the classical and the quantum world (see a discussion about Bohr's views on this point in [11, 77]). And the two problems also have something in common: both point to the need for finding a link between the classical and the quantum domains.

Along the history of quantum mechanics, the classical limit has been approached from many different perspectives, such as those given by the Ehrenfest theorem [58], the Wigner transform [112] and the deformation theory [16, 17]. Traditionally, the problem was conceived as a matter of intertheory relation: classical mechanics should be obtained from quantum mechanics by means of the application of a mathematical limit, in a way analogous to how classical equations of motion are obtained from special relativity. However, this approach has been reconsidered over the past decades: at present it is recognized that the classical limit also involves some kind of physical process, which transforms quantum states in such a way that they finally can be interpreted as classical states. This process is now known as *quantum decoherence*.

One of the main features of quantum mechanics is the superposition principle, which leads to the phenomenon of quantum interference, without classical analogue. Decoherence is viewed as a process that cancels interference and selects the candidates to classical states. That cancellation has been traditionally conceived in terms of the transformation of a pure state into a mixture without interference terms. From a geometrical viewpoint, the state of a quantum system passes from the frontier of the convex set of states (pure states) to its interior (mixture states) ([19]; for a recent approach see [74]). On this basis, decoherence was studied in open and closed systems. Schematically, three periods can be identified in the historical development in the general program of decoherence (see [38]):

- **First period**: Several authors [52, 104–106] studied the approach to equilibrium of quantum systems through the behavior of the so-called *collective observables*, are the observables i.e. accessible from the macroscopic viewpoint. This approach was based on the methods traditionally used to describe irreversible processes. The aim was to understand how classical macroscopic properties emerge from the quantum microscopic evolution. For this purpose, a coarse-grained state $\rho_G(t)$ is defined, which carries all the macroscopic information of the system, and it is shown that, under certain definite conditions, $\rho_G(t)$ decoheres in its eigenbasis in a decoherence time t_D, and reaches equilibrium after a relaxation time t_R. The main problem of this period was the fact that the decoherence time t_D computed with these primitive formalisms proved to be too long when compared with experimental results [94].
- **Second period**: The interest turns out to be focused on open systems. An open quantum system S is considered in interaction with its environment E, and the time evolution of the reduced state $\rho_S(t) = Tr_E(\rho(t))$ is studied. According to the *environment-induced decoherence* (EID) approach [113–116, 118, 122], decoherence is the result of the interaction between the system S and the environment E. It is shown that, under certain definite conditions,

the states of E become orthogonal in a very short decoherence time t_D and, as a consequence, interference disappears from the state $\rho_S(t)$ of S: it is said that $\rho_S(t)$ decoheres in the so-called "preferred" or "pointer" basis. This solves the problem of the first period; moreover, the formalism counts with many successful applications [69].

Under the assumption that quantum systems are never isolated and interact significantly with their environment [113], the EID approach was initially conceived to study the measurement problem, but it was immediately extrapolated to the case of the classical limit. Nevertheless, this approach was questioned regarding its capacity of accounting for the emergence of classicality due to some conceptual difficulties (which will be discussed below).

- **Third period**: Although the EID approach is still the most widespread view, other perspectives have been proposed to deal with cases that are beyond the application domain of EID, in particular, the case of closed systems [28, 29, 55, 56, 90, 97]. Some of them were designed specifically to describe processes that do not dissipate energy to the environment [25, 60, 63, 65–67, 78, 99, 102]. A formalism particularly devoted to describe decoherence in closed systems is the self-induced decoherence approach, according to which a closed system with continuous spectrum may decohere due to destructive interference [31–37, 41, 43–47].

The classical limit based on the EID approach describes a classical world that emerges from the interaction between subsystems of a quantum system: this approach is a *bottom-up view*, since it begins by the analysis of the subsystems of the whole system and of their interactions. In this chapter, we will propose an inverse perspective, a *top-down view*, which begins by analyzing the whole closed system and its evolution, and on this basis identifies the degrees of freedom that will behave classically. From this perspective, decoherence is not a yes-or-no process, but a phenomenon *relative* to the decomposition of the closed system selected in each case. In turn, the classical limit based on this top-down view is more general than that resulting from the traditional EID approach, and dissolves some of its conceptual difficulties.

In order to develop our presentation, this chapter is organized as follows: In Section 2, the conceptual basis of the EID approach will be recalled, and in Section 3 its conceptual challenges will be pointed out. Section 4 will be devoted to show that reduced states are a kind of coarse-grained states of closed composite systems, and that it is for this reason that they cancel the correlations between subsystems. In Section 5, the closed-system approach to decoherence will be introduced by arguing that the loss of coherence in a system can be studied in terms of the internal structure of the expectation values of its observables. On this basis, Section 6 will show that the EID approach can be reformulated from the open-system perspective, and Section 7 will explain how its conceptual difficulties are solved or dissolved from the new approach to decoherence. Finally, Section 8 will be devoted to consider how the closed-system approach fits into a general top-down view of quantum mechanics, which underlies different conceptual and interpretive perspectives on the theory.

2. Environment-induced Decoherence: An Open-System Approach

Let us consider an open system S represented by an Hilbert space $\mathcal{H}_S$, whose initial state $\rho_S(0)$ belongs to the Liouville space $\mathcal{L}_S = \mathcal{H}_S \otimes \mathcal{H}_S$, and that is in interaction with an environment E represented by an Hilbert space $\mathcal{H}_E$, whose initial state $\rho_E(0)$ belongs to the Liouville space $\mathcal{L}_E = \mathcal{H}_E \otimes \mathcal{H}_E$. The initial state $\rho_U(0)$ of the whole composite system $U = S \cup E$ is obtained as $\rho_U(0) = \rho_S(0) \otimes \rho_E(0)$. Therefore, at the initial time $t = 0$, it is possible to recover the initial states of S and of E from the initial state of the closed system U by means of a partial trace:

$$\rho_S(0) = Tr_E(\rho_U(0)) \quad \rho_E(0) = Tr_S(\rho_U(0)). \tag{1}$$

The evolution of the total system is governed by the Hamiltonian $H_U = H_S + H_E + H_{SE}$, where H_S is the self-Hamiltonian of the system

S, H_E is the self-Hamiltonian of the environment E, and H_{SE} is the interaction Hamiltonian. The time evolution of the closed system U, represented by $\rho_U(t)$, can be computed by means of the Liouville–von Neumann equation in terms of the Hamiltonian H_U.

Since partial trace recovers the initial state of S from the initial state of U, from the open-system perspective, it is usual to assume an additional hypothesis: S preserves its identity as a quantum system during the time evolution, and its state $\rho_S(t)$ can be computed in a manner analogous to that used in the initial time (Eq. (1)):

$$\rho_S(t) = Tr_E(\rho_U(t)). \tag{2}$$

The EID formalism proves that, in many physically relevant models, the non-diagonal terms of the reduced state $\rho_S(t)$ rapidly tend to vanish after an extremely short decoherence time t_D:

$$\rho_S(t) \to \rho_S^d(t), \tag{3}$$

where $\rho_S^d(t)$ is diagonal in the preferred basis. In this way, it is said that the system S decoheres as a consequence of its interaction with the large number of degrees of freedom of the environment E.

It is important to stress that, according to the EID approach, decoherence amounts not merely to the diagonalization of the reduced state, but to the diagonalization of the reduced state in the preferred basis, that is, the basis that defines what observables behave classically. Although the precise definition and the strategies to compute this preferred basis have been the subject of much discussion, the treatment of this point is beyond the limits of the present article [34, 36, 44, 79, 118, 123, 124].

3. Conceptual Challenges of the Open-System Approach

The theory of decoherence in its EID version has become the "new orthodoxy" in the quantum physicists community [26, 80]. In fact, decoherence is studied in many areas of physics, and has gained a great relevance in quantum computation, where the aim is to take advantage of superpositions and, therefore, to avoid classicality.

Despite its great success, from a conceptual viewpoint there are still some challenges that must be faced if one wants to offer a self-consistent view of the classical limit. We will focus on three of them:

1. *The closed-system problem*: EID cannot be applied to closed systems, in particular to the universe as a whole.
2. *The defining-system problem*: EID does not provide a criterion to decide where to place the "cut" between the proper system and the environment.
3. *The problem of the emergence of the classical world*: Under certain conditions, EID cannot define univocally the classically behaving systems that emerge from the quantum domain.

3.1. The closed-system problem

According to the authors of the EID approach, the aim of the program "*is to describe the consequences of the 'openness' of quantum systems to their environments and to study the emergence of the effective classicality of some of the quantum states and of the associated observables*" ([120], p. 1793). Therefore, the split of the universe into the degrees of freedom that are of direct interest to the observer — the system — and the remaining degrees of freedom — the environment — is absolutely essential to understand "*the quantum origin of the classical world*" ([96], p. 77). Zurek even claims that the prejudice that seriously delayed the solution of the problem of the emergence of classicality consists in the fact that the role of the "openness" of quantum systems was ignored for a very long time [96, 122].

If decoherence explains the emergence of classicality, but only open systems can decohere, the question is: what about closed systems, in particular, the universe as a whole? [98]. In the literature, several models can be found that describe decoherence in systems without an environment, understood in the traditional way. For instance:

- In the Casati–Prosen model [30], decoherence is manifested by the vanishing of the interference pattern on a screen located in a closed cavity. Independently of the details of the model, this is the case in which it is not possible to consider that the phenomenon is due to the interaction with an external environment [33].

- In the case of the systems studied by the self-induced decoherence approach, the loss of coherence is attributed to a very generic coarse-graining, and not to the intervention of an environment external to the system (see [37] and references therein).
- In the systems studied by Gambini and collaborators [65–67], the authors analyze the influence of an extra term in the evolution equation, which comes from quantum gravity considerations. The term responsible for decoherence does not result from the interaction with an environment, but expresses a coarse-graining due to time-uncertainty.
- Some authors describe decoherence in the Heisenberg representation [78, 99]. In this formalism, the loss of coherence, treated by means of the Bogoliubov transformation, is due to the dynamics of the system itself.

Given the peculiar features of the EID approach, these cases are beyond its application scope. In fact, according to Zurek [119], since a closed system evolves deterministically forever, the issue of its classicality cannot even be posed.

3.2. The defining-system problem

Since EID does not apply directly to closed systems, in these cases "internal environments" are defined, the closed system is partitioned into some degrees of freedom that represent the system of interest, and the remaining degrees of freedom that play the role of the environment. For example, in the cosmological context, long wavelength modes are usually considered as the system, and short wavelength modes are conceived as the environment [87]. However, this is not the only way to introduce the split into the closed system. In a more recent study of the fluctuations generated during the inflationary period of the cosmic evolution, it is supposed that the tensor and the scalar fluctuations interact with each other, and the tensor fluctuations act as an environment that causes the loss of coherence of the scalar fluctuations, whose classicality is so justified [62].

This cosmological case is only an example of the fact that, although the EID approach studies the correlations between system and environment and also between different subsystems [95], there is a conceptual difficulty in the definition of the systems involved in the phenomenon of decoherence: the approach does not supply a general criterion to discriminate between system and environment. In general, the classically behaving degrees of freedom are assumed in advance: the application of the EID formalism does not predict which observables will manifest a classical behavior but only confirms a previous assumption. This problem is acknowledged by Zurek himself: "*one issue which has been often taken for granted is looming big, as a foundation of the whole decoherence program. It is the question of what are the systems which play such a crucial role in all the discussions of the emergent classicality. This issue was raised earlier, but the progress to date has been slow at best. Moreover, replacing systems with, say, coarse grainings does not seem to help at all, we have at least tangible evidence of the objectivity of the existence of systems, while coarse-grainings are completely in the eye of the observer.*" ([121], p. 338; [120]).

3.3. The problem of the emergence of the classical world

According to the EID approach, decoherence explains the transition from quantum to classical [117], that is, the emergence of the classical world from the quantum realm. This classical world must be objective as decoherence itself: it should not be confined to "the eye of the observer". However, as indicated above, the EID approach provides no criterion to distinguish the system of interest from its environment. As a consequence, in order to apply the EID formalism to a closed system U, the decomposition between system and environment can be introduced in many different ways: $U = S_1 \cup E_1$, $U = S_1 \cup E_1, \ldots, U = S_n \cup E_n$. Since there is no privileged decomposition, this situation leads to one of the following three cases:

- If none of the systems S_i arising from the different decompositions decoheres, then classicality does not emerge.
- If none of the systems S_i arising from the different decompositions decoheres except one, say, S_k, then only this subsystem S_k decoheres and becomes classical.
- If more than one of the systems S_i arising from the different decompositions decohere, but their union does not, then the classicality emerging from the underlying quantum domain is not univocally determined.

A concrete example of this last case is proposed by Castagnino, Fortin and Lombardi [39] (see discussion in [86]): a generalized spin-bath model of $m+n$ spin-1/2 particles, where the m particles interact with each other and the n particles also interact with each other, but the particles of the m group do not interact with those of the n group. The study of the model shows that there are definite conditions under which all the particles decohere, but neither the system composed of the m group nor the system composed of the n group decoheres.

This kind of cases poses a conceptual challenge to the EID approach: if classicality is conceived as an objective property, the fact that a system behaves classically or not cannot depend on the way in which the observer decides to split the original closed system. In other words, this situation challenges the spirit of the original EID proposal, according to which decoherence provides the basis of a classical limit that explains the objective emergence of the classical world.

4. About the Reference of the Reduced State

The conceptual difficulties of the EID approach derive precisely from its open-system perspective. Therefore, it seems reasonable to reconsider the status of open systems and, in particular, of what supposedly represents their behavior: the reduced state.

In classical statistical mechanics, the problem of irreversibility consists in accounting for an irreversible approach to equilibrium in systems ruled by time-reversal invariant laws [64, 81]. The standard answer in the Gibbsian framework relies on coarse-graining: whereas the statistical state of the system, represented by a density function, evolves obeying the Liouville theorem, the evolutions of coarse-grained states are not constrained by the theorem and, under definite conditions of instability, may approach a definite limit for $t \rightarrow \infty$. Of course, there are deep disagreements about the interpretation of the irreversibility so obtained. But, independently of them, nobody ignores the difference between the statistical state, whose evolution is ruled by the dynamical postulate of the theory, and the coarse-grained state, which may tend to a final stable state.

The situation in quantum mechanics is quite different: the distinction between the different kinds of states appearing in the quantum context is usually not sufficiently emphasized. For instance, sometimes it is said that the dynamical postulate of quantum mechanics only applies to closed systems, whereas reduced operators actually represent quantum states of open systems. Nevertheless, we are not informed about the dynamical *law* for reduced states; in fact, the evolution of open systems always depends, in the final analysis, on the unitary evolution of the whole closed system of which the open system is a part. It is also admitted that, whereas the states of closed systems embody quantum correlations, reduced states may cancel those correlations and, as a consequence, in certain cases they cannot be used for computations. However, in spite of these relevant differences, the states of closed and open systems are usually treated on equal footing.

Although a minority, some authors have conceived reduced states as coarse-grained states (see, for example, [4, 68, 91]). Nevertheless, in general the claim does not go beyond pointing out the operation of tracing over the degrees of freedom of the environment. However, it can be formally proved that the reduced state provides a description that can be understood by means of a generalized conception of coarse-graining [61]: from being originally conceived as *the quantum state of the open system*, it turns out to be *a coarse-grained state of the closed system*.

In its traditional classical form, coarse-graining is based on the partition of a phase space into discrete and disjoint cells: this mathematical procedure defines a projector Π that cancels some components of the state vector ρ [89]: only certain components are retained in the coarse-grained description $\rho_{cg} = \Pi\rho$. If this idea is generalized, coarse-graining can be conceived as a projection that cancels some components of a vector representing a state.

Let us recall that the reduced state ρ_1 of S_1 is defined as the density operator by means of which the expectation values of all the observables belonging to S_1 can be computed. Precisely, if $\mathcal{H}_1$ and $\mathcal{H}_2$ are the Hilbert spaces of S_1 and S_2 respectively, $\mathcal{H} = \mathcal{H}_1 \otimes \mathcal{H}_2$ is the Hilbert space of $S = S_1 \cup S_2$, $O_1 \in \mathcal{H}_1 \otimes \mathcal{H}_1$ is an observable of S_1, I_2 is the identity in $\mathcal{H}_2 \otimes \mathcal{H}_2$, and $\rho \in \mathcal{H} \otimes \mathcal{H}$ is the state of S, then the reduced state of S_1 is defined as the density operator ρ_1 such that

$$\forall (O = O_1 \otimes I_2) \in \mathcal{H} \otimes \mathcal{H}, \quad \langle O \rangle_\rho = \langle O_1 \rangle_{\rho_1}. \tag{4}$$

Although for dimensional reasons the reduced state ρ_1 cannot be expressed as a direct projection $\Pi\rho$ of the quantum state $\rho \in \mathcal{H} \otimes \mathcal{H}$, the expectation value $\langle O_1 \rangle_{\rho_1}$ can also be expressed as the expectation value of $O = O_1 \otimes I_2$ in a coarse-grained state $\rho_{cg} \in \mathcal{H} \otimes \mathcal{H}$:

$$\langle O_1 \rangle_{\rho_1} = \langle O \rangle_{\rho_{cg}}. \tag{5}$$

The density operator ρ_{cg} represents a coarse-grained state because it can be obtained as $\rho_{cg} = \Pi\rho$, where the projector Π performs the following operation:

$$\Pi\rho = (Tr_{(2)}\,\rho) \otimes \widetilde{\delta}_2 = \rho_1 \otimes \widetilde{\delta}_2, \tag{6}$$

where $\widetilde{\delta}_2 \in \mathcal{H}_2 \otimes \mathcal{H}_2$ is a normalized identity operator with coefficients $\widetilde{\delta}_{2\alpha\beta} = \delta_{\alpha\beta} / \sum_\gamma \delta_{\gamma\gamma}$.

It is quite clear that ρ_{cg}, although belongs to $\mathcal{H} \otimes \mathcal{H}$, is not the quantum state of S: it is *a coarse-grained state of the closed system* that disregards certain information of its quantum state. However, ρ_{cg} supplies the same information about the open system S_1 as the reduced state ρ_1, but now from the viewpoint of the composite system S. Therefore, the reduced density operator ρ_1 can also be conceived as a kind of coarse-grained state of S that disregards certain degrees of freedom considered as irrelevant.

Once the reduced state is viewed as a coarse-grained state, it becomes clear that its non-unitary evolution neither restricts the application of the quantum dynamical postulate nor requires a new dynamical postulate: it turns out to be a situation analogous to the familiar case of classical instability, in which it is completely natural to obtain irreversible coarse-grained evolutions from the underlying reversible dynamics of the unstable system, with no need of restrictions or reformulations of the classical dynamical laws [20]. An author who has emphasized the analogy between the classical statistical case and the quantum case is Omnès [92, 93], by repeatedly claiming that decoherence is a particular case of the phenomenon of irreversibility. Now the claim can be endowed with a more precise meaning: as in the case of classical instability, in which the coarse-grained state approaches a final state in spite of the reversible evolution of the statistical state, in environment-induced decoherence the reduced state approaches a diagonal reduced state, in spite of the fact that the quantum state indefinitely follows its unitary evolution.

5. A Closed-System Approach to Decoherence Based on Expectation Values

If the description given by the reduced state can be recovered from the perspective of the closed system, it is not surprising that, by contrast to the EID approach, decoherence can be accounted for from a closed-system view. In fact, by following the path opened by Zeh [69], it is possible to study the loss of coherence in a system by considering the internal structure of the expectation values of its observables. From this perspective, decoherence is conceived in terms of the vanishing of the interference terms of the expectation values of certain observables of interest [38, 42].

In line with these remarks, the closed-system approach to decoherence relies on a strategy based on two assumptions:

- The object of this study is always the *closed system*, which is considered from the viewpoint of some relevant observables.

Therefore, the state used to describe decoherence is not the reduced state of an open system but a coarse-grained state of the closed system.
- The evolution relevant to decoherence is the evolution of the *expectation values* of the observables of the closed system.

In the rest of this section, the closed-system approach will be developed in the light of these two points.

5.1. Expectation values and closed systems

The usual presentations of quantum mechanics place the concept of state in the center of the scene: the description of a system is given by the system's state and its time evolution. However, states in quantum mechanics do not supply the value of the observables of the system, as in the classical case; by contrast, they are the theoretical means for computing the expectation values of all the observables of the quantum system. And such expectations values amount to the information that can be empirically obtained from the system, that is, the information that can be measured [15]. In turn, given the expectation values of all the observables of a system, it is possible to compute the system's state. Therefore, those expectation values provide a complete description of the system and its time evolution, without resorting to the state. This fact is what supports the expectation-value perspective.

This perspective centered on the observables is in resonance with the algebraic formalism of quantum mechanics [70], according to which a quantum system is mathematically characterized by the space $\mathcal{O}$ of the self-adjoint elements of an algebra of operators representing observables, and states are represented by functionals on $\mathcal{O}$. In this theoretical framework, the observables are the basic elements of the theory, and states are secondary elements, defined in terms of the basic ones. Therefore, in order to study the subsystems of a system, it is necessary to consider the spaces of observables corresponding to those subsystems.

Let us consider a closed system U partitioned as $U = S_1 \cup S_2$. If $\mathcal{O}_U$ is the space of observables of U, and $\mathcal{O}_1$ and $\mathcal{O}_2$ are the spaces of observables of S_1 and S_2 respectively, then $\mathcal{O}_U = \mathcal{O}_1 \otimes \mathcal{O}_2$. If ρ_U is the state of U, the reduced states of S_1 and S_2 can be computed by means of partial traces as $\rho_1 = Tr_2 \rho_U$ and $\rho_2 = Tr_1 \rho_U$ respectively. With these three states, the expectation values of all the observables of the corresponding systems can be computed:

$$\langle O_U \rangle_{\rho_U} = Tr(\rho_U O_U) \quad \langle O_1 \rangle_{\rho_1} = Tr(\rho_1 O_1) \quad \langle O_2 \rangle_{\rho_2} = Tr(\rho_2 O_2), \tag{7}$$

where $O_U \in \mathcal{O}_U$, $O_1 \in \mathcal{O}_1$, and $O_2 \in \mathcal{O}_2$. But, as it is well known, there are always observables of the composite system U that are not observables of the subsystems. For instance, the expectation values of the observables of the form $O_1 \otimes O_2 \in \mathcal{O}_U$ cannot be computed in terms of the subsystems S_1 and S_2. For this reason, from the perspective centered on the observables, the viewpoint given by the closed system has conceptual priority: any partition into subsystems gives a view that is unavoidably partial, to the extent that it cannot capture the information of all the observables of the composite system.

In the context of the closed-system approach, by contrast, the expectation values of the subsystems' observables can be computed in terms of the state and the observables of the closed composite system. The strategy consists in considering the observables O_{U1} and O_{U2} of U that have the form $O_{U1} = O_1 \otimes I_2 \in \mathcal{O}_U$ and $O_{U2} = O_2 \otimes I_1 \in \mathcal{O}_U$. Then, the expectation values of the observables of the subsystems can be computed as

$$\langle O_{U1} \rangle_{\rho_U} = Tr(\rho_U O_{U1}) = Tr(\rho_U (O_1 \otimes I_2)) = Tr(\rho_1 O_1) = \langle O_1 \rangle_{\rho_1}, \tag{8}$$

$$\langle O_{U2} \rangle_{\rho_U} = Tr(\rho_U O_{U2}) = Tr(\rho_U (O_2 \otimes I_1)) = Tr(\rho_2 O_2) = \langle O_2 \rangle_{\rho_2}. \tag{9}$$

But the closed-system approach is completely general, since not restricted to the study of partitions of the closed system U into subsystems. It allows us to consider any subset $\mathcal{O}_R \subset \mathcal{O}_U$, where the $O_R \in \mathcal{O}_R$ do not need to be of the form $O_1 \otimes I_2$: the O_R are the observables considered relevant in the particular situation of interest.

5.2. Decoherence from the closed-system approach

On the basis of the above explanation, in the present framework a closed system U will not be split into a system of interest and its environment, but will be partitioned into relevant and irrelevant observables: $\mathcal{O}_U = \mathcal{O}_R \cup \mathcal{O}_I$.

Let us consider the expectation value of a generic relevant observable $O_R \in \mathcal{O}_R$ when the system U is in the state ρ_U:

$$\langle O_R \rangle_{\rho_U} = Tr(\rho_U O_R) = \sum_i (o_{Rii})(\rho_{Uii}) + \sum_{i \neq j} (o_{Rij})(\rho_{Uij}), \quad (10)$$

where the (ρ_{Uii}) and the (o_{Rii}) are the diagonal components, and the (ρ_{Uij}) and the (o_{Rij}) are the non-diagonal components of ρ_U and O_R, respectively, in the preferred basis. The second sum of Eq. (10) represents the specifically quantum interference terms of the expectation value. If those terms vanished, the expectation value would adopt the structure of a classical expectation value:

$$\sum_i (o_{Rii})(\rho_{Uii}), \quad (11)$$

where the (o_{Rii}) might be interpreted as possible values and the (ρ_{Uii}) might play the role of probabilities, as positive numbers that are less than one and sum to one.

In the light of this idea, from an expectation-value perspective, there is decoherence for the relevant observables when the expectation values $\langle O_R \rangle_{\rho_U}$ tend, in an extremely short decoherence time t_D, to $\sum_i (o_{Rii})(\rho^d_{Uii})$, where the (o_{Rii}) are the diagonal components of the relevant observables O_R in the preferred basis, and $0 \leq \rho^d_{Uii} \leq 1$, $\sum_i \rho^d_{Uii} = 1$. The (ρ^d_{Uii}) can be conceived as the components of a kind of coarse-grained state ρ^d_G, diagonal in the preferred basis. In other words, $\langle O_R \rangle_{\rho_U(t)}$ converges, after t_D, to a value that can be computed as if the system were in a state represented by a diagonal density operator $\rho^d_G(t)$:

$$\langle O_R \rangle_{\rho_U(t)} \to \sum_i (o_{Rii})(\rho^d_{Uii}) = \langle O_R \rangle_{\rho^d_G(t)}, \quad (12)$$

in such a way that

$$\forall t \gg t_D, \quad \langle O_R \rangle_{\rho_U(t)} = \langle O_R \rangle_{\rho_G^d(t)} \quad \text{and}$$
$$\rho_G^d(t) \text{ is diagonal in the preferred basis.} \tag{13}$$

Summing up, the decoherence of certain relevant observables of the whole closed system amounts to the fact that the expectation values of those observables tend very rapidly to certain values that can be computed as their expectation values in a state that is diagonal in the preferred basis. Therefore, the phenomenon of decoherence can be explained in three general steps [37, 38, 42]:

➢ **Step 1:** The space $\mathcal{O}_R$ of relevant observables is defined. Regarding this step, the present proposal agrees with all the other approaches to decoherence, which always select a set of relevant observables in terms of which the time behavior of the system is described: gross observables (van Kampen), macroscopic observables of the apparatus (Daneri), relevant observables (Omnès), observables of the open system (environment-induced decoherence), van Hove observables (self-induced decoherence), etc.

➢ **Step 2:** The expectation value $\langle O_R \rangle_{\rho_U(t)}$, for any $O_R \in \mathcal{O}_R$, is obtained. This step can be performed in two different but equivalent ways:

- $\langle O_R \rangle_{\rho_U(t)}$ is directly computed as the expectation value of O_R in the unitarily evolving state $\rho_U(t)$.
- A coarse-grained state $\rho_G(t)$, such that $\langle O_R \rangle_{\rho_U(t)} = \langle O_R \rangle_{\rho_G(t)}$ for any $O_R \in \mathcal{O}_R$ is defined, and its non-unitary evolution (governed by a master equation) is computed.

➢ **Step 3:** It is proved that $\langle O_R \rangle_{\rho_U(t)} = \langle O_R \rangle_{\rho_G(t)}$ reaches a value $\langle O_R \rangle_{\rho_G^d(t)}$:

$$\langle O_R \rangle_{\rho_U(t)} = \langle O_R \rangle_{\rho_G(t)} \xrightarrow{t \gg t_D} \langle O_R \rangle_{\rho_G^d(t)}, \tag{14}$$

where $\rho_G^d(t)$ remains diagonal in the preferred basis for all times $t \gg t_D$.

This means that, although the off-diagonal terms of $\rho_U(t)$ never vanish through the unitary evolution, it might be said that the system decoheres *from the observational point of view* given by any relevant observable $O_R \in \mathcal{O}_R$.

6. Environment-induced Decoherence from the Closed-System Approach

As explained above, the EID approach proves that, in many physical models of systems in interaction with their environments, the non-diagonal terms of the reduced state $\rho_S(t)$ rapidly tend to vanish after an extremely short decoherence time t_D (see Eq. (3)): the reduced state approaches a state ρ_S^d that is diagonal in the preferred basis.

In the traditional presentation, Steps 1–3 are not explicit. However, EID can be rephrased in such a way that it can be viewed from a closed-system viewpoint.

➢ **Step 1:** In this case, the whole composite system U, whose space of observables is $\mathcal{O}_U$, is partitioned into a system S, represented by the space of observables $\mathcal{O}_S$, and an environment E, represented by the space of observables $\mathcal{O}_E$. Therefore, the members of the space $\mathcal{O}_R \subset \mathcal{O}_U$ of relevant observables have the form,

$$O_{US} = O_S \otimes I_E \in \mathcal{O}_R, \tag{15}$$

where $O_S \in \mathcal{O}_S$ is a generic observable of the system S and $I_E \in \mathcal{O}_E$ is the identity operator corresponding to the environment E.

➢ **Step 2:** The expectation value of any $O_{US} \in \mathcal{O}_R$ can be computed as (see Eq. (8)):

$$\begin{aligned}\langle O_{US}\rangle_{\rho_U} &= Tr(\rho_U O_{US})\\ &= Tr(\rho_U(O_S \otimes I_E)) = Tr(\rho_S O_S) = \langle O_S\rangle_{\rho_S}.\end{aligned} \tag{16}$$

This means that $\langle O_{US}\rangle_{\rho_U}$ can also be computed as the expectation value of any observable $O_S \in \mathcal{O}_S$ in the reduced state ρ_S of S.

➢ **Step 3:** The evolution of the reduced state $\rho_S(t)$ (see Eq. (3)) has its counterpart in the evolution of the expectation values.

Therefore,

$$\langle O_{US}\rangle_{\rho_U(t)} = \langle O_S\rangle_{\rho_S(t)} \to \langle O_S\rangle_{\rho_S^d(t)} = \langle O_{US}\rangle_{\rho_G^d(t)}. \qquad (17)$$

In other words, this step is equivalent to the diagonalization of the reduced state.

This means that all the results obtained by the EID approach can also be obtained by the closed-system approach. However, the closed-system approach is more general, since makes it possible to consider decoherence of completely generic sets of observables, that is, sets of observables considered relevant but that do not define systems. In particular, it can be used to describe the phenomenon of decoherence in cases in which the EID approach cannot be applied.

7. Solving the Conceptual Challenges of the Open-System Approach

As discussed in Section 3, the EID approach faces some challenges which, although not serious in the technical application of the formalism, undermine the conceptual understanding of the phenomenon of decoherence. These difficulties derive precisely from what was considered as the main advantage of the approach: its open-system perspective. For this reason, it is not surprising that they are solved or dissolved from a closed-system view.

7.1. The closed-system problem

The generality of the closed-system approach makes it particularly pertinent to cases that are beyond the scope of the EID approach. For instance, it can be applied to the formalism of self-induced decoherence, which was specifically designed to account for decoherence in closed systems [38, 42]. In fact, from this view, if the closed system has a continuous spectrum of energy, it decoheres in the basis of the energy from the viewpoint of almost all its observables, with the exception of the observables that are not experimentally accessible [44]. Precisely due to its closed-system perspective, the self-induced decoherence approach has been successfully used to describe closed-system models, such as the model of a flat Roberson–Walker universe [43] and the Casati–Prosen model [33].

It is interesting to recall certain presentations of the EID approach that suggest a close relationship between decoherence and dissipation: since decoherence is a consequence of the interaction between an open system and its environment, it should be accompanied by other manifestations of openness, such as the dissipation of energy into the environment. Precisely for this reason, the non-dissipative approaches to decoherence were proposed as alternative or rival to the orthodox EID approach [25, 60, 63, 78, 99, 102]. Schlosshauer clearly stresses that energy dissipation is not a condition for decoherence [101]: the loss of energy from the system is a classical effect, leading to thermal equilibrium in the relaxation time, whereas decoherence is a purely quantum effect that takes place in the decoherence time, many orders of magnitude shorter than the relaxation time. From the closed-system approach, since decoherence is not due to the interaction of a system with its environment, the possibility of confusing decoherence and dissipation vanishes [40].

7.2. The defining-system problem

The closed-system approach implies, by its own nature, the dissolution of the "looming big" defining-system problem, that is, the problem that there is no criterion to distinguish between the system and the environment. In fact, the splitting of the closed system into an open subsystem and an environment is just a way of selecting the relevant observables of the closed system. Since there are many different sets of relevant observables depending on the observational viewpoint adopted, the same closed system can be decomposed in many different ways: each decomposition represents a decision about which degrees of freedom are relevant and which can be disregarded in any case. But since there is no privileged or "essential" decomposition, there is no need of an unequivocal criterion to decide where to place the cut between "the" system and "the" environment. If all the ways of selecting the relevant observables of the closed system are equally legitimate, decoherence is a phenomenon *relative* to which observables of the whole closed system are considered relevant and which are disregarded in each case [39, 86, 88].

These considerations cast new light on the relationship between decoherence and energy dissipation. To the extent that decoherence is a relative phenomenon, no flow of a non-relative quantity from the open system to the environment can account for it. In particular, although energy dissipation and decoherence are in general easily distinguished because of their different timescales, the very reason for their difference is that energy dissipation is not a relative phenomenon but results from the effective flow of a physical entity, whereas decoherence is relative to the observational partition of the whole closed system selected in each situation.

It is worth insisting on the difference between the open-system and the closed-system approaches by emphasizing the difference between the concepts of subjective and relative. The open-system approach relies on the cut between system and environment. However, since there is no univocal criterion to decide where to place the cut, the decision rests with the observer, that is, turns out to be a *subjective* matter. From the closed-system perspective, by contrast, the discrimination between relevant and irrelevant observables does not result from an intrinsic property of a system. The only system univocally defined is the closed system; the selection of a set of relevant observables is a selection of a kind of observational reference frame *relative* to which decoherence is evaluated: for certain sets of relevant observables the interference terms of the expectation values vanish and not for others. This flexibility is what endows the closed-system approach with the capability of studying the behavior of any set of observables, with no need of conceiving one or some of them as the privileged ones.

7.3. The problem of the emergence of the classical world

If classicality is conceived as an objective property of systems, the fact that a system behaves classically or not, cannot depend on the way in which the observer decides to split the original closed system into a system of interest and its environment. The conceptual difficulty is even more serious due to an already pointed out result: in certain situations the fact that classicality emerges in an open system or not, depends on what composite system that open subsystem is embedded in. More precisely, given two partitions of a closed system U, $U = S_1 \cup E_1$ and $U = S_2 \cup E_2$, it may be the case that S_1 and S_2 decohere and behave classically, but $S_1 \cup S_2$ does not decohere, so classicality does not emerge in it.

Although admitting that the relative nature of decoherence dissolves the defining-system problem, such a relativity does not apply to the problem of the emergence of the classical world: even if emergent, classicality is not conceived as a relative property; a given system behaves as classical or not and the theory should account for this. Nevertheless, the closed-system approach solves the problem due to its focus on observables. Given the closed system U, saying that it decoheres from the perspective of the relevant observables $O_R \in \mathcal{O}_R$ amounts to saying that the interference terms of the expectation values of all the observables belonging to $\mathcal{O}_R$ vanish with the (unitary) time-evolution of U, and this is not a relative fact. As a consequence, the observables of U could be considered one by one to see whether they decohere or not, and then the set $\mathcal{O}_{cl}$ of all the classically behaving observables of U could be defined, with no ambiguity or relativity.

According to the explanation of the emergence of the classical world given by the closed-system approach, strictly speaking, classicality is not a property of systems: thinking in systems that become classical leads to the already mentioned difficulties. The difficulties can be overcome once it is recognized that *classicality is a property of observables.* The emergent classical world is, then, the world described by the observables that behave classically with respect to their expectation values.

8. A Top-down View of Quantum Mechanics

The advent of the EID approach was received with great enthusiasm in the scientific community. Many authors considered that decoherence supplies the right answer to the measurement problem and the classical limit of quantum mechanics. For instance, under the assumption that the only legitimate requirement for a physical theory is the explanation of our perceptions (the "appearances"), d'Espagnat ([53], p. 136) says that "*for macroscopic systems, the appearances are those of a classical world* [...] *decoherence explains the just mentioned appearances, and this is the most important result*". In his book on foundations of quantum mechanics, Auletta ([8], p. 289) makes a stronger claim: "*decoherence is able to solve practically all the problems of measurement*". From a similar perspective, Anderson ([5], p. 492) asserts that "*the word 'decoherence'* [...] *describes the process that used to be called 'collapse of the wave function'*". By adopting a more cautious stance, some only claim that decoherence has been successful in reducing the unobserved coherent superposition of macrostates into a mixture of effectively classical states [48]. The idea of the power of decoherence for supplying the final account of the classical limit is still in the air in the physics community; for instance, in a very recent article, Crull [49] claims that decoherence is able to tackle many conceptual problems of quantum physics by itself, with no need of interpretation (see criticism in [107]).

By contrast, many authors, mainly coming from the philosophy of physics, advanced serious warnings about the capability of decoherence for solving those interpretive problems (see, for instance, [9, 26, 73]). In particular, it has been stressed that the diagonalization of the reduced state of the system of interest does not imply that the whole composite system acquires a definite property: "*I do not believe that either detailed theoretical calculations or recent experimental results show that decoherence has resolved the difficulties associated with quantum measurement theory*" ([1], p. 135). Therefore, it has been considered that the only interpretive virtue of decoherence is to identify the preferred basis that defines the observables of classical-like behavior. For example, Schlosshauer thinks that "*it is reasonable to anticipate that decoherence embedded in some additional interpretive structure could lead to a complete and consistent derivation of the classical world from quantum-mechanical principles*" ([100], p. 1287). In a similar vein, Elby ([59], p. 364) claims that "*decoherence cannot help modal, relative state, or many-world interpretations fend off general metaphysical criticisms. The value of decoherence lies in its ability to pick out a special basis.*" In fact, the theory of decoherence has been frequently used in the many-world interpretation to solve the problem of the preferred basis [27, 110, 111], considered the main difficulty of Everett's proposal [103]. In the context of the de Broglie–Bohm mechanics, decoherence has been appealed to in order to argue that classical mechanics emerges as a sub-dynamics in the total configuration space of the trajectory theory corresponding to the system-plus-environment [22, 75]. Decoherence has also been integrated into the framework of modal interpretations [54, 85]; moreover, Bacciagaluppi and Hemmo [10] have suggested that the definition of the preferred basis given by decoherence would allow modal interpretations to overcome some well-known criticisms [2, 3].

The main criticism to the EID approach regarding the solution of the measurement problem is that, even in the case that the open system decoheres in a given basis, the whole closed system is still in a superposition and, therefore, the observables defined by that basis cannot behave classically. This correct criticism is disregarded by those who insist on endowing decoherence with the capability of solving interpretive problems: they continue to conceive open systems as quantum systems of the same nature as closed systems, and their reduced states as legitimate quantum states. Therefore, when they select certain relevant degrees of freedom to build the reduced state, they think that they are identifying an objectively defined open system, with its objective properties. This open-system approach is *a bottom-up view*, which leads to beginning by the open systems, and to considering their interaction only after that.

In the previous sections, we have shown that this approach involves different difficulties, which ultimately derive from ignoring that reduced states are a kind of coarse-grained states defined as the result of particular needs. This means that decoherence is relative to the degrees of freedom selected in each case. Then, if it is supposed that decoherence identifies the systems that behave classically, it is not even clear that the privileged basis is univocally picked up by decoherence: as stressed above, there are different factorizations that may lead to classicality in non-consistent ways.

The closed-system approach, by contrast, is *a top-down view* that begins by studying the whole closed system. Instead of resorting to reduced states, it focuses on the information of interest by selecting the relevant observables of the closed system. This strategy is more general in the sense that it can be applied to cases not covered by the open-system approach. By denying that decoherence is a process that only open systems develop, the closed-system approach dissolves the difficulties of the EID approach. In particular, since what decoheres or not are the observables of the closed system, there is a univocal way to define the set of classically behaving observables. Therefore, the emergence of classicality, although manifested through expectation values, is a well-defined objective phenomenon.

This top-down view of decoherence and the classical limit fits into the general framework of a *top-down view of quantum mechanics*, according to which the only legitimate quantum systems are the unitary evolving closed systems. This view finds a significant affinity with the so called "quantum structure studies", dealing with the different ways in which a quantum system can be decomposed into subsystems (quantum structures) [7, 71, 76]. From this general perspective, entanglement is also relative to the partition of the closed system into parts, which are not characterized as subsystems [13, 14, 108]. More precisely, from a generalized perspective, entanglement is not a relationship between systems or states, but between sets of observables [72]. Therefore, the concept of quantum correlation can also be generalized is such a way to be relativized to the subalgebras of the algebra of observables of the closed system [18]. In this way, this top-down view dissolves the so-called "puzzles" about quantum entanglement [57], derived from the lack of a univocal criterion to introduce a decomposition into the closed system. The same perspective leads to consider locality as relative.

In the interpretive framework, the modal-Hamiltonian interpretation [6, 83, 84] also takes a top-down closed-system perspective. Given the closed system $S = S_1 \cup S_2$, in the case that there is no interaction between S_1 and S_2 and, so, their time-evolutions are governed by the Schrödinger equation, there is no obstacle to consider them legitimate quantum systems, in particular, subsystems of the composite system S. However, if S_1 and S_2 do not follow the unitary evolutions ruled by the dynamical law of quantum mechanics, they are not viewed as subsystems of S but as mere "parts" of it. Those parts are not quantum systems because they lack independent identity: they are conceived as the result of conventional partitions of the whole quantum system S. This conception not only agrees with the well-known holism of quantum mechanics, but also leads to a reinterpretation of indistinguishability, according to which it is not a relation between individual particles but an internal symmetry of a non-individual and indivisible whole [50, 51, 82].

All these works show that the view that endows closed systems with ontological priority is gaining ground in the quantum foundations community. The top-down view of decoherence is one of its different manifestations.

Acknowledgments

This publication was made possible through the support of Grant No. 57919 from the John Templeton Foundation.

References

[1] Adler, S. (2003). Why decoherence has not solved the measurement problem: A response to P. W. Anderson. *Studies in History and Philosophy of Modern Physics*, **34**: 135–142.

[2] Albert, D. and Loewer, B. (1990). Wanted dead or alive: two attempts to solve Schrödinger's paradox. *Proceedings of the 1990 Biennial Meeting of the Philosophy of Science Association*, **1**: 277–285.

[3] Albert, D. and Loewer, B. (1993). Non-ideal measurements. *Foundations of Physics Letters*, **6**: 297–305.

[4] Anastopoulos, C. (2002). Frequently asked questions about decoherence. *International Journal of Theoretical Physics*, **41**: 1573–1590.

[5] Anderson, P. W. (2001). Science: A 'dappled world' or a 'seamless web'? *Studies in History and Philosophy of Modern Physics*, **34**: 487–494.

[6] Ardenghi, J. S., Castagnino, M. and Lombardi, O. (2009). Quantum mechanics: modal interpretation and Galilean transformations. *Foundations of Physics*, **39**: 1023–1045.

[7] Arsenijeviæ, M., Jekniæ-Dugiæ, J. and Dugiæ, M. (2016). A top-down versus a bottom-up hidden-variables description of the Stern–Gerlach experiment. *This Volume.*

[8] Auletta, G. (2000). *Foundations and Interpretation of Quantum Mechanics.* Singapore: World Scientific.

[9] Bacciagaluppi, G. (2012). The role of decoherence in quantum mechanics, in E. N. Zalta (ed.), *The Stanford Encyclopedia of Philosophy* (Winter 2012 Edition), (http://plato.stanford.edu/archives/win2012/entries/qm-decoherence/).

[10] Bacciagaluppi, G. and Hemmo, M. (1996). Modal interpretations, decoherence and measurements. *Studies in History and Philosophy of Modern Physics*, **27**: 239–277.

[11] Bai, T. and Stachel, J. (2016). Bohr's diaphragms. *This Volume.*

[12] Ballentine, L. E. (1990). *Quantum Mechanics.* New York: Prentice Hall.

[13] Barnum, H., Knill, E., Ortiz, G., Somma, R. and Viola, L. (2003). Generalizations of entanglement based on coherent states and convex sets. *Physical Review A*, **68**: #032308.

[14] Barnum, H., Knill, E., Ortiz, G., Somma, R. and Viola, L. (2004). A subsystem-independent generalization of entanglement. *Physical Review Letters*, **92**: #107902.

[15] Baumgarten, C. (2016). Minkowski spacetime and QED from ontology of time. *This Volume.*

[16] Bayen, F., Flato, M., Fronsdal, C., Lichnerowicz, A. and Sternheimer, D. (1977). Quantum mechanics as a deformation of classical mechanics. *Letters in Mathematical Physics*, **1**: 521–530.

[17] Bayen, F., Flato, M., Fronsdal, C., Lichnerowicz, A. and Sternheimer, D. (1978). Deformation theory and quantization I-II. *Annals of Physics* (NY), **111**: 61–110, 111–151.

[18] Bellomo, G., Majtey, A. P., Plastino, A. R. and Plastino, A. (2014). Quantum correlations from classically correlated states. *Physica A*, **405**: 260–266.

[19] Bengtsson, I. and Yczkowski, K. Z. (2006). *Geometry of Quantum States: An Introduction to Quantum Entanglement.* Cambridge: Cambridge University Press.

[20] Berkovitz, J., Frigg, R. and Kronz, F. (2006). The ergodic hierarchy, randomness and Hamiltonian chaos. *Studies in History and Philosophy of Modern Physics*, **37**: 661–691.

[21] Bohm, D. (1951). *Quantum Theory.* Englewood Cliffs NJ: Prentice-Hall.

[22] Bohm, D. and Hiley, B. J. (1993). *The Undivided Universe.* London: Routledge.

[23] Bohr, N. (1920). Über die Serienspektra der Element. *Zeitschrift für Physik*, **2**: 423–478.

[24] Bokulich, A. (2014). Bohr's correspondence principle, in E. N. Zalta (ed.), *The Stanford Encyclopedia of Philosophy* (Spring 2014 Edition), (http://plato.stanford.edu/archives/spr2014/entries/bohr-correspondence/).

[25] Bonifacio, R., Olivares, S., Tombesi, P. and Vitali, D. (2000). Model-independent approach to nondissipative decoherence. *Physical Review A*, **61**: #053802.

[26] Bub, J. (1997). *Interpreting the Quantum World.* Cambridge: Cambridge University Press.

[27] Butterfield, J. (2002). Some worlds of quantum theory, in R. Russell, P. Clayton, K. Wegter-McNelly and J. Polkinghorne (eds.), *Quantum Physics and Divine Action*, pp. 111–140, Vatican: Vatican Observatory Publications.

[28] Casati, G. and Chirikov, B. (1995a). Comment on "Decoherence, chaos, and the second law". *Physical Review Letters*, **75**: 350.

[29] Casati, G. and Chirikov, B. (1995b). Quantum chaos: unexpected complexity. *Physica D*, **86**: 220–237.

[30] Casati, G. and Prosen, T. (2005). Quantum chaos and the double-slit experiment. *Physical Review A*, **72**: #032111.

[31] Castagnino, M. (1999). The classical regime of a quantum universe obtained through a functional method. *International Journal of Theoretical Physics*, **38**: 1333–1348.

[32] Castagnino, M. (2004). The classical-statistical limit of quantum mechanics. *Physica A*, **335**: 511–517.

[33] Castagnino, M. (2006). The equilibrium limit of the Casati–Prosen model. *Physics Letters A*, **357**: 97–100.

[34] Castagnino, M. and Fortin, S. (2011a). New bases for a general definition for the moving preferred basis. *Modern Physics Letters A*, **26**: 2365–2373.

[35] Castagnino, M. and Fortin, S. (2011b). Predicting decoherence in discrete models. *International Journal of Theoretical Physics*, **50**: 2259–2267.

[36] Castagnino, M. and Fortin, S. (2012). Non-Hermitian Hamiltonians in decoherence and equilibrium theory. *Journal of Physics A: Mathematical and Theoretical*, **45**: #444009 (http://dx.doi.org/10.1088/1751-8113/45/44/444009).

[37] Castagnino, M. and Fortin, S. (2013). Formal features of a general theoretical framework for decoherence in open and closed systems. *International Journal of Theoretical Physics*, **52**: 1379–1398.

[38] Castagnino, M., Fortin, S., Laura, R. and Lombardi, O. (2008). A general theoretical framework for decoherence in open and closed systems. *Classical and Quantum Gravity*, **25**: #154002.

[39] Castagnino, M., Fortin, S. and Lombardi, O. (2010a). Suppression of decoherence in a generalization of the spin-bath model. *Journal of Physics A: Mathematical and Theoretical*, **43**: #065304.

[40] Castagnino, M., Fortin, S. and Lombardi, O. (2010b). Is the decoherence of a system the result of its interaction with the environment? *Modern Physics Letters A*, **25**: 1431–1439.

[41] Castagnino, M. and Gadella, M. (2006). The problem of the classical limit of quantum mechanics and the role of self-induced decoherence. *Foundations of Physics*, **36**: 920–952.

[42] Castagnino, M., Laura, R. and Lombardi, O. (2007). A general conceptual framework for decoherence in closed and open systems. *Philosophy of Science*, **74**: 968–980.

[43] Castagnino, M. and Lombardi, O. (2003). The self-induced approach to decoherence in cosmology. *International Journal of Theoretical Physics*, **42**: 1281–1299.

[44] Castagnino, M. and Lombardi, O. (2004). Self-induced decoherence: a new approach. *Studies in History and Philosophy of Modern Physics*, **35**: 73–107.

[45] Castagnino, M. and Lombardi, O. (2005a). Self-induced decoherence and the classical limit of quantum mechanics. *Philosophy of Science*, **72**: 764–776.

[46] Castagnino, M. and Lombardi, O. (2005b). Decoherence time in self-induced decoherence. *Physical Review A*, **72**: #012102.

[47] Castagnino, M. and Ordoñez, A. (2004). Algebraic formulation of quantum decoherence. *International Journal of Theoretical Physics*, **43**: 695–719.

[48] Cohen, E. (2016). Quantum to classical transitions via weak measurements and postselection. *This Volume.*

[49] Crull, E. (2015). Less interpretation and more decoherence in quantum gravity and inflationary cosmology. *Foundations of Physics*, **45**: 1019–1045.

[50] Da Costa, N. and Lombardi, O. (2014). Quantum mechanics: ontology without individuals. *Foundations of Physics*, **44**: 1246–1257.

[51] Da Costa, N., Lombardi, O. and Lastiri, M. (2014). A modal ontology of properties for quantum mechanics. *Synthese*, **190**: 3671–3693.

[52] Daneri, A., Loinger, A. and Prosperi, G. (1962). Quantum theory of measurement and ergodicity conditions. *Nuclear Physics*, **33**: 297–319.

[53] d'Espagnat, B. (2000). A note on measurement. *Physics Letters A*, **282**: 133–137.

[54] Dieks, D. (1989). Resolution of the measurement problem through decoherence of the quantum state. *Physics Letters A*, **142**: 439–444.

[55] Diosi, L. (1987). A universal master equation for the gravitational violation of quantum mechanics. *Physics Letters A*, **120**: 377–381.

[56] Diosi, L. (1989). Models for universal reduction of macroscopic quantum fluctuations. *Physical Review A*, **40**: 1165–1174.

[57] Earman, J. (2014). Some puzzles and unresolved issues about quantum entanglement. *Erkenntnis*, **80**: 303–337.

[58] Ehrenfest, P. (1927). Bemerkung über die angenäherte Gültigkeit der klassischen Mechanik innerhalb der Quantenmechanik. *Zeitschrift für Physik*, **45**: 455–457.

[59] Elby, A. (1994). The 'decoherence' approach to the measurement problem in quantum mechanics. *Proceedings of the 1994 Biennial Meeting of the Philosophy of Science Association*, **1**: 355–365.

[60] Ford, G. and O'Connell, R. (2001). Decoherence without dissipation. *Physics Letters A*, **286**: 87–90.

[61] Fortin, S. and Lombardi, O. (2014). Partial traces in decoherence and in interpretation: What do reduced states refer to? *Foundations of Physics*, **44**: 426–446.

[62] Franco, M. and Calzetta, E. (2011). Decoherence in the cosmic background radiation. *Classical and Quantum Gravity*, **28**: 145024.

[63] Frasca, M. (2003). General theorems on decoherence in the thermodynamic limit. *Physics Letters A*, **308**: 135–139.

[64] Frigg, R. (2007). A field guide to recent work on the foundations of thermodynamics and statistical mechanics, in D. Rickles (ed.), *The Ashgate Companion to the New Philosophy of Physics*, pp. 99–196, London: Ashgate.

[65] Gambini, R., Porto, R. and Pullin, J. (2007). Fundamental decoherence from quantum gravity: a pedagogical review. *General Relativity and Gravitation*, **39**: 1143–1156.

[66] Gambini, R. and Pullin, J. (2007). Relational physics with real rods and clocks and the measurement problem of quantum mechanics. *Foundations of Physics*, **37**: 1074–1092.

[67] Gambini, R. and Pullin, J. (2010). Modern space-time and undecidability, in V. Petkov (ed.), *Fundamental Theories of Physics (Minkowski Spacetime: A Hundred Years Later)*, pp. 149–162, Heidelberg: Springer.

[68] Gell-Mann, M. and Hartle, J. B. (1993). Classical equations for quantum systems. *Physical Review D*, **47**: 3345–3382.

[69] Joos, E., Zeh, H. D., Kiefer, C., Giulini, D., Kupsch, J. and Stamatescu, I. O. (2003). *Decoherence and the Appearance of a Classical World in Quantum Theory*. Berlin: Springer Verlag.

[70] Haag, R. (1993). *Local Quantum Physics (Fields, Particles, Algebras)*. Berlin: Springer Verlag.

[71] Harshman, N. (2011). Symmetry and natural quantum structures for three-particles in one-dimension. *This Volume*.

[72] Harshman, N. and Ranade, K. (2011). Observables can be tailored to change the entanglement of any pure state. *Physical Review A*, **84**: #012303.

[73] Healey, R. (1995). Dissipating the quantum measurement problem. *Topoi*, **14**: 55–65.

[74] Holik, F. and Plastino, A. (2011). Convex polytopes and quantum separability. *Physical Review A*, **84**: #062327.

[75] Holland, P. (2016). The quantum state as spatial displacement. *This Volume*.

[76] Jekniæ-Dugiæ, J., Arsenijeviæ, M. and Dugiæ, M. (2013). *Quantum Structures: A View of the Quantum World*. Saarbrücken: Lambert Academic Publishing.

[77] Kastner, R. E. (2016). Beyond complementarity. *This Volume*.

[78] Kiefer, C. and Polarski, D. (2009). Why do cosmological perturbations look classical to us? *Advanced Science Letters*, **2**: 164–173.

[79] Knill, E., Laflamme, R., Barnum, H., Dalvit, D., Dziarmaga, J., Gubernatis, J., Gurvits, L., Ortiz, G., Viola, L. and Zurek, W. (2002). Quantum information processing — a hands-on primer. *Los Alamos Science*, **27**: 2–37.

[80] Leggett, A. (1987). Reflections on the quantum measurement paradox, in B. J. Hiley and F. D. Peat (eds.), *Quantum Implications*, pp. 85–104, London: Routledge and Kegan Paul.

[81] Lombardi, O. (2003). El problema de la ergodicidad en mecánica estadística. *Crítica. Revista Hispanoamericana de Filosofía*, **35**: 3–41.

[82] Lombardi, O. and Dieks, D. (2013). Modal interpretations of quantum mechanics, in E. N. Zalta (ed.), *The Stanford Encyclopedia of Philosophy* (Fall 2013 Edition), (http://plato.stanford.edu/archives/fall2013/entries/qm-modal/).

[83] Lombardi, O. and Castagnino, M. (2008). A modal-Hamiltonian interpretation of quantum mechanics. *Studies in History and Philosophy of Modern Physics*, **39**: 380–443.

[84] Lombardi, O., Castagnino, M. and Ardenghi, J. S. (2010). The modal-Hamiltonian interpretation and the Galilean covariance of quantum

mechanics. *Studies in History and Philosophy of Modern Physics*, **41**: 93–103.

[85] Lombardi, O. and Dieks, D. (2016). Particles in a quantum ontology of properties, in T. Bigaj and C. Wüthrich (eds.), *Metaphysics in Contemporary Physics*, pp. 123–143, Leiden: Brill-Rodopi.

[86] Lombardi, O., Fortin, S. and Castagnino, M. (2012). The problem of identifying the system and the environment in the phenomenon of decoherence, in H. W. de Regt, S. Hartmann and S. Okasha (eds.), *Philosophical Issues in the Sciences Volume* 3, pp. 161–174, Berlin: Springer.

[87] Lombardo, F. and Mazzitelli, D. (1996). Coarse graining and decoherence in quantum field theory. *Physical Review D*, **53**: #2001.

[88] Lychkovskiy, O. (2013). Dependence of decoherence-assisted classicality on the way a system is partitioned into subsystems. *Physical Review A*, **87**: #022112.

[89] Mackey, M. C. (1989). The dynamic origin of increasing entropy. *Review of Modern Physics*, **61**: 981–1015.

[90] Milburn, G. J. (1991). Intrinsic decoherence in quantum mechanics. *Physical Review A*, **44**: 5401–5406.

[91] Omnès, R. (1994). *The Interpretation of Quantum Mechanics*. Princeton: Princeton University Press.

[92] Omnès, R. (2001). Decoherence: An irreversible process. *Los Alamos National Laboratory*, arXiv:quant-ph/0106006.

[93] Omnès, R. (2002). Decoherence, irreversibility and the selection by decoherence of quantum states with definite probabilities. *Physical Review A*, **65**: 052119.

[94] Omnès, R. (2005). Results and problems in decoherence theory. *Brazilian Journal of Physics*, 35: 207–210.

[95] Paz, J. P. and Roncaglia, A. J. (2009). Redundancy of classical and quantum correlations during decoherence. *Physical Review A*, **80**: 042111.

[96] Paz, J. P. and Zurek, W. H. (2002). Environment-induced decoherence and the transition from quantum to classical, in D. Heiss (ed.), *Fundamentals of Quantum Information: Quantum Computation, Communication, Decoherence and All That*, pp. 77–148, Heidelberg-Berlin: Springer.

[97] Penrose, R. (1995). *Shadows of the Mind*. Oxford: Oxford University Press.

[98] Pessoa Jr., O. (1998). Can the decoherence approach help to solve the measurement problem? *Synthese*, **113**: 323–346.

[99] Polarski, D. and Starobinsky, A. A. (1996). Semiclassicality and decoherence of cosmological perturbations. *Classical and Quantum Gravity*, **13**: 377–392.

[100] Schlosshauer, M. (2004). Decoherence, the measurement problem, and interpretations of quantum mechanics. *Reviews of Modern Physics*, **76**: 1267–1305.
[101] Schlosshauer, M. (2007). *Decoherence and the Quantum-to-Classical Transition*. Berlin: Springer.
[102] Sicardi Shifino, A., Abal, G., Siri, R., Romanelli, A. and Donangelo, R. (2003). Intrinsic decoherence and irreversibility in a quasiperiodic kicked rotor. *Los Alamos National Laboratory*, arXiv:quant-ph/0308162.
[103] Stapp, H. P. (2002). The basis problem in Many-Worlds theories. *Canadian Journal of Physics*, **80**: 1043–1052.
[104] Van Hove, L. (1957). The approach to equilibrium in quantum statistics: A perturbation treatment to general order. *Physica*, **23**: 441–480.
[105] Van Hove, L. (1959). The ergodic behaviour of quantum many-body systems. *Physica*, **25**: 268–276.
[106] Van Kampen, N. G. (1954). Quantum statistics of irreversible processes. *Physica*, **20**: 603–622.
[107] Vassallo, A. and Esfeld, M. (2015). On the importance of interpretation in quantum physics: a reply to Elise Crull. *Foundations of Physics*, **45**: 1533–1536.
[108] Viola, L. and Barnum, H. (2010). Entanglement and subsystems, entanglement beyond subsystems, and all that, in A. Bokulich and G. Jaeger (eds.), *Philosophy of Quantum Information and Entanglement*, pp. 16–43, Cambridge: Cambridge University Press.
[109] Von Neumann, J. (1932). *Mathematische Grundlagen der Quantenmechanik*. Berlin: Springer.
[110] Wallace, D. (2002). Worlds in Everett interpretation. *Studies in History and Philosophy of Modern Physics*, **33**: 637–661.
[111] Wallace, D. (2003). Everett and structure. *Studies in History and Philosophy of Modern Physics*, **34**: 87–105.
[112] Wigner, E. (1932). On the quantum correction for thermodynamic equilibrium. *Physical Review*, **40**: 749–759.
[113] Zeh, H. D. (1970). On the interpretation of measurement in quantum theory. *Foundations of Physics*, **1**: 69–76.
[114] Zeh, H. D. (1971). On the irreversibility of time and observation in quantum theory, in B. d'Espagnat (ed.), *Foundations of Quantum Mechanics*, pp. 69–76, New York: Academic Press.
[115] Zeh, H. D. (1973). Toward a quantum theory of observation. *Foundations of Physics*, **3**: 109–116.
[116] Zurek, W. H. (1982). Environment-induced superselection rules. *Physical Review D*, **26**: 1862–1880.

[117] Zurek, W. H. (1991). Decoherence and the transition from quantum to classical. *Physics Today*, **44**: 36–44.

[118] Zurek, W. H. (1993). Preferred states, predictability, classicality and the environment-induced decoherence. *Progress of Theoretical Physics*, **89**: 281–312.

[119] Zurek, W. H. (1994). Preferred sets of states, predictability, classicality and environment-induced decoherence, in J. J. Halliwell, J. Pérez-Mercader and W. H. Zurek (eds.), *Physical Origins of Time Asymmetry*, pp. 175–212, Cambridge: Cambridge University Press.

[120] Zurek, W. H. (1998). Decoherence, einselection, and the existential interpretation. *Philosophical Transactions of the Royal Society A*, **356**: 1793–1820.

[121] Zurek, W. H. (2000). Decoherence and einselection, in P. Blanchard, D. Giulini, E. Joos, C. Kiefer and I.-O. Stamatescu (eds.), *Decoherence: Theoretical, Experimental, and Conceptual Problems*, pp. 309–342, Berlin-Heidelberg: Springer-Verlag.

[122] Zurek, W. H. (2003). Decoherence, einselection, and the quantum origins of the classical. *Reviews of Modern Physics*, **75**: 715–776.

[123] Zurek, W. H. (2009). Quantum Darwinism. *Nature Physics*, **5**: 181–188.

[124] Zurek, W. H., Habib, S. and Paz, J. P. (1993). Coherent states via decoherence. *Physical Review Letters*, **70**: 1187–1190.

Chapter 15

A Top-down versus a Bottom-up Hidden-variables Description of the Stern–Gerlach Experiment

M. Arsenijević*, J. Jeknić-Dugić† and M. Dugić*

*Department of Physics, Faculty of Science, Kragujevac, Serbia

†Department of Physics, Faculty of Science, Niš, Serbia

We employ the Stern–Gerlach experiment to highlight the basics of a minimalist, non-interpretational top-down approach to quantum foundations. Certain benefits of the "quantum structural studies" (QSS) highlighted here are detected and discussed. While the top-down approach can be described without making any reference to the fundamental structure of a closed system, the hidden variables (HV) theory *á la* Bohm proves to be more subtle than it is typically regarded.

1. Introduction

Reductionist, i.e. bottom-up, thinking is prominent in physics: features and dynamics of the structural components (subsystems), all the way down to the elementary particles, are assumed to exhaust the description of the features and dynamics of the whole. Huge success of application of physical theories seem to weaken the possible merits of the question: what might be wrong with the physical reductionism?

However, there are indications that this "wrong" is rather subtle considering that reductionism may not be the "whole story". On one hand, efficient physical description of many-particle systems is lacking [2]. On the other hand, even the orthodox "quantum" systems may hide subtleties in their structural description, given entanglement and other features of "quantum wholeness". It may be more realistic to allow native variables for quantum composites.

In the context of the universally valid and complete quantum mechanics not extended by any additional rules or (e.g. interpretational) assumptions [28]:

> "*Without further physical assumption, no partition has an ontologically superior status with respect to any other.*"

as well as [13]:

> "*However, for many macroscopic systems, and in particular for the universe as a whole, there may be no natural split into distinguished subsystems and the rest, and another way of identifying the naturally decoherent variables is required.*"

The apparent lack of preferred structure (partition into subsystems) of a closed quantum system yields a top-down approach to quantum structures, cf. e.g. [17] (and references therein).

In the context of the universally valid but incomplete quantum theory, the things may look the opposite. For example, in the de Broglie–Bohm quantum theory, the particles are assumed to define the physically fundamental (ontic) structure of the Universe [5]. This kind of structuralism is bottom-up that *supplements* the standard quantum mechanical formalism.

Hence, the foundational character of the quantum structure studies (QSS) and particularly of the topical questions highlighting this volume that can be shortly expressed as "How components relate to a composite?" Scientific relevance of this question stands even without any reference to "applications", since majority of the working physicists agree that [16]:

> "*But our present [quantum mechanical] formalism is not purely epistemological; it is a peculiar mixture describing in part realities of Nature, in part incomplete human information about Nature — all scrambled up by Heisenberg and Bohr into an omelette that nobody has seen how to unscramble. Yet we think that the unscrambling is a prerequisite for any further advance in basic physical theory. For, if we cannot separate the subjective and objective*

aspects of the formalism, we cannot know what we are talking about; it is just that simple."

In this chapter, we contrast a top-down with a bottom-up (hidden-variables (HV)) description of the illustrious Stern–Gerlach experiment. Contrasting each other, these descriptions highlight the QSS as an useful tool in the foundations and interpretation of quantum theory that is our main goal. Certain ramifications of the observations made in this chapter will be elaborated elsewhere.

2. Outlines of a Top-down Approach to Quantum Structures

This volume covers the different top-down approaches and interpretations of quantum structures, see e.g. [9]. In this section, we briefly overview perhaps the simplest one, which equates "quantum structure" with the "tensor product structure [of the composite system's Hilbert space]" [17] (and references therein).

In the universally valid and complete quantum theory,[a] every set of the *linearly independent and commuting* degrees of freedom, $\{q_i\}$, defines a tensor-product structure for the system's Hilbert space:

$$\mathcal{H} = \otimes_i \mathcal{H}_i, \tag{1}$$

where $\mathcal{H}_i$ concerns the ith degree of freedom.[b]

By "alternative degrees of freedom", we assume the sets of *arbitrary* degrees of freedom, which are mutually linked via the invertible linear canonical transformations (LTS), e.g.

$$\xi_m = \sum_i \kappa_{mi} q_i, \tag{2}$$

and analogously for the conjugated momentums (if these exist). For the unitary matrix (κ_{mi}), applies the constraint $\sum_l \kappa_{li}\kappa_{lm} = \delta_{im}$ for real κs.

[a]To be precise, by "universally valid and complete quantum theory", we have in mind the standard non-relativistic formalism not extended or amended by any additional assumptions or interpretational elements.

[b]Notice that a factor-space $\mathcal{H}_i$ may refer to a particle's spin as a vector observable.

Then the two sets of the degrees of freedom, $\{q_i\}$ and $\{\xi_m\}$, define a pair of structures of the composite system.

To illustrate, consider the paradigmatic example of the hydrogen atom, which is defined as a pair of quantum particles "electron" and "proton" via the respective position observables $\vec{r}_e, \vec{r}_p$. However, this structure, $e + p \equiv \{\vec{r}_e, \vec{r}_p\}$, is not the only one possible. Rather, it is typically regarded the alternative atomic "center of mass + relative particle" ($CM + R$) structure that is defined by the respective position observables, $\vec{R}_{CM}$ and $\vec{\rho}_R$. The structural transition [21]

$$e + p \rightarrow CM + R \tag{3}$$

is due to the (invertible) linear canonical transformations:

$$\vec{R}_{CM} = \frac{m_e \vec{r}_e + m_p \vec{r}_p}{m_e + m_p}, \quad \vec{\rho}_R = \vec{r}_e - \vec{r}_p, \tag{4}$$

and with the tensor re-factorization:

$$\mathcal{H}_e \otimes \mathcal{H}_p = \mathcal{H}_{CM} \otimes \mathcal{H}_R. \tag{5}$$

All kinds and types of the LTS-induced structures are of interest in the context of the universally valid and complete quantum theory. For closed quantum systems (subjected to the Schrödinger law), there is no privileged (preferred) structure, Section 1. However, for open quantum systems, it is often conjectured, e.g. [12], and sometimes justified [1, 21, 24] existence of a preferred structure (decomposition into subsystems) due to the environmental influence. Sometimes, the preferred structure is postulated [31] or expected to exist due to the additional symmetry-based requirements, see e.g. [14].

In general, *locality* is clearly defined for every possible subsystem of a composite (total) system: tensor-factorization of the composite system's Hilbert space uniquely defines *local observables* for every subsystem. For example, any observable A of the atomic CM system is defined $A_{CM} \otimes I_R$ and, according to Eq. (4), is a "collective" observable regarding the atomic $e + p$ structure. Equivalently, any observable B for the atomic proton is local for the $e + p$ structure in the form of $B_p \otimes I_e$ and is a "collective" observable for the atomic $CM + R$ structure. Therefore, as long as a measurement of the $A_{CM} \otimes I_R$ observable can induce nonlocal effects for the $e + p$ structure, it never — by definition — induces any influence on the atomic R system.

It is essential to note that the tensor-product nonlocality is more general than nonlocality often regarded in the context of interpretation of Bell inequalities. An action exerted on e.g. the atomic CM system *only partially interrupts the atomic electron and proton.* This is a direct consequence of Eq. (4): according to Eq. (4), to say that both electron and proton are simultaneously influenced by an action *implies* that the atomic CM and R systems are also *both* simultaneously influenced by that action. Therefore, a local action on the atomic CM or R systems does not influence either the electron or the proton, except partially, and *vice versa.* Hence, we can conclude that the concept of locality introduced here is a prerequisite for the locality versus nonlocality studies regarding the Bell inequalities, which assume a fixed bipartite structure of a composite system.

At this point, this version of the top-down approach to quantum structures directly tackles the so-called Tsirelson's problem [25] and also naturally calls for the analysis in the context of the so-called Categorical quantum theory, e.g. [6], that here will not be considered. The secondary role of "quantum subsystems", which lack independent individuality, is exhibited by the so-called "parallel occurrence of decoherence" on the purely formal level in the context of the standard environment-induced decoherence theory [7]. Simultaneous unfolding of the mutually irreducible decoherence processes regarding different (mutually irreducible) structures of a single closed system [7] is the price that must be paid as long as quantum theory is considered to be universal and complete in a non-interpretational context.

3. The Stern–Gerlach Experiment in the Universally Valid and Complete Quantum Mechanics

Now we are prepared to give a top-down description of the Stern–Gerlach experiment for arbitrary atom not carrying a net electric charge.

Consider an atom as a set of electrons and the atomic nucleus, which is typically considered as a point-like (non-structured) quantum particle. The atom's Hamiltonian:

$$H = \sum_i \frac{\vec{p}_{ei}^2}{2m_e} + \frac{\vec{p}_{\rm nucl}^2}{2m_{\rm nucl}} - k\sum_i \frac{Ze^2}{|\vec{x}_{ei} - \vec{x}_{\rm nucl}|} + V \tag{6}$$

with the obvious notation and the V stands for all the (weaker) Coulomb- and spin-interactions between the particles that can be treated as the gross perturbation term. Equation (6) defines the following tensor-factorization of the atomic Hilbert state-space:

$$\mathcal{H} = \mathcal{H}_e \otimes \mathcal{H}_n \otimes \mathcal{H}_{\rm spin}, \tag{7}$$

where the atomic spin state-space, $\mathcal{H}_{\rm spin}$, is assumed to be isomorphic to the single-electron's spin-1/2 space and the remaining factor spaces concern the standard "orbital" (spatial) degrees of freedom of the electrons (e) and of the atomic nucleus (n) systems.

However, the standard theoretical model [5] of the Stern–Gerlach experiment regards the alternative "center-of-mass + internal (relative)" atomic structure for the spatial degrees of freedom. This structural change,[c] $e + n \to CM + R$, induces refactorization of the Hilbert state-space

$$\mathcal{H} = \mathcal{H}_{CM} \otimes \mathcal{H}_R \otimes \mathcal{H}_{\rm spin} \tag{8}$$

as well as the alternative form of the atomic Hamiltonian, $H = \frac{\vec{P}_{CM}^2}{2M} + \sum_i \vec{p}_{Ri}^2/2\mu_i + V_{\rm Coul} + V$ where the μ_is represent the "reduced masses" while $V_{\rm Coul}$ is the nucleus-induced classical Coulomb field for the R system's degrees of freedom [17, 21]. Neglecting the weak term V and bearing in mind Eq. (3), the Hamiltonian exhibits the variables separation:

$$H = \frac{\vec{P}_{CM}^2}{2M} \otimes I_R \otimes I_{\rm spin} + I_{CM} \otimes H_R \otimes I_{\rm spin}, \tag{9}$$

where the internal atomic energy $H_R = \sum_i \vec{p}_{Ri}^2/2\mu_i + V_{\rm Coul}$.

[c] A generalization of Eq. (4).

Placing the atom in a sufficiently strong magnetic field along the z-axis is modelled by the interaction term for the *CM + spin* system [5]:

$$H_{CM+\text{spin}} = -\vec{\mu}_{\text{spin}} \cdot \vec{B}(Z_{CM}), \tag{10}$$

where $\vec{\mu}_{\text{spin}} = -\mu\vec{S}$ is the atomic-spin magnetic dipole; if we apply the standard approximation $\vec{\mu}_{\text{spin}} \approx \vec{\mu}_e$, then the constant $\mu = \mu_B$ is the standard "Bohr magneton" and S is the spin of the electron.

From Eqs. (9) and (10), it directly follows that the unitary dynamics of the atom induces quantum entanglement between the atomic *CM* and "spin" systems, while the absence of interaction between the R system with the *CM* and the "spin" system allows for the "separation" of the R system's state from the rest in the form of:

$$|\Psi\rangle = \left(\sum_i c_i|i\rangle_{CM} \otimes |i\rangle_{\text{spin}}\right) \otimes |\chi\rangle_R, \tag{11}$$

in an instant of time t; for the spatial $i = +, -$, [indicating the above and below the XY-plane], the spin states $i = \downarrow, \uparrow$, respectively, and $c_i = 2^{-1/2}$.

Bearing in mind Section 2, the physical meaning of Eq. (11) is rather obvious:

> ($\mathcal{S}$) *Magnetic field* locally *acts on the atomic CM system, while leaving the atomic internal (the R) system intact.*

To the extent that Eq. (11) captures the phenomenology, the same can be told for the statement $\mathcal{S}$.

This trivial observation carries a nontrivial content regarding the atomic $e + n$ structure. Actually, bearing in mind Section 2, the $\mathcal{S}$ statement straightforwardly implies the following observation:

> ($\mathcal{O}$) *Neither the electrons (e) nor the atomic nucleus (n) are influenced by the magnetic field.*

The statement $\mathcal{O}$ is easily proved as a repetition of the arguments of Section 2: Assume that the magnetic field "sees" both the atomic electrons and the atomic nucleus. Then an influence exerted on the electrons positions, $\vec{r}_i$, *and* on the nucleus position, $\vec{r}_n$, in analogy with Eq. (4), directly gives rise to the conclusion that *both* the atomic CM and R systems are influenced by the magnetic field — in contradiction with Eq. (11) i.e. with $\mathcal{S}$.

Therefore, equal status of every (physically reasonable) structure of a composite system *implies* the purely local influence of the magnetic field. Furthermore, this influence cannot be precisely described for the atomic $e + n$ structure as, due to $\mathcal{S}$ and in analogy with Eq. (4), both e and n are only "partially" seen by the magnetic field.

4. A HV Description of the Stern–Gerlach Experiment

Influence of the magnetic field on the atomic e and n systems cannot be imagined in the universally valid and complete quantum theory — the statement $\mathcal{O}$ of Section 3. Hence, let us assume the following:

> ($\mathcal{S}'$) *Magnetic field locally acts on the atomic e and p systems, independently.*

This directly leads to the conclusion that there are certain HV in the description of the atom in the Stern–Gerlach experiment. That is, a "non-hidden" influence of the magnetic field implies the above statement $\mathcal{S}$, i.e. the $\mathcal{O}$ of the previous section, not to be correct.

It is natural to assume that the electrons-system and the nucleus positions, $\vec{r}_i$ and $\vec{r}_n$, respectively, can play the roles of "HV". Assuming the ontic status of those variables like in the Bohm's theory [5], it is possible qualitatively to devise a scenario in which the standard description, Eq. (11), applies and still to describe the influence of the field on *both* the e and n atomic systems. For simplicity, consider the hydrogen atom described by Eq. (11). Then it is easy to imagine that the external magnetic field $\vec{B}$ drives both the electron and the proton so as to have the dynamical change of the $CM + spin$ system as described by Eq. (11), while the relative distance between the e and p is determined by the probability density $|\chi(\vec{\rho}_R)|^2$; $\vec{\rho}_R = \vec{r}_e - \vec{r}_p$, cf. Section 2 for the notation.

Needless to say, in this picture, the alternative atomic structure $CM + R$ is artificial — simply a mathematical artifact without any physical meaning. Just like in classical physics, this structure can be used to ease mathematical manipulation, while all the physically relevant results must be expressed in the terms of the fundamental degrees of freedom, of $\vec{r}_e$ and $\vec{r}_p$.

The magnetic field Eq. (10), which is of the form $\vec{B}(Z_{CM} \otimes I_{\text{spin}})$ for the $CM + R$ structure, can represent an interaction term for the atomic $e+p$ structure: $\vec{B}((m_e z_e \otimes I_p + m_p I_e \otimes z_p)/M)$; $M = m_e + m_p$. This interaction would produce entanglement for the e and p even if there were not the Coulomb interaction in the $e + p$ structure. That is, regarding the atomic $e + p$ structure, Eq. (11) takes the form[d]:

$$\sum_i c_i |\psi_i\rangle_{e+p} \otimes |\chi_i\rangle_{\text{spin}}$$
$$= |\Psi\rangle = \sum_i c_i \left(\sum_\alpha d_\alpha^i |\alpha\rangle_e \otimes |\alpha\rangle_p \right) \otimes |\chi_i\rangle_{\text{spin}}, \tag{12}$$

as an example of Entanglement Relativity [8, 17] (and references therein).

By definition, Eq. (12) is *insensitive* to the fundamental *(subquantum)* influence of the magnetic field on the atomic e and p subsystems and can be assumed to give rise to the subquantum (HV) probability density with purely classical correlations [that are induced by the Coulomb and/or the external magnetic field], of the general form of:

$$\mu(\lambda) = \sum_i p_i f_{1i}(\vec{r}_e) f_{2i}(\vec{r}_p). \tag{13}$$

If normalized, the local probability densities f_is imply $\sum_i p_i = 1$, while $\int \mu(\lambda) d\lambda = 1$.

Hence, a HV theory *á la* Bohm can be summarized by the following observation while taking over the meaning of "locality"

---

[d]The adiabatic cut of the electron from the proton due to $m_e/M \ll 1$ would not change our argument even if numerically justified, see Ref. [17] for more details.

from Section 3:

> *($\mathcal{O}'$) Both the electrons* (e) *and the atomic nucleus* (n) *systems are locally influenced by the magnetic field so as that cannot be observed on the quantum level of* Eq. (12).

5. Discussion

QSS of Section 2 are *minimalist* in that they neither extend nor interpret the standard quantum mechanical formalism. In this sense, QSS introduce a non-standard methodology in the standard quantum theory with the following benefits.

First, the very basic QSS concept of *locality*, Section 2, is a prerequisite of the locality (i.e. of "local causality") *á la* Bell [27] and is compatible with the "non-locality" in the HV theories *á la* Bohm [27] — cf. ($\mathcal{O}'$), Section 4. Therefore, different contents of "locality" are naturally linked through the QSS.

Second, on the *ontological level*, QSS as per Section 2 is in sharp contrast with the reductionistic interpretations of quantum theory *á la* Bohm [5, 27]. Distinguishing between these two approaches to quantum theory may be seen as an amendment to the tests of the Bell inequalities [27]: one should simultaneously consider more than one composite-system's structure, which leads to the following observation.

Third, all the structural considerations are by definition *contextual*. Regarding the atom as a whole, all observables of every single atomic subsystem are the atomic observables too. However, measurements that are local relative to one atomic structure need not imply any information regarding an alternative atomic structure. Capturing the atoms impinging on a screen directly reveals the atomic CM position on the screen but does not provide any information regarding the atomic R system. Thus, the Stern–Gerlach experiment is local relative to the $CM + R$ structure and does not provide much information regarding the atomic $e + p$ structure. Therefore, the concept of contextuality is also extended, i.e. in a sense generalized: it does not assume exclusively a single (fixed) structure of a composite system and also allows for the mutually compatible observables, cf. e.g. Eq. (4).

Fourth, while Eqs. (12) and (13) are compatible, they are still limited from the point of view of the QSS. Simultaneous consideration of the physically fundamental and artificial structures pose specific constraints on the HV theories as per Section 3. To see this, consider the asymptotic limit of the hydrogen atom — a pair of the free non-interacting electron and proton — out of any external field. The "natural" choice of the classical density probability

$$\mu(\lambda) = f_1(\vec{r}_e) f_2(\vec{r}_p) \tag{14}$$

implies a conflict with the standard task of introducing the artificial $CM + R$ structure, Eq. (4). Actually, as a classical counterpart of the quantum correlations relativity [8, 17], typically:

$$f_1(\vec{r}_e) f_2(\vec{r}_p) = \mu(\lambda) = \sum_i q_i g_{1i}(\vec{R}_{CM}) g_{2i}(\vec{\rho}_R) \neq G_1(\vec{R}_{CM}) G_2(\vec{\rho}_R), \tag{15}$$

which is in contrast with the standard task of "variables separation" (and integrability) in classical mechanics. That is, mutually separated CM and R systems are typically described by the RHS of Eq. (15).

Conversely, the inverse of Eq. (14):

$$G_1(\vec{R}_{CM}) G_2(\vec{\rho}_R) = \mu'(\lambda) = \sum_i p_i f_{1i}(\vec{r}_e) f_{2i}(\vec{r}_p) \neq f_1(\vec{r}_e) f_2(\vec{r}_p), \tag{16}$$

challenges the above-distinguished "natural" choice of Eq. (14).

Thereby, a choice for the probability density $\mu(\lambda)$ becomes not as free as it may seem in the standard HV theories. Despite the fact that the alternative structures are artificial, the simple *mathematical* considerations as per Eqs. (13)–(16) confine the considerations in a nontrivial way yet fully to be explored. In the context of the Stern–Gerlach experiment, this choice should be accompanied with Eqs. (11) and (12) while bearing in mind that Eq. (11) describes the *operationally accessible* [28] situation.

It is worth repeating that structure of a composite system is conditional in the universally valid and complete quantum theory, Section 2. As distinct from the prevailing reductionistic wisdom, the QSS of Section 2 highlight existence of a preferred structure of an *open* quantum system as a *relational* concept. There is nothing in the formalism as well as in the physical ontological basis that could suggest the ontologically superior status of any structure of a closed composite quantum system. Thus, the QSS as per Section 2 directly challenge the standard physical conceptualization embodied in the concept of "elementary particles" [17, 27]: *in the non-relativistic domain, quantum formalism can properly work even without assuming the ontological status of elementary particles.*

This brings us to the following mathematical forms of the basic task of the QSS. On one hand, it is the so-called Tsirelson's problem [25]: whether or not a composite system's state space is built via the tensor-product structures as assumed in QSS? And the corollary: what if the tensor-product structures have an alternative — what is the meaning of "locality" in that new, hypothetical discourse? As another natural frame for the QSS, we emphasize the Category studies [6], in which *correlations* are primary and "*correlata*" (subsystems) are secondary. This is a natural framework for the emerging approach of "there are no particles" to quantum foundations [17, 29].

From the purely operational point of view, QSS elevate the following seemingly "philosophical" issue: How can we be sure in a concrete physical situation which degrees of freedom of a composite system have been targeted by our apparatus(es)? The natural assumption that acquiring information about a composite system is limited to a "small" set of the system's degrees of freedom *implies* locality as per Section 2. On one hand, measurement of the proton's position in the hydrogen atom gives a close value for the atomic CM position but introduces entanglement for the CM and R systems, cf. Eq. (4), that in principle can be experimentally tested. Hence, non-equivalence of measurements of the proton and the CM system's positions that instantiates locality of Section 2. On the other hand, targeting *both* the e and p atomic systems also provides an information regarding the positions of the formal systems $\vec{R}_{CM}$ and $\vec{\rho}_R$ [contextual measurements with only linear increase in uncertainties for $\vec{R}_{CM}$ and $\vec{\rho}_R$]. Hence, the "total" measurements — of the pair $\vec{r}_e, \vec{r}_p$ (or of the pair $\vec{R}_{CM}, \vec{\rho}_R$) — on the atom are practically equivalent.

Now the point strongly to be stressed is that quantum uncertainty, i.e. the non-increase of information regarding the related conjugate momentums, may appear as a kind of *limitation* in the physical process of *measurement* (that is understood as the process of acquiring information), not necessarily of the fundamental physical description of the atomic e and p subsystems. In effect, the quantum part of the theory, Eq. (12), may be *not a fundamental* but an emergent description of the quantum system called "atom" *due to the process of acquiring information* now crying for explanation and called "measurement". In our opinion, the dichotomy of "particles + quantum field" [5] is a suitable phrase, not a satisfying explanation yet. This position is in contrast to that of Bell's [4], which discards foundational importance of the quantum measurement process. To this end, setting "measurement" as a purely operational concept [10, 22] makes the two competing theories (Bohm [5] versus Qubism [10]) virtually indistinguishable everywhere except possibly on the operationally-inaccessible ontological level.

Operationally again, QSS are sometimes reduced to the topic of experimental accessibility of the composite system's observables [28]. While "accessibility" assumes the preferred tensor-product structure, which may be environment-induced [1, 12, 21, 24], there appear the following subtleties. On one hand, not all observables of a subsystem pertaining to a preferred structure of the composite system are on the equal physical footing. To this end, the symmetry of the open system's effective Hamiltonian [15] may be decisive-not only a choice between the non-commuting e.g. "position" and "momentum" (or energy) observables should be made, but also a choice between the commuting observables defining "object" (e.g. the Descartes versus spherical degrees of freedom) are of interest [30]. On the other hand, the environmental influence is typically approximate and hence one may expect some "emergent" degrees of freedom [26].

This brings us to the following answers to the topical questions of this volume. The familiar classical center-of-mass degrees of freedom are environment-induced, relationally realistic and operationally accessible degrees of freedom, which do not require any ontologically superior subquantum degrees of freedom. Hence, the transition from quantum to classical regards the specific, *local* degrees of freedom that, we believe, may be at the core of the other aspects of the problem of "transition from quantum to classical" [12]. In other words, a part of the problem may reside on the assumption that the problem must be solved in terms of the "fundamental" (ontological) structure of a composite system. Unfortunately, we are not aware nor do we offer an elaborated experimental proposal in this regard.

While QSS challenge [20] the interpretations *á la* Everett [23, 26], the discourses regarding the non-universally valid quantum theory [11] as well as "the problem of time" [3, 18, 19] (and references therein) are in order. To this end, the work is in progress and the results will be presented elsewhere.

Acknowledgments

We benefited much from discussions with Ruth Kastner. Financial support from Ministry of Science Serbia Grant No. 171028 is acknowledged, as well as for MD by the ICTP — SEENET-MTP project PRJ-09 Cosmology and Strings in frame of the SEENET-MTP Network.

References

[1] Arsenijević, M., Jeknić-Dugić, J. and Dugić, M. (2013). Asymptotic dynamics of the alternate degrees of freedom for a two-mode system: an analytically solvable model. *Chinese Physics B*, **22**: 020302.

[2] Auyang, S. Y. (1998). *Foundations of Complex-system Theories*, Cambridge University Press: Cambridge.

[3] Baumgarten, C. (2016). Minkowski spacetime and QED from ontology of time, forthcoming in Kastner, R. E., Jeknić-Dugić, J. and Jaroszkiewicz, G. (eds.), *Quantum Structural Studies*. Singapore: World Scientific Publishers.

[4] Bell, J. S. (1990). Against "Measurement", in *Sixty-Two Years of Uncertainty Historical, Philosophical, and Physical Inquiries into the Foundations of Quantum Mechanics*, Arthur I. Miller (ed.), NATO ASI Series Volume 226, pp. 17–31.

[5] Bohm, D. (1951). *Quantum Theory*, Prentice Hall, New York.

[6] Coecke, B. and Kissinger, A. Categorical quantum mechanics I: Causal quantum processes, e-print arXiv arXiv:1510.05468 [quant-ph].

[7] Dugić, M. and Jeknić-Dugić, J. (2012). Parallel decoherence in composite quantum systems. *Pramana*, **79**: 199.

[8] Dugić, M., Arsenijević, M. and Jeknić-Dugić, J. (2014). Quantum correlations relativity. *Science China Physics, Mechanics & Astronomy*, **56**: 732.

[9] Fortin, S. and Lombardi, O. (2016). A top-down view of the classical limit of quantum mechanics, forthcoming in Kastner, R. E., Jeknić-Dugić, J., Jaroszkiewicz, G. (eds.), *Quantum Structural Studies*. Singapore: World Scientific Publishers.

[10] Fuchs, C. A., David Mermin, N. and Schack, R. (2014). An introduction to QBism with an application to the locality of quantum mechanics. *American Journal of Physics*, **82**: 749.

[11] Ghirardi, G. C., Rimini, A. and Weber, T. (1986). Unified dynamics for microscopic and macroscopic systems. *Physics Review D*, **34**: 470.

[12] Giulini, D., Joos, E., Kiefer, C., Kupsch, J., Stamatescu, I.-O. and Zeh, H. D. (1996). *Decoherence and the appearance of a classical world in quantum theory*, Springer: Berlin.

[13] Halliwell, J. (2010). Emergence of hydrodynamic behaviour. In *Many Worlds? Everett, Quantum Theory, and Reality*, Saunders, S., Barrett, J., Kent, A., Wallace, D. (eds.), Oxford University Press, Oxford, pp. 99–117.

[14] Harshman, N. L. (2016). Symmetry and Natural Quantum Structures for Three-Particles in One-Dimension, forthcoming in Kastner, R. E., Jeknić-Dugić, J., Jaroszkiewicz, G. (eds.), *Quantum Structural Studies*. Singapore: World Scientific Publishers.

[15] Harshman, N. L. and Wickramasekara, S. (2007). Galilean and dynamical in-variance of entanglement in particle scattering. *Physcis Review Letters*, **98**: 080406.

[16] Jaynes, E. T. (1990). In *Complexity, Entropy, and the Physics of Information*, Zurek, W. H. (ed.), Addison-Wesley, p. 381.

[17] Jeknić-Dugić, J., Arsenijević, M. and Dugić, M. (2013). *Quantum Structures. A View of the Quantum World*, LAP Lambert Scientific Publishing, Saarbrücken, also available at e-print arXiv:1306.5471 [quant-ph].

[18] Jeknić-Dugić, J., Arsenijević, M. and Dugić, M. (2014). A local-time-induced pointer basis. *Proceedings of the Royal Society of London. Series A*, **470**: 20140283.
[19] Jeknić-Dugić, J., Arsenijević, M. and Dugić, M. (2016). Dynamical emergence of Markovianity in local time scheme. *Proceedings of the Royal Society of London. Series A* **472**: 20160041.
[20] Jeknić-Dugić, J., Dugić, M. and Francom, A. (2010). Quantum structures of a model-universe: questioning the everett interpretation of quantum mechanics. *International Journal of Theoretical Physics*, **53**: 169.
[21] Jeknić-Dugić, J., Dugić, M., Francom, A. and Arsenijević, M. (2014). Quantum structures of the hydrogen atom. *Open Access Library Journal*, **1**: e501.
[22] Nielsen, M. A. and Chuang, I. L. (2000). *Quantum Computation and Quantum Information*, Cambridge University Press: Cambridge.
[23] Saunders, S., Barrett, J., Kent, A. and Wallace, D. (eds.) (2010). *Many Worlds? Everett, Quantum Theory, and Reality*, Oxford University Press, Oxford.
[24] Stokes, A., Kurcz, A., Spiller, T. P. and Beige, A. (2012). Extending the validity range of quantum optical master equations. *Physics Review A*, **85**: 053805.
[25] Tsirelson, B. S. (1993). Some results and problems on quantum Bell-type inequalities. *Hadronic Journal Supplement*, **8**: 329345.
[26] Wallace, D. (2012). *Emergent Multiverse: Quantum Theory According to the Everett Interpretation*, Oxford University Press, Oxford.
[27] Wiseman, H. M. (2014). The two Bell's theorems of John Bell. *Journal of Physics A: Mathematical and Theoretical*, **47**: 424001.
[28] Zanardi, P. (2001). Virtual quantum subsystems. *Physical Review Letters*, **87**: 077901.
[29] Zeh, H. D. (1993). There are no Quantum Jumps, nor are there Particles. *Physics Letters A*, **172**: 189.
[30] Zurek, W. H. (1993). Preferred states, predictability, classicality and the environment-induced decoherence. *Progress of Theoretical and Experimental Physics*, **89**: 281.
[31] Zurek, W. H. (2003). Decoherence, einselection, and the quantum origins of the classical, *Review of Modern Physics*, **75**: 715.

Index